A SYSTEMS-BASED APPROACH TO POLICYMAKING

A SYSTEMS-BASED APPROACH TO POLICYMAKING

Edited by

Kenyon B. De Greene
University of Southern California

Kluwer Academic Publishers
Boston/Dordrecht/London

Distributors for North America:
Kluwer Academic Publishers
101 Philip Drive
Assinippi Park
Norwell, Massachusetts 02061 USA

Distributors for all other countries:
Kluwer Academic Publishers Group
Distribution Centre
Post Office Box 322
3300 AH Dordrecht, THE NETHERLANDS

Library of Congress Cataloging-in-Publication Data

A Systems-based approach to policymaking / edited by Kenyon B. De Greene.
p. cm.
Includes bibliographical references and index.
ISBN 0-7923-9325-2
1. Policy sciences. 2. System analysis. I. De Greene, Kenyon B.
H97.S96 1993
320'.6--dc20 93-538
CIP

Printed on acid-free paper.

Printed in the United States of America

CONTENTS

ABOUT THE EDITOR AND CONTRIBUTORS

Kenyon B. De Greene is a Professor of Systems Management at the Institute of Safety and Systems Management, University of Southern California, Los Angeles, California. Dr. De Greene previously held positions at System Development, Northrop, and Aerospace Corporations. He is the editor and chief contributing author of *Systems Psychology* and the author of *Sociotechnical Systems: Factors in Analysis, Design, and Management* and of *The Adaptive Organization: Anticipation and Management of Crisis* . He is the author of some 90 articles on systems theory, systems management, sociotechnical systems, and human factors/ergonomics. He has visited all seven continents and some 185 countries.

Peter M. Allen is Scientific Director of the International Ecotechnology Research Centre, Cranfield Institute of Technology, Bedford, England. Previously, Dr. Allen was a Senior Research Fellow with Nobel Laureate Professor Ilya Prigogine at the Free University of Brussels, where he continued to advance the dissipative-structure theory developed by Prigogine. He is the author of numerous articles and is the associate editor of several journals in the fields of evolution, urban modeling, and environmental science. He is the editor (with W. Schieve) of *Self-Organization and Dissipative Structures.*

Harold Chestnut was a director of corporate planning at General Electric Corporation. Dr. Chestnut is the author of *Systems Engineering Tools* and *Systems Engineering Methods,* and of numerous articles; he is the editor of several proceedings. He has been president of the Institute of Electrical and Electronics Engineers and of the International Federation for Automatic Control. He founded SWIIS in Schenectady, New York, a foundation devoted to applying systems engineering methodology to the solution of problems of international stability.

Yehezkel Dror is Professor of Political Science and Public Administration at the Hebrew University in Jerusalem. Dr. Dror has recently completed a two-year leave at the European Institute of Public Administration, Maastricht, Holland, where he worked on policy problems of Western and Eastern Europe. He is the author of some 100 books and articles, including *Public Policymaking Reexamined* and *Policymaking Under Uncertainty.* He has held distinguished appointments at policy-research organizations in a number of countries, where, among other things, he has conducted intensive workshops for top-level policymakers.

Fred E. Emery is a main co-founder (with Eric Trist) of the theory, methodology, and application of sociotechnical systems. Dr. Emery is based at the Australian National University, Canberra, and he has held many positions in Europe, the United States, and Canada. He is the author of numerous articles and analytic reports and of the books *Toward a Social Ecology* (with Trist), *A Choice of Futures* (with M. Emery), and *Futures We Are In* . He is the editor of *Systems Thinking*. He has organized many policy studies with the Australian government.

Jay W. Forrester is Germeshausen Professor at the Sloan School of Management, Massachusetts Institute of Technology, Cambridge, Massachusetts. Dr. Forrester is the originator of the system dynamics theory, computer-simulation methodology, and approach to policymaking. His is the inventor of the random-access computer core-memory. He is the author of *Industrial Dynamics, Urban Dynamics, World Dynamics, and Principles of Systems.* He and his students have designed a number of very influential global models, including *The Limits to Growth* and *The National Economic Model* .

Willis W. Harman is President of the Institute of Noetic Sciences in Sausalito, California. Previously, Dr. Harman founded the futures-research group at SRI, International, where he led numerous studies of emerging complex situations. He is the author of *An Incomplete Guide to the Future*, *Global Mind Change: The Promise of the Last Decade of the Twentieth Century*, and (with John Hormann) *Creative Work: The Constructive Role of Business in a Transforming Society*, and of numerous articles on futures research and social learning.

Harold A Linstone is University Professor of Systems Science at Portland State University, Portland, Oregon, where he has served as Director of the Systems Science Ph.D. Program and Futures Research Institute. Earlier, Dr. Linstone's industrial positions included the Associate Directorship of Corporate Planning at Lockheed Corporation. He is Editor-in-Chief of *Technological Forecasting and Social Change*, which he founded in 1969. He is the co-editor of the books *The Delphi Method, Futures Research: New Directions,* and *Technological Substitution.* He is also the author of numerous articles and of the book *Multiple Perspective for Decisionmaking.*

Jacques G. Richardson headed UNESCO's Science and Society Section for 13 years. He was associate publisher for *La Recherche.* He is the author of numerous articles and the editor of *Models of Reality: Shaping Thought and Action.* He now operates his own consultancy in Paris, specializing in communication problems based on natural and engineering science, economics, and management.

PREFACE

This book is based on a number of systems concepts, of which the following are emphasized here:

- The interacting systems of society and the environment are dynamic and evolutionary
- Evolution of these systems carries them through stages of differential stability and instability, continuity and discontinuity
- Associated with evolution and instability is structural change that is essentially irreversible
- The present is a stage of world transformation that may not have been equaled for decades or even centuries
- Policies and decisions must match the times, in the present case the stage of world transformation

The time 11:59:59 PM, approximately, on December 31, 2000 has an important symbolic meaning. It marks the end of a minute, the end of an hour, the end of a day, the end of a year, the end of a decade, the end of a century, and the end of a millennium. The time and date provide a convenient yardstick against which we can evaluate the evolution of our thinking and the adequacy of our assumptions, mental models, paradigms, and policies. Will the beginning turn out to be appropriately different from the end? We hope that this book is helpful in such evaluation.

This is a *new-paradigm* book, which both presents and advances the new way of thinking about the systems of science, technology, society, economics, politics, and the environment, and actively calls for the replacement of the worn-out cognitive/sociotechnical paradigm. The book mirrors the systems theory discussed within it in the sense that it represents a fluctuation or nucleation that, under the far-from-equilibrium and unstable conditions of the present time, confronts the still dominant paradigm. We simply see no reason for reconfirming the *status quo* at a time when novelty and innovation are so sorely needed.

As the title suggests, the book deals with two components. The first component is the systems approach, which consists of systems thinking, systems theories, and systems models. The second component consists of the policymaking situation and the policymakers themselves. The interface between the two components is, however, neither conceptually trivial nor transparent. Myriad books and articles have dealt with areas like "models in the policy process" and "decision-support systems." In most of this large body of literature, it has been assumed that the policy or policy/decision component is more or less fixed. Thus, if the fit between the systems component and the policy component is poor, it is usually assumed that it is the former that needs redoing, perhaps because models convey too much jargon or are not tuned to the busy schedules of policymakers and decisionmakers. Systems theories and models can certainly be improved, but this book also places great emphasis on deficiencies in the structure of the policy situation and in policymakers themselves.

Accordingly, some readers may find some portions of this book controversial and the arguments forcefully presented. Nevertheless, from ideas in conflict, from "cognitive dissonance," can emerge fresh and creative new ideas. We believe that the book can play quite a useful role as a catalyst to individual, organizational, and social learning.

The authors provide the results of many, many man-years of working with complex systems in the contexts of systems analysis, systems design, and policymaking and decisionmaking. Some are the originators of great schools of systems thinking.

The book should be of interest to a wide variety of readers. With the collapse of Communism and the end of the East-West Cold War, and the associated greatly lessened emphasis on defense, many students, who otherwise would have gone into hardware engineering, will now want to devote their interests and talents to the increasingly challenging problems of society and the environment. With the number and variety of such problems becoming more evident day by day, in spite of the application of familiar policies, many faculty and students in the disciplines of mainstream behavioral and social science may want to try alternative interdisciplinary or transdisciplinary approaches. The reading audience will also include working policy analysts and advisors who may have experienced the exhaustion of traditional approaches, systems developers and designers who need a broader context for their work, and the systems community in general.

The book should find use as a core-curriculum specific or ancillary text in upper-division and graduate courses in colleges and universities, in the growing number of "summer universities," and in think tanks and policy institutes and programs.

Except for the first and introductory chapter, all chapters are arranged alphabetically by author. Instructors may wish to follow the order of chapters as presented. However, different flexible groupings can be made to fit the needs of the instructor and students. One such grouping follows:

I. Introduction and Systems Theory (Chapters 1, 2, and 4)
II. The Systems Movement: Its History and Critics (Chapter 9)
III. Policymaking: Its Nature and Improvement (Chapters 5 and 6)
IV. Systems Models and Designs for Policymaking (Chapters 3 and 7)
V. World Transformation and Systems Thinking for Policymaking (Chapter 8)
VI. Case Histories of Policymaking and Decisionmaking for Large-Scale Systems (Chapter 10)

Acknowledgments

I should like to acknowledge my gratitude to the contributors for the scholarship and insight of their chapters and for their patience in revising the original drafts. I owe a special intellectual debt to Harold Chestnut, Fred E, Emery, and Jay W Forrester, whose brilliant ideas over the years stimulated the growth of my own understanding of the dynamics of complex systems. In the same category is Ilya Prigogine, whose pioneering works are widely cited in the present volume. I owe an additional kind of debt to Harold Chestnut, who, through his Supplementary Ways for Increasing International Stability (SWIIS) activity, brought together people from all over the world to work on the war/peace problem. The main thrust of SWIIS over the past 12 years has been to utilize the expertise of systems engineering, systems science, and conflict resolution, both in the understanding of the underlying dynamics that can lead to war or peace, and in the development of practical means for reducing the likelihood of war.

I should like to thank my wife Maria and my son Ken for their valuable help in preparing the final manuscript, particularly the figures. I wish to thank the many anonymous reviewers of the draft materials for their helpful suggestions and for otherwise stimulating us to complete this work. Finally, I extend my thanks to the staff of the publisher.

KENYON BRENTON DE GREENE
Woodland Hills, California

A SYSTEMS-BASED APPROACH TO POLICYMAKING

CHAPTER ONE

POLICY CHALLENGES IN A WORLD OF NONLINEARITY AND STRUCTURAL CHANGE

by Kenyon B. De Greene

INTRODUCTION AND MAIN ASSUMPTIONS AND PROBLEM SETS

This book is about thinking, doing, and the consequences of thinking and doing at a time of *transformational* change in the nations, societies, and natural systems of the world. The analysis and nature of such intensive global change, and recommendations for policymakers in the face of change, are presented in considerable detail throughout the various chapters.

A "systems-based approach" refers to systems theories, philosophies, and models and to the concepts and constructs that are the building blocks of the former. Policymaking, briefly, refers to means leading to a governing course of action. There is, however, no simple, mechanical bridge nor simple rational transfer of knowledge between the two. There is rather a complex "policy reality," and we shall examine these concerns throughout the book.

Scientists and engineers have aided policymakers and decisionmakers since antiquity. Examples include the pyramids of Egypt and the other six Wonders of the Ancient World; bridges and aqueducts; and refinements in the accuracy of artillery. Systems-based support began during and just after World War II. For example, the new field of operational (operations) research was defined as "the scientific method applied to decisionmaking." However, operational researchers and their descendants, systems analysts, soon experienced some major difficulties with implementing this concept; and these difficulties have been magnified as analysts and their clients have dealt with more and more complex systems.

These major difficulties remain as major concerns in this book. First, human beings are not rational actors in the abstract sense of the term. Computer scientist and psychologist Herbert Simon defined the term "bounded rationality;" but Simon referred mostly to information-processing limits, and rationality is even more limited by the interplay of cognitive and emotional factors that characterizes all people. Second, the interrelationship between analyst and client was frequently strained and nonproductive because of differences in cognitive models, use of language, job situations, and outlooks. Third, the impressive developments in the systems engineering of weapons and space systems ("hard systems") were not readily transferable to organizational, social, and sociotechnical systems ("soft systems"). Finally, the very nature of the scientific method is being rethought. Epistemologically, some of the problems of systems study are approached from the perspective of "inquiring systems," such as Aristotelian, Hegelian, and Singerian. Very good reviews of this historical experience are provided by Checkland (1981), Linstone (1984), and Miser and Quade (1985). A review of the "systems quandary" is given by Linstone in Chapter 9 of this book.

Let us now state a number of assumptions about complex systems on which this book is based.

1. The earth with its inhabitants consists of a system of interlocking subsystems, and, continuing the hierarchy of organization, each subsystem consists of sub-subsystems, and so on.
2. The world system is a dynamic *field* (the *world system-field of forces*), and interactions at a more microlevel generate qualitatively new properties at a more macrolevel.
3. The world system-field consists of two major subsystems: the *world societal-field* and the *world ecosystem.*
4. A change anywhere in the world system-field may trigger explosive changes throughout the entire field; as a corollary: human decisions and actions may be surprisingly incapable of being restricted to local effects.
5. The world system-field must be viewed as evolutionary, and evolution does not proceed at a constant rate.
6. No system nor environment can be perpetually stabilized, and over time they drift or are pushed to the boundaries of stability.
7. Structural continuity is limited in time and space, and in the longer range the natural evolutionary tendency is toward structural change.
8. The world system-field is differentially sensitive in time and space to fluctuations and perturbations. As a corollary: because of time lags and amplification factors, human actions may show effects that are displaced in time and space and that are surprisingly much stronger than anticipated.
9. All people have mental or cognitive (or better cognitive/ emotional) models, but these models sometimes bear little resemblance to reality. No model is or can be purely "rational."
10. Systems, environments, and individual and collective cognitive models must *co*evolve.

These assumptions are pursued further in Chapters 2 and 7, and especially in Chapter 4.

Systems theories and models have not yet had the impact on societal policymaking and decisionmaking that many scholars had hoped for and expected, partially for the reasons mentioned above. The lack of fit between the systems approach and policymaking concerning these complex systems became evident in the early 1970s. Then, as now, scholars pointed out the essential interrelatedness of things and the dangers of policies based on simplistic notions of cause and effect. Critics of the 1970s-era models focused on apparent computer-programming errors and on the specificity of some predictions. Modelers, then as today, were often depicted as purveyors of gloom and doom. Reviewers in the popular media today are quick to point out how "wrong" and overly pessimistic were the earlier forecasts. Lost was the major message of the models: that new ways were needed to look at increasing complexity, and that seemingly straightforward decisions quite often led to the opposite results. Nevertheless, many of the earlier predictions have been borne out. Some of the original *Limits to Grow*th modelers (Meadows, Meadows, and Randers, 1992) provide a very interesting comparison of then (the year 1972) with later (the year 1992). In June 1992 at the time of the United Nations (UN) Conference on Environment and Development, held in Rio de Janeiro, these models and other environmentalist and sustainable-development approaches were condemned by pro-growth, pro-business, pro-industry, and pro-*status quo* interests from both the right and the left perspectives and from both the Northern developed and Southern underdeveloped worlds. In these and other conflict situations, each side can marshal impressive amounts of evidence. Which side is "right"? Which side is rational?

Meanwhile, according to many observers, world society and the natural environment have continued to evolve (or to devolve) and to show signs of incipient, probably irreversible structural change. It may be no overstatement to say that most subsystems in the world system-field are in a state of disequilibrium (implying possible controllability and reversibility) or nonequilibrium (implying very difficult control at best and imminent structural change). At the same time, many man-made and natural systems may both be becoming increasingly unstable and losing their resiliency.

This discussion is continued, with some differences in emphasis, in Chapters 2, 3, 4, 7, and 8.

The 1970s-era models dealt with problems like human population growth, growing enough food to feed the world, energy, and world pollution. Although there have been some major advances, these problems have not gone away; most simply, they were allowed to sleep during the later 1970s and 1980s. Now the world faces an intersecting set of newly emergent difficulties: debt at every level from individual to international, trade deficits, defense spending that no longer accounts for the major new threats to national and international security, crime, drug abuse, deterioration of health services, homelessness, deterioration of transportation systems and of infrastructure, urban and rural decay, deterioration of education, massive transfers of wealth from the Have Nots to the Haves, polarization of the rich and the poor at both class and national levels, disintegration of nations on at least five continents, and destruction of the natural environment of worldwide proportions. Many of these ubiquitous problems appear to be accelerating. The structure and process of national and international policymaking and decisionmaking, expressed as acts of omission and commission, may have exacerbated these difficulties and may actually encourage and subsidize further destruction.

NEW THINKING FOR NEW TIMES

Dealing with situations like these, as mentioned earlier, goes far beyond coming up with another systems model and then hoping that policymakers will use it. Here the important concept of paradigm shift enters in. History has shown that certain epochs are characterized by a dominant paradigm, the dimensions of which can variously be identified as religious, social, technological, economic, macropsychological, political, and so forth. The paradigm is not simply a set of rules and procedures or means of approach. The paradigm is a holistic thing that represents the *élan* or spirit of the times and the ways challenges and opportunities are perceived and met. The paradigm expresses the collective consciousness of the society or group in society (e.g., scientists), that during periods of stability or "normal times" is not easily amenable to "rational" appeals for change. Paradigm shifts occur during periods of societal flux and crisis when the knowns and knowables no longer fit together nicely. Unfortunately, during epochs of crisis, there is no time to develop policies and programs that may take decades to implement and to show positive results.

In the past, theories and models entered into the policy process, often with major impacts, because they *fitted into and reflected the dominant paradigm*. Consider, for example, the two or so decades immediately following World War II, the epoch that we shall associate later with the rising leg of the fourth Kondratiev cycle. The 1950s and 1960s were times of hope and promise. Technology could seemingly handle any difficulty that came along. The war had been fought and won by the Allies, and just about everywhere nations were miraculously arising from the ashes. Technology had sent men to the moon and had brought them back safely. Affluence was seen as growing and permanent, and the standard of living would continue to improve for individuals and generations. Great depressions were a thing of the past, and even business cycles could be eliminated by finetuning the economy. The world was a place of simple equilibria. Goods and bads and rights and wrongs could be clearly delineated. The world looked linear--responses were proportionate to stimuli and policies could match this linearity. Cost/effectiveness, benefit/cost, optimization, and econometric models seemed to be quite appropriate. In much of science, logical positivism--detached observation of existing facts--played a dominant role.

More appropriate for today, we argue, are systems theories, models, concepts, and constructs that deal with complex systems possessing many interacting subsystems, variables, and forces that in turn can be described as social, technological, economic, ecological, etc. These systems are strongly nonlinear and possess numerous positive- and negative-feedback loops. They may be surprisingly sensitive to internal fluctuations and to external fluctuations and perturbations. These systems usually have hidden critical thresholds that when passed yield results that are surprising in their nature and magnitude. The systems evolve or are driven into disequilibrium or nonequilibrium states, and attempts to restore them to equilibrium may produce results ranging from wild oscillations to self-organized new structures. These states and results may not necessarily be to most people's liking.

The applicable theories go by names like cybernetics, system dynamics, stability theory, evolutionary theory, thermodynamic theory, bifurcation theory, catastrophe theory, dissipative-structure theory, attractor theory, and resilience theory. The developing field of deterministic chaos (related to attractor and stability theory) may provide insights into both the limits to modeling and entirely new sources of information about system state. These theories may variously complement or oppose one another. Some theories may be more relevant at a given stage of societal evolution, and hence for policymaking, than are others. Most importantly, however, all these theories and models are general purpose. They apply to physical, biological, and social systems and can be applied to different hierarchical levels of organization and to specific kinds of problem subsets. Thus, the focus of this book is *transdisciplinary*, not just multidisciplinary or interdisciplinary. See Chapters 2 and 4 for more detail.

The Newtonian Paradigm

Let us now return to the examination in more detail of one major feature of the old paradigm or collective way of thinking, the so-called Newtonian Paradigm. The Newtonian Paradigm has received considerable reevaluation and criticism among members of the scientific community. See especially Jantsch (1980); Prigogine and Stengers (1984).

The name comes of course from Isaac Newton (1642-1727), one of the greatest scientists of all time. The importance of the paradigm far transcends Newton's discoveries in mechanics, and even the limitations within physics, to include the impact not only in science as a whole but throughout society itself. The Newtonian paradigm has shaped to a great extent the way most people in Western and Westernized societies think about Nature and society, the way measurements are made and data collected, and the ways policies are formulated.

Newton's laws of motion, apparently universal laws of dynamics, deal with a single force, gravity, which can explain movements of both the stars and of free-falling bodies. Newtonian science shows the universe to be lawful, possessed of natural order, deterministic, and dominated by equilibrium. It is a mechanical world ruled by deductive logic and mathematics: the highly rationalistic and mechanistic world view. It is a reversible world in which time equals minus-time, and velocity equals minus-velocity. Initial conditions determine a trajectory, and a change can return the system to these initial conditions. Once the instantaneous state of the system is determined, deductive logic can yield predictions of past and future trajectories. What has been changed can be exactly undone by another change. With exact prediction, the behavior of Nature can be exactly controlled.

Although there were always critics, Newtonian science was first seriously threatened by developments in thermodynamics in the 19th Century; for example, the work of Sadi Carnot, Ludwig Boltzmann, and others who identified both probabilistic effects and irreversible processes. At the present time in history of science, the laws of quantum mechanics have replaced those of classical mechanics at the microscopic level, and relativistic physics has replaced Newtonian physics at the level of the universe. Nevertheless, the concept of deterministic, static, reversible trajectories remains the core of physics (Prigogine and Stengers, 1984).

A derived concept is logical positivism which emphasizes the methods and procedures of science and, in some versions, the separation of theory and fact. As interpreted by Auguste Compte (1798-1857), positivism rejects speculation about or search for ultimate origins. In experimental and empirical study, strict objectivity denies any reference to the observer. The observation and measurement of the system is not only dispassionate, but the processes of observation and measurement do not tamper with the system. The Heisenberg indeterminacy principle and the role of the observer in relativity dealt blows to positivism in its most extreme form.

Unfortunately, the behavioral and social sciences, intrinsically most related to systems management functions like policymaking, appear to have been influenced by the Newtonian paradigm in the most lasting way. The waves of change represented by thermodynamics, statistical mechanics, and relativity in physics had little impact on these sciences. For example, the 19th-Century physicist/physiologist fathers of experimental psychology in Germany, Hermann von Helmholtz, Gustav Fechner, Ernst Weber, and Wilhelm Wundt continued the Newtonian tradition and laid a firm basis in positivism. The Americans John B. Watson (behaviorism) and Frederick W. Taylor (scientific management) continued in this tradition. In economics Adam Smith, David Ricardo, Léon Walras, Vilfredo Pareto, and others set the stage for the equilibrium-centered thinking that dominates the discipline to this day. Chapter 2 extends this discussion.

The New Paradigm

A number of scholars, however, are now recognizing the limits to Newtonian rationality. Interestingly, the development of a new paradigm stems in part from reactions against high technology: computers, information systems, artificial intelligence (AI), and expert systems. For example, rules and procedures can rigidify a system, and the design of files and displays can channel thinking into the most obvious paths and can inhibit creativity.

Denning (1989, 1990) and also Dreyfus and Dreyfus (1986) note that, in spite of the increase in the number of applications (e.g., to robotics, chemical-structure determination, and speech recognition), little theoretical progress has been made in AI. Lack of progress can be traced to models that were established in psychology and neurology, as noted above, in the early years of this century or in the 19th Century. Denning states that the failure of AI to produce machines with truly human cognitive capabilities after 40 years of research is not due to lack of attention. It is a failure of the philosophy of rationalism which is deeply rooted in Western thinking. Similar failures have occurred throughout the behavioral and social sciences. In each case there has been a search for context-free, meaningless elements or facts, and the attempt to combine these elements into systems governed by formal laws. Dreyfus and Dreyfus state that over 25 years of AI research has lived up to very few of its promises, and that there is no evidence that AI will ever be capable of doing so. Human expertise and holistic knowledge are basically irreducible to rules. To regard people as only information processors and abstract calculating machines is to provide a limited and incorrect model of humanity. Searle (1990) comments that AI has been strongly influenced by behavioristic theory from psychology. AI is seen as a last gasp of an outmoded tradition, especially in terms of viewing the mind as a computer program independent of the biological reality of the brain.

In short: theories, models, philosophies, and methods stemming from an earlier era of scientific thought and developed for simpler, mostly physical systems are largely inapplicable for a mind, a society, an economy, or an ecosystem. Unfortunately, by far the usual practice is for people to apply the simplest possible interpretations to complex situations. Table 1.1 contrasts the old and the new paradigm. See also De Greene (1990e, 1991a, 1991b); Kuhn (1970); Simonton (1988); and Chapters 2, 8, and 9.

WHAT IS POLICYMAKING?

In the most general sense, policymaking consists of a general set of concepts, assumptions, frames-of-reference, mental models, and activities directed toward some perceived problem area.

Definitions of policymaking, and contrasts of policymaking with related activities like strategic planning, are provided in detail in Chapters 5 and 6. Policymakers usually have groups of policy *advisors* who possess expertise in important specialty areas. For example, environmental policymaking can be supported by specialists in atmospheric physics or chemistry, and defense policymakers are supported by experts on specific weapons systems. Unfortunately, specialized disciplinary knowledge may be too isolated or fragmented to be integrated into a holistic picture of the policy situation. Policy *analysts* may sometimes have a better understanding of interrelationships and impacts than do the advisors. *Decisionmakers* operate within a policy context in relatively structured situations. They make yes/no choices or choices among a fairly limited and well defined set of alternatives.

Policymaking can also be *shaped* indirectly. Here public perceptions, values, attitudes, and behaviors can play an important role. Changing values about wilderness, old-growth forests, protection of species, and the harmful effects of many practices and products are changing both national and international government and business policies toward the natural environment.

Finally, as noted above, the evolutionary *context* for policymaking must be recognized. Policies must coevolve with the given system and external environment. A policy that fits one time and place and set of interacting forces may not be relevant to another time and place with its own set of interacting forces. The momentum, inertia, and time lags that characterize a system very often lead to the implementation of obsolete policies in radically changed settings.

Table 1.1. The Old and the New Paradigm

Old Paradigm	New Paradigm
Rationality	Nonrationality
Linearity	Nonlinearity
Simple causality	Mutual causality
Equilibrium	Nonequilibrium
Reversibility	Irreversibility
Determinism	Stochasticity/determinism
Optimization	Satisfaction
Certainty	Uncertainty
Noise and fluctuations suppressed	Opportunity seen in noise and fluctuations
Averages always dominate	Exceptions dominate near critical thresholds
Asymptotic or Lyapounov stability	Structural stability
Structural constancy	Evolution/structural change
Analysis/reduction	Holism/synthesis
Convergent thinking	Divergent thinking
Legal and financial emphasis	People and technology emphasis
Assumed predictability	Predictability severely limited by instability, structural change, and chaos

Policy Reality

This area for concern was summarized earlier and is discussed more fully in Chapter 5. For our purposes here there are two aspects. First, is the well known lack of fit between theorists/modelers and policymakers. For example, policymakers usually live in a world of time pressures, power, influence, conflict, bluff, bargaining, and concessions. It may be that there is very little that can be done to change policy in the short term. The policymaker has invested his/her time and effort to obtain a job that meets his/her deepest underlying psychological needs. The policymaker may view new ideas, including systems theories and models, as a threat to the stability of a position that represents years of investment and commitment. Yet, on occasion, systems support may be coopted as just one more weapon against the adversary. A policy almost always follows the values of those responsible for its conceptualization and implementation.

It is worth noting that the words "policy" and "police" both stem from the same Greek word, *politeia.*

The second aspect deals with the question as to whether a policy is "good" or "bad." There is no easy way to answer this question for reasons that are widely discussed throughout this book. Suffice it to say that all policies are value-oriented, that values change over time, and that knowledge is never perfect at any given time. Policies are often made in the face of ignorance about the real dynamics of the system, and this ignorance may stem from the fact that the requisite knowledge lies (far) in the future.

In Brief Summary: Some Characteristics of Policymakers and Policy Situations. Following are some generalized points that may or may not apply in varying degrees to any given policymaker or policy situation:

- Policymakers may be parochial in background and experience
- Policymakers may lack mental models that are appropriate to the times and to the complexity of things
- Policymakers tend to discount, placing much lower values on things that are distant in time and space
- Policymakers must usually oversimplify greatly in order to stay in power
- Policymakers have traditionally had difficulty communicating with experts, whose expertise indeed may be narrow, specialized, and highly disciplinary
- Policymakers usually emphasize the short term, with attendant searches for quick fixes
- Policymakers have very strong psychological needs for power and tend to enact policies and practices that ensure their retention of power
- Policymakers usually have backgrounds in the military, in the law, or in the civil service, areas that seek confirmation in rules and procedures rather than seeking novelty and innovation
- Policymakers tend to seek economic and technological solutions to societal and environmental problems
- Policymakers do not usually consider the side effects or externalities of their efforts
- The many levels of policymaking and the many vested interests and pressure groups render making integrated or unified policy most difficult
- Policymaking is strongly adversarial and generative of conflict
- The very object of policymaking may be diffuse, obscure, and hard to define

COGNITIVE/EMOTIONAL MODELS OF POLICYMAKERS AND DECISIONMAKERS

It is sometimes said that men are rational and women emotional, and that business is rational whereas concern for the natural environment is emotional. Most readers today would probably reject such oversimplified, generalized, and erroneous statements. As discussed earlier, except as an abstraction, there can be no purely rational person, model, policy, or decision. The detached, objective person who carefully evaluates the pros and cons of an issue, assesses and weighs the risks, and makes a rational choice among alternatives is a myth.

Figure 1.1 expands the familiar information-decision-action-feedback model. Note the interactions among the Feelings and Emotions, Perception, and Decision blocks. Note also that uncertainty characterizes not only information input as is recognized by the decision-analysis and decision-support schools. Even more important is consequences-of-action uncertainty, an area of study that has been poorly defined. For example, will an action lead to crossing a critical threshold, precipitating undesirable structural change?

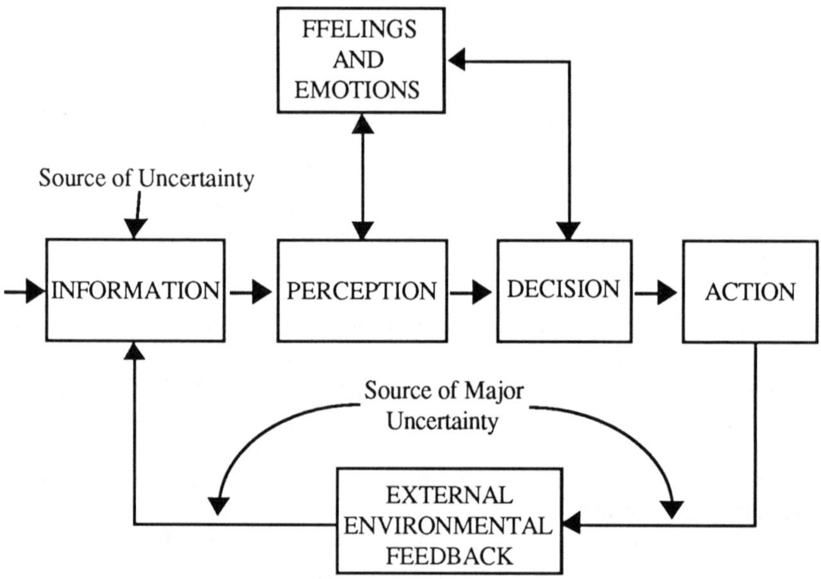

Figure 1.1. Basic information-decision-action-feedback model modified to show the importance of the emotions and of unexpected sources of uncertainty.

The cognitive subsystem of a person is not independent of the emotional subsystem. At the level of physiology, the nervous, endocrine, immune, cardiovascular, and respiratory sub-subsystems mutually support and reinforce one another. At the level of individual psychology, cognition and emotion are inextricable aspects of personality. Both stem from the interaction of genetic, congenital, and postpartum forces. The personality of an adult is an outgrowth of the kinds of (including the lack of) interactions with parents, siblings, members of the extended family, schoolmates and playmates, nation(s), neighborhood(s), social- and natural-environment(s), races, cultures and religions, status or class, values, wealth, epoch of upbringing, sheer luck, and so on. Each person possesses various needs, and the dynamics of personality are shaped by the gratification or frustration of these needs and by conflict among needs. Personality has both conscious and unconscious (below the level of awareness) aspects. In the language of systems, the development of personality reflects many past bifurcation points.

Consider now some studies that illustrate these points. Kull (1988) points out that strategic policymakers tend to be quite resistant to changes in their belief structures, and that their seemingly rational thinking may mask deep, subconscious emotional conflicts. Kull interviewed high-level US and Soviet defense policymakers and identified a number of inconsistencies in these managers' responses and an overall attempt to maintain consistency. There was a strong tendency to ignore the realities of nuclear weaponry and therefore the vulnerability of the two nations. The policymakers transferred prenuclear military logic to the different nuclear situations. When Kull noted the inconsistencies, the policymakers' response was an elaborate and highly convoluted hierarchy of rationalizations. Many of these rationalizations camouflaged deep-seated conflicts and emotional needs for power, prestige, status, and the alleviation of fear and anxiety. The emotional block to recognizing the reality of the nuclear situation arose because recognition was perceived as interfering with the gratification of significant personal needs.

Tolcott, Marvin, and Lehner (1989) also observed a need for consistency in the behavior of defense-system personnel. Military analysts tend to make consistent interpretations and explanations in military and intelligence-analysis situations. Analysts developed a cognitive model (a single basic truth) from the information first available, then account for update information in terms of this mental model. They place very high confidence in an initial hypothesis, regardless of the nature of the hypothesis. They emphasize familiar kinds of information disproportionately in decisionmaking. For the most part it seems that analysts believe that a consistent explanation is probably the correct explanation. Such decisionmaking behavior can be inappropriate and dangerous in evolving situations.

Do Policymakers and Decisionmakers Necessarily Think Systemically?

Forrester (1990); Greenberger, Crenson, and Crissey (1976); and Lee, Ball, and Tabors (1990), as well as authors cited earlier, have discussed the lack of *rapprochement* between systems managers on the one hand and modelers and technical experts on the other. See also Chapters 7 and 9. We emphasize here that we should not regard the policymaker or decisionmaker as *fixed* and the correct basis for judging utility. The following discussion focuses on apparent systemic limitations in the mental models of many societal managers (De Greene, 1987).

Linearity/Monotonicity/Reversibility. When a linear function is plotted on a graph, it appears as a straight line; and when a monotonic function like an exponential function is plotted, it does not go up and down. Linear thinkers believe that a policy action produces a proportionate effect. If a small, perhaps incremental action produces a desired result, then more of the same action must yield even better results. Action and results are closely and clearly related, and other factors possibly contributing to the situation are not considered. An action taken in the present works the way it did in the past and the way it will in the future. Usually more of a given policy is better because it produces the more or less of a behavior that is desired in the results. Even if the results turn out to be undesirable, the results can be reversed or undone by applying a counteractive policy. Linear thinkers tend not to understand diminishing returns, saturation effects, and the need for substitution.

Direction of Causality. Nonsystemic thinkers believe that cause and effect are directly, closely, and linearly related. Cause yields effect unidirectionally, and effect does not modify cause. Cause-effects can be isolated from context, and a given cause-effect can be handled apart from other cause-effects. The linear thinker does not easily comprehend mutual causality, a situation of realworld complex systems characterized by numerous positive- and negative-feedback loops that shift in dominance and control.

Determinism. People who do not think systemically believe that when a policy action is taken, the results are inevitable. Contrariwise, under other conditions, if a policy action is not taken, then the results are inevitable. Chance and uncertainty are thought to be ruled out by careful planning. The real world, however, is chancy and uncertain. Indeed, stochasticity and determinism (chance and necessity, novelty and confirmation) underlie systems evolution (e.g., Jantsch, 1980; Nicolis and Prigogine, 1977, 1989).

Equilibrium. Nature and society are often viewed in terms of a simple balance between forces. A system can be knocked out of balance, but sooner or later balance will return or can easily be restored. There is usually an ideal balance for a given situation, and a system in disequilibrium should return to this level. Even if equilibrium levels are perceived to move upward or downward, these changes are seen in terms of numbers, not in terms of system states or control mechanisms. The world provides limitless resources and opportunities, the exploitation of which will always restore balance. However, living systems often display two or more equilibrium configurations (e.g., Rosser, 1991; Thom, 1975); Chapters 2 and 4 of this book. Human activities can cause a system to flip to a configuration wherein resources suddenly and surprisingly become scarce or disappear. Or a system may evolve or be pushed to a state far from equilibrium in which it is hypersensitive to fluctuations and perturbations and to unexpected structural change. See Jantsch (1980); Nicolis and Prigogine (1977, 1989); Chapters 2 and 4 of this book.

Complexity. Systems managers may not believe that everything is connected to everything else; complexity is in the eyes of the beholder. The most significant factors can be easily identified and other factors minimized or ignored. The relative importance of factors remains constant over time. The really important factors can almost always be measured. Even if complexity has to be acknowledged in some situations, careful selection and control of variables, aggregation, and averaging make this complexity manageable. Parts of the whole can be distributed to specialists, and the specialists' findings can be put back together into an intelligible whole. In realworld systems, however, complexity has several dimensions: the number of interacting elements, the number of different interactions, and the strength of the interactions. More profoundly, complexity is a function of the number of hierarchical levels of organization of matter/energy and information with their emergent new properties. A complex system is capable of more varied kinds of behavior than is a simple system; but a complex system is also more fragile, especially in the face of environmental change. Complex systems are best suited to predictable environments. There appear to be limits to the complexity of ecosystems and societal systems, and beyond some critical threshold more complexity means more instability. See May (1974); Nicolis and Prigogine (1977, 1989); Tainter (1988).

Structural Change. Nonsystemic thinkers believe that things do not change very much, but if change does occur it is almost always gradual, continuous or continual, quantitative, predictable, and reversible. Qualitative change in form, pattern, or structure is rarely perceived as an important consideration. Structure, function, and behavior are seen to have a continuity that permits what is happening today to be projected meaningfully into the past and future. In actuality, systems evolution is punctuated by frequent discontinuities. We acknowledge here the great value of tentative or *working* projections of things like population growth and changes in greenhouse gases--within a holistic context. Nevertheless, this is a tricky area with which to deal and an area to which we shall frequently return.

Is Systems Thinking Easily Developed?

The answer to the above question appears to be *no*. Kitchener and also Salner (cited in De Greene, 1987) provide more answers. Kitchener distinguished among three levels of cognitive processing; (1) basic information processing exemplified by perceiving and memorizing; (2) metacognition which involves monitoring, evaluating, and correcting basic information processing; and (3) epistemic cognition, which functions in the evaluation of thought processes themselves, the determination of how solutions come to be true, the evaluation of whether reasoning leads to a correct solution, and the generation of knowledge about the limitations of one's own cognitive abilities and strategies. Cognitive theory and research have dwelt almost exclusively on the first two levels.

Salner tried to teach systems theory and methodology to graduate students. She found that some students, otherwise mature and intelligent, could not grasp and apply systems concepts. Theory could not be understood until students had reached a high level of development. This level requires an awareness of context in the recognition of truth and value. Importantly, this level is emergent and is not reducible to the lower levels of cognition, nor to recognizable particulars of psychosocial and personality development. The high level of cognitive awareness evolves as a time-based and nonlinear structural reorganization toward greater complexity. Unfortunately, there are no quick fixes for separating epistemic cognitive development from the overall development of the student. Systems thinking, *when it does develop*, emerges from a long period of personal experiences and interactions and from the self-testing of one's own mental models.

Chapters 4, 5, 7, and 8 discuss other features of cognitive models that we propose to be of value in aiding the development of systems thinking.

A SIMPLE MATHEMATICAL MODEL OF NONLINEARITY

The *logistic function* is one of the "workhorses" of systems science and of technological forecasting. Its differential equation, most simply stated (e.g., Marchetti, 1991, p. 414) is:

$$dN/dt = a \cdot N \cdot (K-N),$$

where N represents the number of individuals or items or a population; t represents time; a is a constant of proportionality; and K is a limit, a carrying capacity, or a saturation level. Figure 1.2 shows the simple logistic function. We shall return to other versions later.

Figure 1.2 provides a great deal of information. First, as plotted, the curve is clearly nonlinear; it is not a straight line. Further, when the N's are squared, the result is a quadratic term, another indication of nonlinearity. Second, the curve reflects the underlying growth dynamics. The leg up to the inflection point represents exponential growth (an even more elementary mathematical "workhorse"), generated by positive-feedback mechanisms including self-amplification and autocatalysis that are ingredients of much structural change. Past the inflection point the rate of change begins to level off and to approach an asymptotic limit, a reflection of negative feedback and the regulation and control of growth. Third, the section of the curve around the inflection point *(K/2) looks linear*. If the plot is that of a stage in the evolution of a society or an industry, policymakers may get a very misleading impression of constancy and stability.

Fourth, an examination of the curve near K may indicate the exhaustion of ideas (De Greene, in press) as well as the saturation of a product or industry in the market. This may suggest an opportunity for the implementation of new ideas, the development of a new-generation product, or the emergence of a new competitor "out of nowhere."

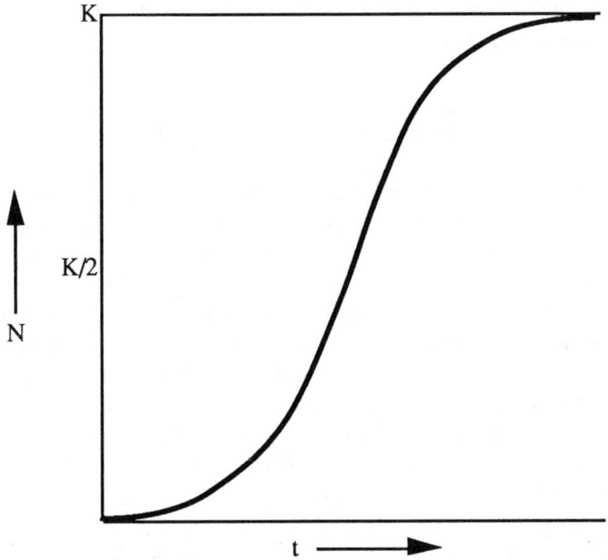

Figure 1.2. The logistic function graph

Further information can be extracted if the function is integrated and parameter estimates are made. In other cases the basic function is subjected to a logarithmic fit that produces a straight line. Technological forecasters (e.g., Marchetti, 1986, 1988, 1991; Nakićenović, 1988) often plot the ratio of N/K as F, the fraction of saturation. Δt then represents the time to go from 10 percent saturation to 90 percent saturation, a value useful in forecasting the time of future saturation of, say, energy use or use of a given mode of transportation (e.g., Vasko, Ayres, and Fontvieille, 1990; and the just-cited authors).

Fifth, the limit or carrying capacity K may or may not be stable. If stable, a system may become locked into a period of prolonged stagnation, or even be forced into a dive of *diminishing returns*. If unstable, the system may shoot through the ceiling or limit, destroying the carrying capacity or initiating a new stage of growth. This behavior is characteristic of evolutionary systems. The result may be families of often overlapping logistic functions or of logistic functions followed by diminishing returns. See Figure 1.4.

This simple model and its interpretations will be expanded later in this chapter. A third mathematical "workhorse" is the Lotka-Volterra predator-prey model discussed briefly in Chapter 8; see also Marchetti (1991, p. 413). Our interpretations emphasize the *qualitative* nature of mathematical models, a point emphasized especially by the catastrophe theorists (e.g., Thom. 1975). Contrariwise, there are situations in which everything that can be is calculated, but little or nothing is known about the overall situation.

STRUCTURAL CHANGE

Structural change is an inherent feature of evolving systems. It is driven by instability. It does not occur evenly across time. Overall, it has evolved in the direction of greater complexity, and the structural change of complex systems is irreversible but destructible.

The present is a time of accelerated evolution, instability, and structural change. Affected at the national and international levels are the economy, business and industry, technology, energy use, political and military systems, agriculture, ecosystems, and the ways people think and behave; and also the oceans and global climate and weather. The planet and its inhabitants are tightly coupled and the overall system is at best metastable, allowing a change anywhere to trigger larger changes that sweep irresistibly across the entire *field.*

In the United States, policies in the 1980s exacerbated latent difficulties leading to increasing unemployment and underemployment, quadrupling of the national debt and the associated persistent internal deficit, imbalances in trade, decline of the dollar, high debt obligations to foreign lenders, and the collapse of whole industries like savings and loans. Manufacturing capabilities and job skills, even high-tech, were lost through improper automation and mechanization, mass firings, and the transfer of plants and jobs to cheap-labor countries. The US changed from being a major creditor nation to becoming the world's largest debtor. In many ways the US has retrogressed to the status of a Third World nation as exports have shifted from manufactured goods to a proportionately greater emphasis on commodities like agricultural and forest products; and as greater numbers of Americans face a decreased standard of living. These may be features of long-lasting economic structural change.

Changes in the former Soviet Union and in Eastern Europe have been even more dramatic. The collapse of Communism has revealed almost incredible civil and industrial incompetence and widespread destruction of the natural environment. But make no mistake! We have hère no simple matter of the evil and inefficiency of Communism and the goodness and efficiency of Western Captalistic Democracy (cf. De Greene, 1990d). For better or worse, 40 to 70 years of Communism provided considerable stability, albeit in many places and especially more recently the stability of stagnation and stultification. Experience is showing that there will be no simple and rapid change to freedom and a market economy. It is estimated that, even if that restructuring were to be feasible, it would still take 15 to 20 years. In the meantime, the republics of the former USSR have drifted farther apart, with actual warfare arising between some (e.g., Armenia and Azerbaijan). Yugoslavia has broken apart, with vicious and destructive war between Serbia and the former Central Government on the one hand and the seceded republics like Croatia or Bosnia-Herzegovina on the other. There is danger of regional war.

In most of the Eastern European countries, unemployment is increasing and will increase much more severely. Whole structures of society are falling apart. It has been estimated that it will take a trillion US dollars and 15-20 years to bring the former East Germany up to West German standards; this commitment has converted Germany internally into a debtor nation. In all these countries, whole industries, communities, and regions will have to be rebuilt if achieving a Western standard of living and a healthful natural environments are the goals. How to accomplish this mammoth task, considering side effects like unemployment and the availability of the necessary human and monetary resources, is not at all immediately evident. Nor is it evident that the Western model *should* be closely emulated at this critical, historical bifurcation point when a perhaps one-time-only choice of future development paths is possible. Choice might best follow the analysis made by Harman in Chapter 8.

The Islamic world is also beset with instability and turbulence. Western policymaking based on, say, "peace and balance of power in the Middle East" is at best an oversimplification. Muslims remember unfavorably the intrusions of Western colonial powers, especially in North Africa, the Near East, and the Middle East. To many Muslims the formation and existence of Israel is regarded as just one of the latest and most humiliating encounters with foreigners. Modernization and technology have produced shock and have not been assimilated well in many places. Added to these unifying factors have been splitting factors such as differences between Shia and Sunni Islam, differences in forms of government, ethnic differences, and historic rivalries. The Islamic world was an unstable societal-field, sensitive to fluctuations and perturbations, even before the 1990-1991 Persian Gulf Crisis and War. Richardson in Chapter 10 provides a much fuller analysis and discussion of the events of and factors contributing to that situation.

The natural environment at local, regional, and global levels also illustrates instability and structural change. The biosphere, atmosphere, and hydrosphere are inextricably interlinked under healthy conditions. A large forested area helps generate its own climate and weather. Deforestation has serious consequences: water that normally sinks into the ground or is transpired by plants into the atmosphere decreases and run-off water increases; soil erosion increases and lakes and reservoirs are silted; desertification and lateritization (conversion of soil to "brick") may increase, and streams may dry up; the remaining top soil blows away; the carbon cycle is disrupted, the amount of carbon dioxide in the atmosphere and greenhouse warming appear to increase, and carbon may become less available as it is precipitated out as limestone in bodies of water; and the delicate ecological web is broken as species are threatened and become extinct.

Ecology and economics have also developed as a mutually causal web. For example, deforestation has been intensified by economic factors. In the western US, Alaska, and western Canada, trees are clear cut and timber exported, mainly to Japan, in attempts to equalize the balance of payments. The hasty acquisitions of the 1980s have also been destructive to the ecosystem as companies accelerate the clear cutting of old-growth forests in order to pay back loans. Because the overall system has been pushed to the limits, pressures mount to make hasty, lose/lose choices as between jobs and old-growth forests.

In the underdeveloped world, loans by commercial banks in the developed world, by regional banks, and by the World Bank, and Capitalism-based austerity pressures by the International Monetary Fund, have accelerated forest destruction and local destabilization. Many of the loans have been for huge projects like transcontinental highways and hydroelectric dams. Critics maintain that these projects have harmed not only forests and wildlife, but also tribal peoples and the rural poor, benefitting mostly rich and powerful elites, who may not even reside in the areas impacted. It is these kinds of situations that can contribute to the often-Marxist insurrections, drug trade, and poverty that now seem endemic. In addition, Third World debt cripples improvements in most of the countries involved and is a serious threat to the stability of the world economy as well as to world peace.

As a further example, the stratospheric ozone layer has turned out to be surprisingly sensitive to the perturbations by chlorofluorocarbons (CFCs). The structural change is the thinning developing into a hole in the layer, first observed over Antarctica in the spring but now reported over the Arctic and mid-latitudes as well. Ozone absorbs most of the ultraviolet (UV) rays from the sun, so that the biological consequences of ozone destruction can be quite severe. The chemical and physical reactions are well known, and the problem in simplified form is discussed in De Greene (1990a). It is estimated that one chlorine atom can destroy up to 100,000 ozone molecules--a most striking example of the nonlinear impact of human activities. In addition, CFCs are also greenhouse gases which could contribute to the warming of the atmosphere.

The CFC problem also illustrates the tradeoffs that are eventually forced onto policymakers when necessary knowledge is absent or ignored. CFCs are important refrigerants, aerosol propellants, cleansers for electronic components, and insulating foams. They are inert in the troposphere and have been considered to be ideal chemicals. But in the stratosphere, where they may remain for decades, they break down under the influence of UV radiation, releasing chlorine which is recycled. Ameliorative policies have been enacted on two occasions. In 1978 the US and several other countries banned the use of CFCs in aerosol propellants. After it had been determined that the problem was much more serious than had originally been recognized, on September 16, 1987 in Montreal, most of the industrialized nations agreed to this policy: freeze the consumption of CFCs at 1986 levels beginning in 1990, and cut back emissions by 50% annually by the end of the century. The most recent evidence suggests an even greater urgency to phasing out the CFCs. Replacement chemicals, themselves apparently greenhouse gases, are under development. However, the policy problem is even more complicated as the industrial North and developing South confront one another as to who gives up what. Many in the South see their future following the path previously set by the North.

Finally, it is clear that policymakers need help in developing a *systemic basis of policymaking*, as opposed to separate economic policies, energy policies, farm policies, military policies, transportation policies, industrial policies, and environmental policies. Next, let us examine the evolutionary epochs and structures into which, systems-wise, effective policies must fit.

THE KONDRATIEV AND HEGEMONIC CYCLES/STRUCTURES AND THE LIFE CYCLE OF THE PRESENT STAGE OF WORLD CIVILIZATION

Cyclicity is ubiquitous in Nature. Consider the birth and death of stars, oscillations of Earth's orbit, glacial epochs, the seasons, animal estrus and menstrual cycles, circadian rhythms, human moods, population rises and crashes in ecosystems, and many more. It is logical, therefore, to expect cyclic behavior in the large system of human society. Consider first "the economy."

Let us start out with the several "business cycles" that have been identified. These are the Kitchin or business cycle, *per se,* of length three-to-four or four-to-seven years. This is the cycle that most economists refer to when they speak of recessions and recoveries. It can be interpreted as reflecting fluctuations in employment and commodity inventories. Many observers recognize the Juglar cycle of 8-10 years, interpreted as representing fixed capital in the durable equipment of production; and Kuznets cycle of 15-25 years' duration, reflecting fixed capital in real estate and structures and the mix of capital and labor. Next comes the Kondratiev (formerly usually Kondratieff) cycle or economic long wave of about 45-60 years. The hegemonic or super-long wave lasts 150 years or more. Finally, we note here the life history of a civilization. This section is devoted to the last three *evolutionary structures.*

Cyclicity does not imply periodicity. Each of the cycles is a more surficial expression of deeper, evolutionary, and structural changes. A modern economy may be viewed as three-tiered. At the top is the behavior recognized by most business people and economists. This is the world of prices, wages, interest rates, money supply, inflation, and hiring and firing, the fluctuations of which describe the business cycle *per se.* The middle tier deals with the means of production and consumption: with factories, equipment, and tools, and of model changes within the prevailing paradigm. The Juglar and Kuznets cycles reflect this intermediate degree of depth. At this level, experts try to understand the meaning of productivity, foreign trade, and competition. This is no longer the world, mostly of the economist. Many other "disciplines" contribute their expertise. For example, industrial engineers help design and layout factories; behavioral and social scientists design work systems and jobs that improve motivation, morale, job satisfaction, and productivity; and human factors specialists and ergonomists improve the fit between users and equipment. The third and lowest tier involves the basic capital, technology, and energy structures of society. It is the world of the exhaustion of capital and of past investments; the wearing out of one technology and replacement by a newer, more efficient technology; and the substitution of one energy resource by another. Indeed, the way of thinking of the society and the prevailing sociotechnical paradigm can themselves become exhausted (De Greene, in press). This is the world of the Kondratiev cycle or economic long wave.

To emphasize the importance of evolution and structural change, I prefer the terms Kondratiev cycle/structure or Kondratiev phenomenon.

The Kondratiev Cycle/Structure

Most of the original impetus for the study of long waves came from the analysis of time-series economic data. In the first quarter of this century, the Dutch workers van Gelderen and de Wolff identified such waves. In 1922 the Russian economist Nikolai Dimitrievich Kondratiev independently made the same discovery. After the revolution, Kondratiev was assigned to study the mechanisms whereby Capitalism contains the seeds for its own destruction. He identified two and one-half long waves; the first wave (Kondratiev Number One) started with the Industrial-Production Revolution. Kondratiev used both monetary and physical time-series data from England, France, Germany, and the United States. Most importantly, Kondratiev went far beyond recording fluctuations in the data. He described social, technological, institutional, trade, military, and other changes that accompany the rise and fall of the cycle. He also made *systemic* interpretations of three levels of equilibrium behavior, interpretations that more or less match the tiers described above. Kondratiev proposed that the major driving force in lowest-tier, third-level equilibrium is the generation, expansion, and replacement of the capital sector, and the radical regrouping of, and changes in, society's productive forces. Unfortunately for Kondratiev, his studies showed that Capitalism also contains the seeds for its own recovery. This was not a popular viewpoint during the Stalin dictatorship, and Kondratiev was discredited and reportedly died in a Siberian *gulag* in the 1930s.

I believe that the Kondratiev phenomenon provides a link between the study of science and the study of history. I consider it to be a basic organizing principle for long-term sociotechnical evolution. There is controversy, however, regarding the very existence of Kondratiev cycles. Mainstream economists almost wholly reject their existence. And mainstream economists have a great deal of power in today's world. Many organizations and many countries have boards or councils of economic advisors. These economists base their interpretations, forecasts, and policy advice on assumptions and methodologies of structural constancy, equilibrium, linearity, and nature of feedback that we should consider *nonsystemic*. Most of their advice centers on the top tier and shorter range discussed above. Policy, of course, can not be separated from the needs and interests of policymakers and policy advisors. Interestingly, the Organization for Economic Cooperation and Development (OECD) in 1981 indicated the need for a structural (long-wave) response to mounting problems of unemployment stemming from the shift from old to new industries. The US and UK delegations made strenuous efforts to make certain that reports remained merely the opinions of a small group of experts, not an official OECD statement with higher political status (citation in De Greene, 1988a).

Similarly, with the collapse of Communism in the USSR and Eastern Europe, it is now acknowledged that the Kondratiev phenomenon is not limited to Capitalist economies. N. D. Kondratiev has been rehabilitated in the USSR/Russia. The year 1992, the 100th anniversary of his birth, marks a major conference on his ideas in Moscow and St. Petersburg. Overall, the Kondratiev phenomenon is a structural/functional/behavioral manifestation of the evolution of modern industrial societies, a manifestation that probably intensifies patterns of boom and bust that were evident in pre-industrial times. Nakićenović (1988), for example, identifies price swings in Britain in the 1580s.

Kondratiev's work was extended in the 1930s by Joseph Schumpeter (1939). Schumpeter emphasized the role of technological innovations and particularly the bunching of innovations during the phase of depression. He also emphasized the process of "creative destruction," whereby an ensemble of technologies both creates new opportunities for economic growth and paves the way for the slowdown of growth and replacement by newer technologies. Following World War II, during the phases of recovery and prosperity, there was not much interest in economic long waves. Indeed, Keynesian economists believed that even the shortest-term business cycles could be eliminated through fine-tuning of the economy. The period 1967-1973 showed a surprising slowdown of growth worldwide. The Organization of Petroleum Exporting Countries (OPEC) oil embargo and sharply rising oil prices also indicated that things had changed. This period marks the peak of Kondratiev Number Four, and the beginning of the downturn into the phases of recession (*not* the business-cycle use of the term) and depression. The problems that became increasingly evident in the 1970s were not simply reversible and did not go away. Rather, according to many thoughtful observers, these problems intensified. Concern about the nature of the change led to a great renewal in interest in Kondratiev cycles.

A good deal of bibliographic material is now available in the form of compendia of individual interpretations, theoretical integrations, overviews and summaries, critical reviews, and forecasts. This literature shows an evolution from mostly an economic emphasis to a more eclectic recognition (much in the spirit of Kondratiev himself) of the interrelationships among economic, technological, social, political, institutional, macropsychological, military, and natural environmental factors. Evidence continues to mount justifying the actuality of Kondratiev cycles. Although there is an implicit recognition of evolution and structural change in much of the writing, the present author has perhaps been most insistent in regarding the

Kondratiev phenomenon as a succession of evolving structures bounded by instabilities.

The following references (and the references they cite) should provide the reader with a global comprehension of the Kondratiev cycle/structure as it is presently understood: Berry (1991); Bianchi, Bruckmann, Delbecke, and Vasko (1985); De Greene (1988a, 1988b, 1989, 1990b, in press); Delbecke (1981); Forrester (1977); Goldstein (1988); Grübler and Nakićenović (1991); Grübler and Nowotny (1990); Kleinknecht (1990); Kondratieff (1926, 1984); Marchetti (1986, 1988); Mensch (1979); Perez (1985); Schumpeter (1939); van Duijn (1983); Vasko (1987); and Vasko, Ayres, and Fontvieille (1990).

Table 1.2 summarizes the four Kondratievs by the four phases. Note the succession or replacement of sources of energy, dominant technologies, and dominant industries. Implied is the sequence: scientific discovery, invention, and innovation (the first successful commercial exploitation of a discovery or invention). There may be time lags of years or decades among the three.

Table 1.2. The Four Kondratiev Cycles

	1st-2nd	2nd-3rd	3rd-4th	4th-5th(?)
Prosperity	1785-1815	1860-1873	1905-1920	1948-1970
Recession	1815-1825	1873-1886	1920-1929	1970-1990?
Depression	1825-1840	1886-1896	1929-1937	1990?-2000?
Dominant new technologies or industries	steampower, textiles	coal, steel, railroads	oil, electricity, chemicals, automobiles	aircraft, electronics, computers, control systems, rockets and missiles
Break between cycles	*	*	*	*
Cycle length	55 years	56 years	41 years	63 years?
Recovery	1840-1860	1896-1905	1937-1948	2000?

In reviewing Table 1.2 note that the actual years may differ somewhat from country to country and indicator to indicator and that different researchers propose somewhat different spans. The table does not reflect the full impact of military technologies (e.g., nuclear). The first three cycles are sometimes called, respectively, the "Industrial Revolution" Kondratiev, the "Railway" or "Bourgeois" Kondratiev, and the "Neo-Mercantilist" Kondratiev. Number Four apparently has no name, and I have called it the "American-Way-of-Life" Kondratiev. The possible Number Five is sometimes called the "Information and Communications" Kondratiev.

Figure 1.3 is an idealized representation of the four Kondratievs, derived from use of a moving-average statistical technique that eliminates the "chaos" of shorter-term fluctuations and cycles. Figure 1.3 shows the rising legs (upswings), the falling legs (downswings), and the dates of the upper and lower turning points.

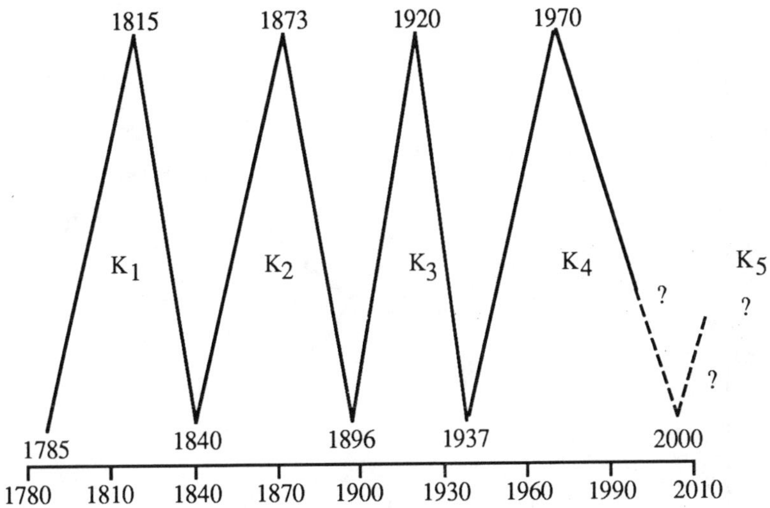

Figure 1.3. Idealized representation of the Kondratiev cycles. The values are not absolute.

Figure 1.4 depicts the evolution of Kondratiev structures as overlapping logistic functions followed by diminishing returns. Momentum carries the system beyond a limit or level of saturation into the domain of diminishing returns. Note that the seeds for a subsequent recovery are sewn in the previous phase of depression (cf. the clustering of innovations mentioned above). Note also the envelope curve, tangent to the individual curves, that indicates the overall evolution of the human sociotechnical system toward sustained exponential or hyperbolic (hyperexponential) growth and greater complexity.

That the moods of a people or nation can differ from time to time has long been recognized. Waves of optimism or pessimism can sweep over countries. We therefore refer to a collective psychology or macropsychology. The Swiss psychiatrist Carl Jung described a collective conscious and unconscious, and other writers have referred to collective perception and collective intelligence. As discussed in Chapter 4, coherent or collective behavior emerges out of interactions at a more micro-level. The Kondratiev phenomenon is also macropsychological. The psychological climate of the times (an "order parameter"), generated from individual interactions, in turn affects individual willingness to try out new ideas and to take risks. Table 1.3 gives the macropsychological attributes of the four phases of the Kondratiev cycle/structure.

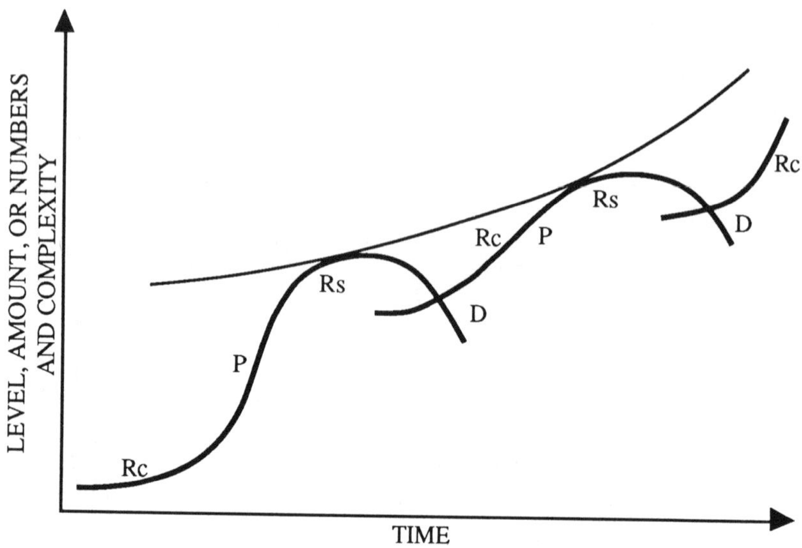

Figure 1.4. The evolution of the Kondratiev cycles/structures shown as overlapping logistic functions with diminishing returns. (Rc = recovery, P = prosperity, Rs = recession, D = depression). The envelope curve shows exponential or hyperbolic systems growth.

In his original studies, Kondratiev recognized that long-wave recoveries were characterized by considerably increased social turmoil, upheavals, and radical changes in the life of society. Kondratiev noted the increase in insurrections, rebellions, wars, coalitions and treaties among nations, and constitutional changes. Subsequent observations confirm that wars involving the great powers are strongly associated with the phases of recovery and prosperity.

Table 1.4 summarizes these relationships. During the phases of recession and depression, nations become increasingly concerned with internal problems and people with raw, day-to-day survival (although leaders may try to redirect peoples' attention outward with the ancient cry: "the enemy is at the gates"). With recovery, confidence and enthusiasm and often a Messianic fervor spread. Economic growth allows increases in international trade and competition for land, other resources, and world markets, and the development and production of more weapons of more sophisticated kinds (arms races may arise).

Table 1.3
Macropsychological Features of the Four Phases of the Kondratiev Cycles/Structure

	Prosperity	Recession	Depression	Recovery
Perception of immediate threat	lowest	rising	falling from a maximum	falling
Perception of opportunity	narrowing	very constrained and selective	broadening	broad
Creativity	oscillating with local high then falling	lowest	rising to a maximum	sustained high
Problem-solving	falling	lowest	rising to a maximum	sustained high
Learning	falling	lowest	rising	high
Anxiety	lowest	mounting to a maximum	decreasing from a maximum	low
Risk taking	falling	lowest	highest	high
Motivation, morale, and job satisfaction	falling	lowest	rising from a minimum	high
Alienation	rising	highest	falling	lowest
Values	cosmopolitan	conservative	economic (parochial)	politicomilitary conflict and cooperation (progressive)

Table 1.4
Occurrence of Major Wars Involving the Great Powers by Phase of the Kondratiev/Cycle Structure

	1st-2nd	2nd-3rd	3rd-4th	4th-5th
Prosperity	1785-1815 French (Revolution), Napoleonic, 1812, Liberation in Spanish America (beginning)	1860-1873 US Civil, Austro-Prussian, Franco-Prussian	1905-1920 Balkans, World War I	1948-1970 Korean, Liberation in European Colonies, Vietnam (all but end)
Recession	1815-1825 Liberation in Spanish America (largely concluded)	1873-1886	1920-1929	1970-1990 Religious and Modernization Wars in the Near/Middle East, Central America, Africa; continued hegemonic suppression
Depression	1825-1840	1886-1896	1929-1937	1990?-2000?
Break between cycles	*	*	*	*
Recovery	1840-1860 Revolution/Set Back in Austria, Hungary, France, Prussia, etc., Franco-Austrian, Crimean, US-Mexican	1896-1905 Spanish-American, Boer, Russo-Japanese	1937-1948 World War II including Spanish Civil and Japanese-Chinese	2000-? ?

The Falling Leg of Kondratiev Number Four. The turning point within Kondratiev Number Four occurred between 1967 and 1973 and for convenience is placed at 1970. By the early 1990s the world economy had moved from recession to depression, about 60 years after the last great depression. Because we are dealing here with structural change, not simply fluctuating macroeconomic variables, the controls that economists have built in to ward against a repetition of the *last* depression may be inappropriate.

Forrester and colleagues (in Bianchi, Bruckmann, Delbecke, and Vasko, 1985) indicate a syndrome of interrelated symptoms characteristic of the past two decades: replacement of labor with capital (increasing capital intensity), rising unemployment, declining utilization of capacity, increased severity of business cycles, deepening depression of capital industries, and declining real return on investment. See also Chapter 7.

A great many industries (manufacturing and service) worldwide are mature or post-mature, saturated, and possess excess capacity. These include steel, aluminum, autos, petroleum refining, steam turbines, farm equipment, traction equipment, transformers, generators, switchgear, shipbuilding, textiles, semiconductors, and commodity chemicals. Nearing saturation are such high-tech products as telecommunications equipment, commercial airliners, and all computers except for the very largest. There is no indication that industries not yet affected are aware of the overcapacity problems, so that they can take anticipatory actions before it is too late (Lee and Nakićenović, in Vasko, Ayres, and Fontvieille, 1990).

Outside production industry, areas like commercial and home real estate show overcapacity in many parts of the industrialized world. In higher education the need for, and real productivity of, Ph.D.s may show saturation and overcapacity (Stephan, 1991). Studies also indicate an overcapacity of academic administrators (Bergmann. 1991).

Mensch (1979) contrasted basic innovations on the one hand and pseudo-innovations and refinements of the basic innovations on the other. The epoch following the early 1970s does appear to be one of stalemate (Mensch's term). There appear to have been comparatively few technological innovations and even fewer widely accepted creative social and institutional designs and approaches to problemsolving. This epoch might well be characterized as one of greater ingrownness, rigidity, insistence on maintenance of the *status quo*, design of cloned products, contrasting of products by packaging rather than by inherent quality features, and greatly diminished problemsolving capabilities.

Baily (cited in De Greene, 1988a) provides data which indicate that declining productivity in the US is strongly associated with declining invention and innovation. Both the number of US patents granted to American inventors and the amount of spending on research and development (R&D) declined significantly since 1972 and 1969, respectively. Stephan (1991) indicates that the quality of people going into science in the 1970s and early 1980s may not have been so high as was the case in the 1950s and 1960s. Much time and energy are wasted as researchers seek "sure bets" from funding agencies, themselves in need of structural change. Simonton (1988), discussing historical fluctuations and clusterings of creative persons per generation within nations, cultures, and civilizations, cites evidence that, if present patterns hold, then American scientific leadership which began in 1920 will end in the year 2000.

The various measures of productivity show fluctuations and mixed and often misleading results. In the major industrial countries, the rate of productivity growth fell in the 1970s from the 1955-1973 level. The picture in the 1980s shows more fluctuation and is probably distorted in the favorable direction by the huge amount of debt financing. Baily (cited in De Greene, 1988a) indicates that in the US the efficiency of the use of capital (quantity of output produced per unit of capital) fell by 3% from 1968 to 1973 and by 7% from 1973 to 1984. The worst performance was in capital-intensive industries like petroleum refining and chemicals. Perhaps most disappointing, if a Fifth Kondratiev recovery (cf. Grübler and Nowotny, 1990) is to be based on such equipment, the computer and electronics revolution has not yielded the kinds of productivity increases that accompanied earlier generations of equipment. This is true even though much of US capital investment in the 1970s and 1980s was for computers and other electronics devices. Other sources also indicate static white-collar productivity. In many private and public organizations the percentage of managerial and clerical workers rose, as did generally the ratio of overhead or nonproduction workers to total employment. Bureaucratic bloat, and in some cases excessive executive salaries unrelated to actual performance, swallowed up much of the investment.

This pattern has been complicated by other changes discussed further elsewhere (De Greene, 1991a, 1991b). In the 1980s a large number of blue-collar jobs were eliminated, and a large number of service jobs were created. But the service jobs were generally lower paying and poorer quality jobs. The very definitions of employment and unemployment were distorted. For example, a person who had been unemployed for months was no longer counted as unemployed, whereas a person who worked only a few hours per week was counted as employed. In the late 1980s large numbers of clerical, staff, managerial, and even executive personnel were laid off. In some cases fired personnel were rehired as contract or contingency workers at much lower salaries and without fringe benefits.

Thus, the true levels of employment, unemployment, and underemployment are imperfectly known and in themselves provide a shaky basis for formulating policy. It does appear that the once touted concept of a carefree "postindustrial society" or "service society" was naive. Creative accounting by governments distorts the record in other areas as well, further compromising the meaning and accuracy of data.

There are few if any panaceas. The roles of flexible manufacturing systems (FMS), computer integrated manufacturing systems (CIM), robotics, and office automation are reviewed elsewhere (De Greene 1999a, 1991b). There is evidence that many US users of FMS do not understand their purposes and indeed apply the FMS incorrectly as just another means of mass production. The strategies of many firms in the US and other Western countries of automation, plant shutdown, large-scale firing, and transferring jobs to low-wage countries may further exacerbate the decline of industry and competitiveness (beginning particularly with the US). Basis skills and collective knowledge can be lost. Further, some technologies, especially large computer information systems and networks, may turn out to be surprisingly both rigid and fragile.

The roles of microelectronics and microprocessors in a possible Fifth Kondratiev recovery have been examined also by Perez (1985) and by Grübler and Nowotny (1990). These technologies may steer a development path away from energy intensity and heavy dependence on expensive materials. Products and services may become more information intensive. There may be a greater shift from economies of scale to economies of scope. Overall designs may become more systemic, partly aided by information networks. Nevertheless, institutional inertia and the momentum of past investments and psychological commitments may preclude early implementation. It is likely that realization of the full potential, in humanistic designs and in designs that do not further degrade the workforce and the natural environment, of the new information technologies must await the "creative destruction" of the present worn-out sociotechnical or technoeconomic paradigm. Here and throughout this book "sociotechnical" (De Greene, 1973, 1982, 1990a, 1991a) and "technoeconomic" (Perez, 1985) indicate the interactions of subsystems to induce or yield a larger system with emergent, qualitatively new properties.

A classical source of instability that may mark the formal inauguration of the depression phase of Kondratiev Number Four is the world stock market. The stock market is overinflated and detached from real growth (cf. our earlier discussion of tiers of the economy). As before October 1929 there is an increased ratio of speculative content to content tied to real production. The waves of acquisitions, mergers, and takeovers; heavy borrowing; and the generation of junk bonds that characterized the 1980s has augmented this kind of instability. The crash of the New York stock exchange in October 1987, and the less noticed but deeper crash of the Tokyo exchange in 1990, may remind us that systems pushed far from equilibrium are especially sensitive to fluctuations and perturbations.

The Hegemonic Cycle/Structure

The hegemonic cycle/structure is associated with the rise and fall of a great power. It may last 150 years or so. Bieshaar and Kleinknecht, and Menshikov and Klimenko, (cited in De Greene, 1988a), identified super-long waves in the statistical analysis of British time-series data. The statistical evidence is essentially consistent with historical analyses and interpretations (e.g., Goldstein, 1988; Kennedy, 1987).

At the beginning of our stage of world civilization starting around the year 1490, Portugal may have been the first hegemonic power. However, emphasis is usually placed on Imperial Spain or the Hapsburg Monarchy (1492-1648). Holland may have been a brief hegemon during its golden age in the mid-17th Century. Britain is considered to have been the hegemon between 1815 and 1914, and the United States the hegemon from the end of World War II to the present. Would-be hegemons include France under Louis XIV and Napoleon Bonaparte, the Second and Third German Reichs, and probably Russia/USSR and Japan.

Kennedy (1987, 1991) has detailed the economic, military, and political dimensions of the rise and fall of the great powers. He refers to "imperial overstretch," of which the US, like Imperial Spain around 1600 and Britain around 1900, is now a victim. The US has inherited a worldwide set of military and economic obligations but is increasingly unable to pay for these obligations. There are numerous reasons for US decline, relative both to other nations and to prior capabilities within the US. Reasons include poor economic competitiveness against Japan and Germany, and perhaps against the European Economic Community, and decay of numerous high-tech and low-tech industries (cf. our earlier discussions). Further, there has been a decline in agriculture; oscillations regarding free trade with many countries and a surge of economic protectionism; an increase in the huge internal budget, national debt, and external trade deficits; and transformation of America within a few years from the world's major creditor to the world's principal debtor country. Most of these features have been mentioned earlier but bear repeating here.

Kennedy notes that, for great powers, there is no easy way to deal with the three-way tension among defense, consumption, and investment. Nevertheless, American policy and practice that reacts to real or phony challenges by massively increased spending on defense, and reacts to the budgetary crisis by eliminating the already inadequate social and environmental expenditures, will provoke eventual political and environmental backlashes. It is interesting to speculate on the real purposes and accomplishments of the US military strikes against Grenada, Tripoli, Panama, and Iraq. The rioting and burning in Los Angeles in April 1992 illustrate dramatically the eventual outcomes of long-neglected, complex problems.

Miraculous cures and quick fixes for deep-seated, long evolving sets of problems are still sought, perhaps as a relief from profound thinking. One is reminded of the belief of 20-25 years ago: if we can send man to the moon, then we can lick poverty, etc. At the end of the short Persian Gulf War of early 1991, in which US high-tech weaponry originally developed for war with the USSR quickly smashed the Iraqi military capability, such hopeful thinking was again popular. Gordon (1991), like others cited in this chapter, points out the steady decline in US economic performance over the past two decades. Both average annual growth of GNP and average annual growth of real capital stock showed this decline. During the 1980s the US ranked last among the seven largest Capitalist economies on measures of investment performance and productivity growth. Some policymakers expected the Persian Gulf War to help "jumpstart" the economy, but the actual impact may be the pushing of the US and other countries deeper into depression.

Attempts to transfer military and defense capabilities to the civil sector have failed in the past. In fact, the decline in US productivity growth can be interpreted to be a consequence of the quasi-military Tayloristic ("scientific management") and Fordian (assembly-line) designs inherited from the past. Gordon (1991) states that US sluggishness stems considerably from the costs of trying to run a modern economy with outmoded kinds of labor discipline, including top-heavy supervisory bureaucracies and coercive treatment of workers. This is of course what humanistic behavioral scientists and sociotechnical systems designers have been saying all along (e.g., De Greene, 1973, 1982). Note here again the symptoms of the worn-out paradigm.

The American System of Management itself may be part of this exhausted paradigm. The American System developed over the epoch 1850-1920 (during Kondratievs Numbers Two and Three). Its key aspects include some of the features just indicated and the design of organizations for efficiency through systematization of communications and control. Yates (1989) states that the transformation of the nature, extent, and role of communications in US firms was essentially complete by 1920. Management control through communications is management as we now think of it. Recent technology may only be reinforcing a system that was firmly established over 70 years ago (cf. De Greene, 1990c, 1991a). And the American System has increasingly become the Western System and the World System.

Note that the above symptoms, and those discussed earlier, are those of both hegemonic decline and Kondratiev decline.

The Life Cycle of the Present Stage of World Civilization

The present stage of world civilization began about 1490. The Dark Ages were ending with the emergence of the Renaissance and somewhat later the Protestant Reformation. The modern nation-state like England, France, and Spain arose out of the unification of smaller kingdoms, duchies, and principalities. The Great Discoveries of Columbus, Magellan, Vasco da Gama, and others opened up the rest of the world to European understanding and settlement and introduced the major structural change of Europeanization. Modern science arose from the observations and interpretations of Copernicus, Kepler, Galileo, Descartes, and Newton. Golden ages of classical European art and music arose. These basic developments have since been accelerated, expanded, intensified, branched (bifurcated), and locally opposed, halted, or reversed to produce the metastable world system-field of today.

In envisioning the end of the present stage of world civilization, we can resort again to two important theoretical concepts used earlier in this chapter and elsewhere in this book: the exhaustion of a paradigm with attendant entry into a domain of diminishing returns and increasingly destructive impacts, and the confrontation with a complexity barrier. Many examples of declining marginal returns have been given earlier. Consider also Simonton (1988).

The origin of systems complexity and limits to the management of complexity have been discussed by several authors (e.g., De Greene, 1982; May, 1974; Nicolis and Prigogine, 1977, 1989; and Prigogine and Stengers, 1984). Tainter (1988), a non-mainstream archeologist, sees societal collapse (a sudden pronounced loss of an established level of sociopolitical complexity) as the result of reaching a point of declining marginal returns in investments in complexity. Tainter's version of complexity includes features we have already discussed: extremes of bureaucratization, spread of cumulative organizational problemsolving and solutions, and increased investment merely to preserve the internal and external *status quo*. Tainter studied past and collapsed societies in Europe, Asia, Africa, and pre-Columbian North America.

In the past, isolation and power vacua allowed individual societies and civilizations to collapse. This is no longer possible, and nations have allowed themselves to become trapped in a competitive spiral. If collapse occurs again, it will be global as world civilization disintegrates as a whole (Tainter, 1988). In the language of Chapter 4, a system pushed or allowed to evolve far from equilibrium is liable to break; the world system-field of forces has coalesced producing great fragility as sensitivity to fluctuations and perturbations has heightened. No system can be stabilized forever.

Fortunately, evidence for an emerging new intellectual, scientific, and values-based paradigm can be seen in many areas. One reads statements from sociologists, economists, computer scientists, students of linguistics, chemists, ecologists, and so on that their work does not fit into the mainstream of their disciplines. Forgers of the new paradigm in apparently disparate disciplines are working on many of the same kinds of problems, problems that involve evolution, stability, and the origin and maintenance of structure. Here again we see an incipiently reconfiguring societal-field which explains the frequent independent discoveries and constructions of theory that are noticeable today. And in some cases researchers in apparently distantly related disciplines are working together. For example, economists and physicists in an area called "complex systems" (a term long used by systems theorists) are trying to overcome the theoretical and practical limitations of neo-classical economics. Rosser (1991) has used chaos theory and catastrophe theory in the study of large economic systems. Complementing these new intellectual approaches are values-based approaches as summarized by Harman in Chapter 8 of this book.

LET THE POLICY FIT THE TIME

Clearly, models and policies that fit periods of relatively slow evolution, relatively greater stability, and relative structural constancy must differ from those in which the opposite conditions obtain. Referring again to the logistic function followed by diminishing returns (in some cases a logistic up is followed by a logistic down), we can note four kinds of qualitative situations. Around the inflection point, the situation looks linear although the underlying dynamics, as discussed earlier, are changing. Just before the linear section is a section of increased exponential growth and just after the linear section is a domain of cybernetic regulation and control of that growth. The regions around the "niche" ceiling where the system has exhausted its capabilities, and at the beginning when system structure is tentative and growth is just getting started, call for a field-theoretic or reconfigurational approach.

In the last nonequilibrium situations, the dynamic, evolving system is in a state of structural instability with its attendant critical thresholds and bifurcation points and is much more susceptible to fluctuations and perturbations. Policies based on information-systems and decision-systems approaches, optimizations, and even cybernetics may strengthen the old system in the short term. However, this may render the system both more rigid and more fragile, may inhibit its necessary evolution, and may prolong its existence well into the domain of diminishing returns until control mechanisms eventually falter, with all that this connotes. Policies should therefore utilize supplementary models (De Greene, 1990b, 1990c, 1991a). An inappropriately timed policy will most likely guarantee the difficult field-theoretic reconfigurational approach. These esoteric concepts are explained more fully in Chapter 4. Figure 1.5 provides a summary.

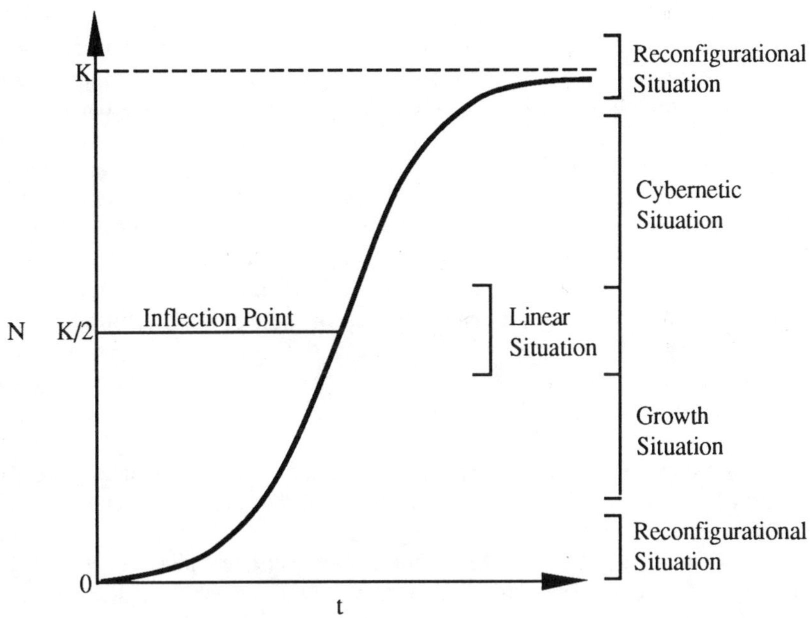

Figure 1.5. Idealized logistic-function representation of supplementary systems models within a life history of a paradigm or cycle/structure such as the Kondratiev, hegemonic, Newtonian, or civilizational. Policy must be anticipatory and not delayed until the given evolutionary epoch. For example, a cybernetic policy should be developed while things still look "normal" but there are early-warning indications of movement to undesirable system states.

In dealing with complex societal systems and ecosystems, the observer may be uncertain as to (1) the equilibrium and resilience states of the system, (2) how far away is the system from the boundaries of stability, and (3) the existence and nature of internal and external fluctuations and external perturbing forces. In such reconfigurational situations, just the slightest breath (called the "butterfly effect" by chaos theorists) may blow the system into a domain of instability or into an unsuspected and very different domain of stability. Figure 1.6 depicts unstable, metastable, stable, and ultrastable systems as epigenetic landscapes or potential wells or basins of attraction (cf. De Greene, 1987). Biologist Conrad Waddington introduced the term "epigenetic landscape" (cf. Jantsch, 1980). "Kinetic potentials" are illustrated in Nicolis and Prigogine (1989). For a discussion of chaotic attractors and uncertainty of capture by a given basin, based on initial conditions, see Grebogi, Ott, and Yorke (1987). We shall not deal here with further technical distinctions. Chapter 2 discusses more complex "landscapes."

Because of these kinds of unknownness, unknowability, or uncertainty, policy actions may variously have no effect, have delayed effects, have counterintuitively the effect opposite to that expected, amplify natural fluctuations and perturbations, or even have the desired effect. On some occasions, by modifying its internal relationships, the system may simply swallow any attempts to change it. On other occasions, in the seemingly identical system, a minuscule push may trigger uncontrollable change throughout.

When the moon, Earth, and the sun are aligned, especially high tides are generated. This is syzygy. Metaphorically, therefore, we can identify a syzygy among the Kondratiev, hegemonic, Newtonian, and world-civilization cycles/structures. The end of the 20th Century may mark the end of Kondratiev Number Four, the end of American hegemony, the end of the utility of the Newtonian paradigm beyond its original application, and perhaps the end of the present stage of world civilization.

Limited, Incorrect, or Failed Past Policies

In this section, let us use the example of energy policy during the falling leg of Kondratiev Number Four. Energy policymaking remains among the most intractable and convoluted of all policymaking.

Following the 1973 oil crisis there were four price discontinuities and a sudden and unexpected, global and systemic structural change. The models that analysts had developed pertained to a completely different and vanishing world. Social, technological, and economic forces were poorly understood; the global dynamics were highly unstable. Nevertheless, decisions were made with little or no previous anticipation and preparation, even though these decisions had long-term, highly uncertain, and quite expensive consequences. Particularly deficient were operations research and econometric models that were based on static or slowly changing situations and on extrapolating past and present trends into the future. The new world was one of nonequilibrium, increasing fluctuation, and structural change although it may have looked linear and structurally stable in the vicinity of the inflection point of the logistic function.

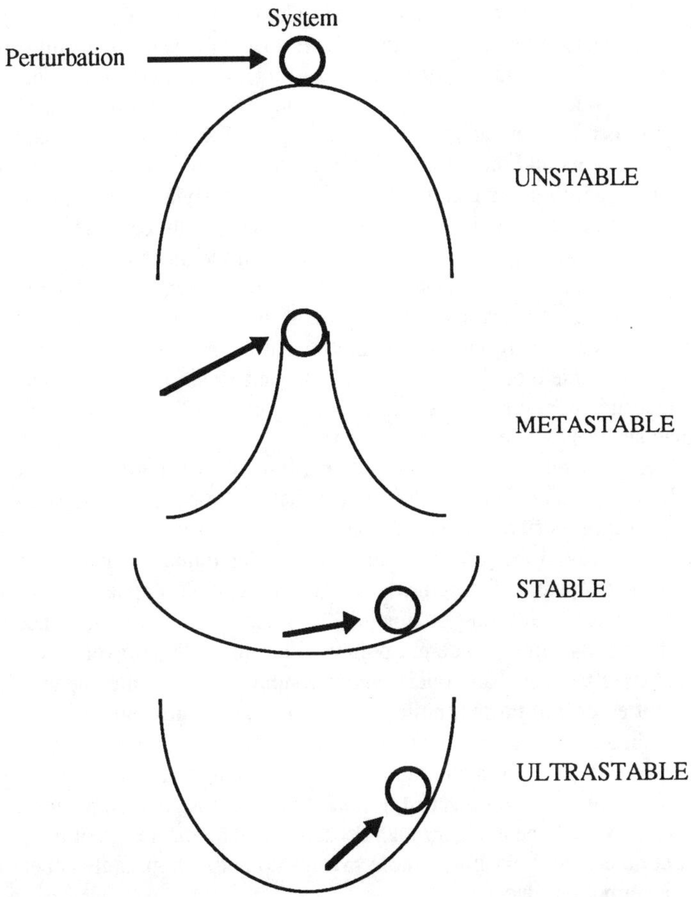

Figure 1.6. Simplified epigenetic landscape (or potential well or basin of attraction) representation of systems stability and instability. Stability increases from the top to the bottom of the figure. The placement of the ball indicates the relationship of the system to equilibrium.

Lee, Ball, and Tabors (1990) provide further insight into the failures. They regard attempts to forecast petroleum prices to be futile because prices do not move as trends but rather as fluctuations and discontinuous spikes. The more the future is approached, the less accurate may be the forecasts. One US Department of Energy forecast of oil prices for 1985 was more accurate in 1977 than was another in 1980. Unfortunately, energy policy indicates that humanity after all may not always be a learning system! In the case summarized here, the very approach to policymaking constrained the collective thought patterns and imaginations. Policymakers and decisionmakers reacted to each discontinuity as if it were the onset of a simple new trend.

Lee, Ball, and Tabors (1990) identify four major reasons why large analytic models have been less than useful in solving the world's energy problems: (1) overemphasis on optimization techniques in spite of the necessary reliance on many qualitative factors and on defective information; (2) limitations associated with an unquestioning faith in the meaningfulness of the underlying assumptions; (3) inadequate attention to the man-model interface, especially with regard to the lack of comprehension by policymakers and decisionmakers of the inner workings of the model; and (4) overreliance on forecasts. Of several recommendations for improvement, three are particularly pertinent to our present discussion: (1) assumptions and the identification of system boundary conditions are more important than is analytical quality; (2) simple models based on sophisticated assumptions are much better than is the opposite, more commonly practiced design; and (3) mathematics and quantitative analysis must not be overemphasized.

Having attended an international, energy-crisis simulation workshop in about 1988, Lee, Ball, and Tabors (1990) noted that participants had apparently learned nothing from experience. The recommendations were a repeat of 1973-1983. Nonsystemic, vested-interest, and conflict-generating thinking ruled the day.

The development of integrated, coherent energy policy appears to be as far away today as it was in 1970 despite much better understanding by many people of systems interrelationships. Coherent policy will require the unraveling of the myriad threads that reflect past bad policies and decisions. For example, the true costs of oil use do not enter into present policy. Total costs, of course, include damage to human health, agriculture, and the natural environment; military expenditures to protect overseas sources (cf. again the Persian Gulf Crisis and War of 1990-1991); and government subsidies to powerful, *established* industries and companies. Although major savings could be made through conservation and the design of more energy-efficient structures and machines, these savings do not appropriately appear in the energy-budget bottom line.

An energy approach in the 1990s in the US or any other country, which stresses power derived from oil, coal, and nuclear-fission sources, and a program of accelerated highway building, appears to be more fitted to the middle of the century than to its end. Similarly, a policy that querulously questions changing environmental parameters, such as the atmospheric buildup of greenhouse gases, while maintaining unquestioning faith in the structure and stability of the economic subsystem, may be obsolete. According to the theory underlying this chapter, the approach fosters the continued domination of confirmation over novelty and the irreversible movement of the system farther from equilibrium and farther into the domain of diminishing returns.

Similar or related interpretations can be made with regard to most other areas; for example, water policy, farm policy, population policy, criminal-justice policy, urban and regional policy, health policy, and military policy. Clearly, new thinking is in order as more and more saturation limits are reached and the surprising and undesirable results of past policies and decisions become increasingly evident.

CONCLUSIONS AND RECOMMENDATIONS--POLICY GUIDANCE

Policymaking and especially the effects of policymaking are not self-evident. The objects of policymaking cannot be treated as static and unchanging. Complex societal systems and ecosystems are inherently dynamic and evolutionary. They can evolve or be driven into domains that are far from equilibrium and in which they are surprisingly sensitive to small changes and forces. No system is perpetually stable, and policies can exacerbate spontaneous instabilities and lead irreversibly to major changes in structure. A policy cannot be rigid and everlasting but must be congruent with the particular system state at the given time.

Complex systems show an overall evolutionary trend toward greater complexity, which in itself may produce greater instability and greater vulnerability. Within this secular trend systems go through various cycles and develop temporary structures. Policies are often not in accord with the phases of the cycles and with the nature and state of the structures. Although the evolution and behavior of realworld systems are characterized by a complementarity of probability and determinism, of chance and necessity, and of novelty and confirmation, policies usually stress the latter of each pair. Thus, the energy of all kinds that is expended on reinforcing the *status quo*)--the maintenance of political, economic, and military power by dominant individuals in key organizations--may drive the system into the domain of diminishing returns and potential failure and collapse.

Policymaking in the 1990s and beyond must be much more systems based--more attuned to the reality of and reasons for complexity, more oriented to the longer range, more aware of the possible deleterious consequences of policies and decisions, and more anticipatory. The new policymaking will place heavy cognitive demands on all stakeholders. In addition, a new rationality must be recognized, a rationality that incorporates the emotional and value investments that all people make in their beliefs. Needed is an epistemic thinking coupled with a search for broader sets of values. Chapters 5, 6, 7, 8, and 9 discuss ways of upgrading systems thinking and the education of policymakers. Chapters 2, 3, and 7 present formal models that are applicable to a variety of policy situations.

Finally, broad frameworks, into which more specific policies would fit, must be developed. In this book these frameworks include the limits to growth, the exhaustion of the present sociotechnical/techno-economic paradigm, the end to war, preservation of the natural environment, and humanistic values and designs. Not every chapter, of course, has the same kind of emphasis. For more on the limits to growth, say, the reader should consult the references at the end of Chapter 7; for more on new values, those at the end of Chapter 8.

The following policy advice is now offered:

1. Retain and expand undeveloped or natural areas. These have considerable, but not unlimited, resiliency and can absorb, buffer, or attenuate otherwise uncontrolled shock effects.
2. Undertake local solutions to local problems and redesign institutions and inhabited areas accordingly.
3. Accelerate the implementation of means for reducing the rate of human population growth with the goal of a net reduction in numbers. Focus on the sustainable *quality* of human life.
4. Dismantle the Machine Model of Humanity, applying social solutions to social problems and psychological solutions to psychological problems, rather than the technological and institutional solutions now largely sought.
5. Create new institutions to replace worn-out, entrenched, and self-serving bureaucracies of all sorts.
6. Redo economic priorities, reducing support for wasteful consumption and for the military; greatly reduce the design, production, and sales of military systems and equipment.
7. Demilitarize the world, starting with more circumscribed regions like Mexico, Central America, and Northern South America. The initiative must come from the nations of the area itself.
8. Develop criteria for evaluation and choice that transcend the purely economic and monetary features that are presently dominant.
9. Encourage the acceleration of value change in constructive, "systems-consistent" directions.
10. Plan for the contingency of a great depression that is likely to end Kondratiev Number Four.

REFERENCES

Bergmann, Barbara R., 1991. "Bloated Administration, Blighted Campuses." *Academe*, 77(6), 12-16.

Berry, Brian J. L., 1991. *Long-Wave Rhythms in Economic Development and Political Behavior*. Baltimore: The Johns Hopkins University Press.

Bianchi, Giuliano, Gerhart Bruckmann, Jos Delbecke, and Tibor Vasko (Eds.), 1985. Long-Waves, Depression, and Innovation: Implications for National and Regional Economic Policy. *Proceedings of the Siena/Florence Meeting (26-30*

October 1983). Laxenburg, Austria: International Institute for Applied Systems Analysis.

Checkland, Peter, 1981. *Systems Thinking, Systems Practice.* New York: Wiley.

De Greene, Kenyon B., 1973. *Sociotechnical Systems: Factors in Analysis, Design, and Management.* Englewood Cliffs, NJ: Prentice-Hall.

De Greene, Kenyon B., 1982. *The Adaptive Organization: Anticipation and Management of Crisis.* New York: Wiley.

De Greene, Kenyon B., 1987. "Cognitive Models of International Decisionmakers and International Stability." *Systems Research,* 4(4), 251-267.

De Greene, Kenyon B., 1988a. "The Kondratiev Phenomenon: A Systems Perspective." *Systems Research,* 5(4), 281-298.

De Greene, Kenyon B., 1988b. "Long-Wave Cycles of Sociotechnical Change and Innovation: A Macropsychological Perspective." *Journal of Occupational Psychology,* 61(1), 7-23.

De Greene, Kenyon B., 1989. "Micro-Macro Interrelations and the Order Parameter Concept in the Field Theory of Societal Systems." *Systems Research,* 6(4), 277-288.

De Greene, Kenyon B., 1990a. "The Turbulent-Field Environment of Sociotechnical Systems: Beyond Metaphor." *Behavioral Science,* 35(1), 49-59.

De Greene, Kenyon B., 1990b. "Supplementary Systems Paradigms for Different Stages of Societal Evolution with Special Reference to War and Peace." *Systems Research,* 7(2), 77-89.

De Greene, Kenyon B., 1990c. "Nonlinear Management in Technologically-Induced Fields." *Systems Research,* 7(3), 159-168.

De Greene, Kenyon B., 1990d. "The Suicide of Western Capitalistic Democracy: Can It Happen? Is It Happening?" *World Futures,* 30(1-2), 17-40.

De Greene, Kenyon B., 1990e. "Contextual Aspects of Human Factors: The Case for Paradigm Shift." *Human Factors Society Bulletin,* 33(9), 1-3.

De Greene, Kenyon B., 1991a. "Rigidity and Fragility of Large Sociotechnical Systems: Advanced Information Technology, the Dominant Coalition, and Paradigm Shift at the End of the 20th Century." *Behavioral Science,* 36(1), 64-79.

De Greene, Kenyon B., 1991b. "Large Technology-Based Systems and the Need for Paradigm Shift." *Technological Forecasting and Social Change,* 39(4), 349-362.

De Greene, Kenyon B., 1992 (in press). "The Growth of Exhaustion." *European Journal of Operational Research.*

Delbecke, Jos, 1981. "Recent Long-Wave Theories: A Critical Survey." *Futures,* 13(4), 246-257.

Denning, Peter J., 1989. "The Science of Computing: Sparse Distributed Memory." *American Scientist,* 77(4), 333-335.

Denning, Peter J., 1990. "Modeling Reality." *American Scientist,* 78(6), 495-498.

Dreyfus, Hubert L., and Stuart E. Dreyfus, 1986. *Mind Over Machine: The Power of Human Intuition and Expertise in the Era of the Computer.* New York: Free Press.

Forrester, Jay W., 1977. "Growth Cycles." *De Economist*, 125(4), 525-543.

Forrester, Jay W., 1990. *The Beginnings of System Dynamics.* Cambridge, MA: Massachusetts Institute of Technology, Sloan School of Management, System Dynamics Group.

Goldstein, Joshua S., 1988. *Long Cycles: Prosperity and War in the Modern Age.* New Haven, CT: Yale University Press.

Gordon, David M., 1991. "Unlike the War, There's No Quick Fix for the Economy." *Los Angeles Times*, Business Section, March 24, pp.D2, D6.

Grebogi, Celso, Edward Ott, and James A. Yorke, 1987. "Chaos, Strange Attractors, and Fractal Basin Boundaries in Nonlinear Dynamics." *Science*, 238(4827), 632-638.

Greenberger, Martin, Matthew A. Crenson, and Brian L. Crissey,1976. *Models in the Policy Process: Public Decision Making in the Computer Era.* New York: Russell Sage.

Grübler, Arnulf, and Nebojša Nakićenović, 1991. "Long Waves, Technology Diffusion, and Substitution." *Review*, XIV(2)(Spring), 313-342.

Grübler, Arnulf, and Helga Nowotny, 1990. "Toward the Fifth Kondratiev Upswing: Elements of an Emerging New Growth Phase and Possible Development Trajectories." *International Journal of Technology Management*, 5(4), 431-471.

Jantsch, Erich, 1980. *The Self-Organizing Universe: Scientific and Human Implications of the Emerging Paradigm of Evolution.* New York: Pergamon.

Kennedy, Paul, 1987. *The Rise and Fall of the Great Powers: Economic Change and Military Conflict from 1500-2000.* New York: Random House.

Kennedy, Paul (Ed.), 1991. *Grand Strategies in War and Peace.* New Haven, CT: Yale University Press.

Kleinknecht, Alfred, 1990. "Schumpeterian Waves of Innovation? Summarizing the Evidence." In Tibor Vasko, Robert U. Ayres, and Louis Fontvieille (Eds.), *Life Cycles and Long Waves*, pp. 41-55. New York: Springer-Verlag.

Kondratieff, Nikolai D., 1926. "Die Langen Wellen der Kunjunktur." *Archiv für Sozialwissenschaft und Sozialpolitik*, LVI, 573-609.

Kondratieff, Nikolai D., 1984. *The Long Wave Cycle.* Translated by Guy Daniels with Introduction by Julian M. Snyder. New York: Richardson and Snyder.

Kuhn, Thomas S., 1970. *The Structure of Scientific Revolutions.* Second edition. Chicago: University of Chicago Press.

Kull, Steven, 1988. *Minds at War: Nuclear Reality and the Inner Conflicts of Defense Policymakers.* New York: Basic Books.

Lee, Thomas H., Ben C. Ball, Jr., and Richard D. Tabors, 1990. *Energy Aftermath.* Boston: Harvard Business School Press.

Linstone, Harold A., 1984. *Multiple Perspectives for Decision Making: Bridging the Gap between Analysis and Action.* New York: North-Holland.

Marchetti, Cesare, 1986. "Stable Roles in Social and Economic Behavior." Unpublished paper. Laxenburg, Austria: International Institute for Applied Systems Analysis.

Marchetti, Cesare, 1988. "Kondratiev Revisited--After One Kondratiev Cycle." Unpublished paper. Laxenburg, Austria: International Institute for Applied Systems Analysis.

Marchetti, Cesare, 1991. "The Future." RR-91-5. Laxenburg, Austria: Internal Institute for Applied Systems Analysis.

May, Robert M., 1974. *Stability and Complexity in Model Ecosystems.* Second edition. Princeton, NJ: Princeton University Press.

Meadows, Donnella H., Dennis L. Meadows, and Jørgen Randers, 1992. *Beyond the Limits: Confronting Global Collapse, Envisioning a Sustainable Future.* Post Mills, VT: Chelsea Green.

Mensch, Gerhard, 1979. *Stalemate in Technology.* Cambridge, MA: Ballinger.

Miser, Hugh J., and Edward S. Quade (Eds.), 1985. *Handbook of Systems Analysis: Overview of Uses, Procedures, Applications, and Practice.* New York: North-Holland.

Nakićenović, Nebojša, 1988. "Dynamics of Change and Long Waves." WP-88-074. Laxenburg, Austria: International Institute for Applied Systems Analysis.

Nicolis, Grégoire, and Ilya Prigogine, 1977. *Self-Organization in Nonequilibrium Systems: From Dissipative Structures to Order through Fluctuations. New York: Wiley.*

Nicolis, Grégoire, and Ilya Prigogine, 1989. *Exploring Complexity: An Introduction.* New York: Freeman.

Perez, Carlota, 1985. "Microelectronics, Long Waves and World Structural Change: New Perspectives for Developing Countries." *World Development*, 13(March), 441-463.

Prigogine, Ilya, and Isabelle Stengers, 1984. *Order Out of Chaos: Man's New Dialog with Nature.* New York: Bantam.

Rosser, John B., Jr., 1991. *From Catastrophe to Chaos: A General Theory of Economic Discontinuities.* Boston: Kluwer.

Schumpeter, Joseph. A., 1939. *Business Cycles: A Theoretical, Historical, and Statistical Analysis of the Capitalist Process.* (2 volumes). New York: McGraw-Hill.

Searle, John R., 1990. "Is the Brain's Mind a Computer Program?" *Scientific American*, 262(1), 26-31.

Simonton, Dean K., 1988. *Scientific Genius: A Psychology of Science.* New York: Cambridge University Press.

Stephan, Paula, 1991. "Research Productivity over the Life Cycle: Evidence for American Scientists." *American Economic Review*, 81(March), 114-132.

Tainter, Joseph A., 1988. *The Collapse of Complex Societies.* New York: Cambridge University Press.

Thom, René, 1975. *Structural Stability and Morphogenesis.* Reading, MA: Benjamin.

Tolcott, Martin A., F. Freeman Marvin, and Paul E. Lehner, 1989. "Expert Decisionmaking in Evolving Situations." *IEEE Transactions on Systems, Man, and Cybernetics*, 19(3), 606-615.

Van Duijn, Jacob J., 1983. *The Long Wave in Economic Life.* Boston: George Allen & Unwin.

Vasko, Tibor (Ed.), 1987. *The Long-Wave Debate.* New York: Springer-Verlag.

Vasko, Tibor, Robert U. Ayres, and Louis Fontvieille (Eds.), 1990. *Life Cycles and Long Waves.* New York: Springer-Verlag.

Yates, JoAnne, 1989. *Control through Communication: The Rise of System in American Management.* Baltimore: The Johns Hopkins University Press.

CHAPTER TWO

POLICY IN A WORLD OF EVOLUTION, LEARNING, AND IGNORANCE

by Peter M. Allen

INTRODUCTION

This chapter expands the discussion of evolution, equilibrium, nonequilibrium, the mechanical or Newtonian paradigm, rationality, and model adequacy first introduced in Chapter 1. This chapter presents several sophisticated models valuable in policy and decision situations. These models provide additional, experimental rigor to many of the ideas presented in this chapter and elsewhere in this book.

In order to make effective policy decisions, a knowledge of the possible futures under the different possible options is required. Unfortunately, until now the conceptual and mathematical basis for such vision has not existed. Complex human systems were represented either by mechanical equations, having fixed mechanisms excluding learning and adaptation, or worse still by assumptions of equilibria and optimality. These ideas were based on the false analogy between an isolated physical system, which can be shown to approach an equilibrium state, and the complex evolution of a socioeconomic system over time.

In fact, the behavior of *open* physical systems offers a much richer and more appropriate set of concepts with which to begin a new reflection on human systems. In this view, *nonequilibrium* phenomena are much more important, for they offer a new understanding of the natural emergence of structure and organization in systems with many interacting individual elements. In this chapter new models of the evolutionary process are presented which show how the dialog between the individual and collective levels generates "ecological" structures. These structures represent a new domain of organization beyond the mechanical, where the identities and behaviors of the actors are mutually interdependent, the system has many possible responses to perturbations, and survival is related to the capacity to change, adapt, and maintain diverse and varied strategies. This view of nonoptimal behaviors, imperfect information, mistaken inferences, and the power of creativity is contrasted with the traditional mechanical representation of human systems. The models discussed here offer a new, quantitative basis for policy exploration and analysis, allowing us to take into account the long-term implications of the system as a whole.

Application of Inappropriate "Scientific" Theories to the Human Domain

Policymaking and decisionmaking are the subject of constant discussion and debate within the political and organizational arenas. Mostly, positions and views are held as a result of personal history and experience, inherited sets of values, and beliefs about how the world works. But of course policymaking and decisionmaking concern the future; that is, they are about actions which should be taken in order that the future will be influenced in a certain way. This in turn implies that we believe that we can know not simply *the* future, but also the nature of different possible futures that our actions might produce. It implies that we understand the processes of change and can influence them according to our objectives.

Such a position seems to be at odds with reality in view of present crises and not to correspond with the level of understanding that really exists concerning social and economic systems. Surprisingly, much of the confusion and disagreement in fact arise from the understandable, but possibly naive, human desire to order, classify, and fit experiences into neat boxes and categories so that reality "makes sense." This seemingly innocent preoccupation implicitly accepts the validity of certain propositions which are not necessarily true and therefore generates confusion in place of the clarity sought.

This desire to "make sense" leads, for example, to the supposition that the future can be predicted and *explained,* implying a belief that events do not occur at random, by happenstance, and by accident, but instead according to some *knowable* pattern or design. Of course, if the future were knowable, then people's actions would make use of this knowledge and thus change the path of the system, raising thereby the question of whether they really *did* know the future. Neglecting this annoying paradox, the scientific view has been that, if the future is produced by the workings of some continuous mechanisms, then the system would have a trajectory along which it would move "naturally." Moreover, by analogy with mechanical systems with frictional forces, it is assumed that the system will move to its new equilibrium state. In this view, *change* is seen as the system's moving from its initial condition to a predictable final state as the result of some imposed action, and it is this assumption that has been taken up by the human sciences, particularly economics, as a basis for theories in the human sciences.

In a further unjustified analogy with physics, human-systems theorists have hypothesized that the supposed movement to equilibrium must also express the increase or decrease of some characteristic function that is analogous to "entropy" in physics; this situation is taken to represent the "utility" of the actors in the system. For producers it is assumed that the move to equilibrium optimizes profits and for consumers that it optimizes their utility, with the whole system's either ascending or descending some "potential field." The implication is that change is not arbitrary but has some overall effect, which in turn implies that time must bring "progress" and that change will be the visible evidence of such progress.

In this unconsciously prevailing view, the ideas of evolution and of God come close to each other, and the complexity of the natural and human systems that surround us seems to offer proof of the miraculous skill of a creator, whether it be God or evolution. And where there is complexity and organization, where before there was none, then surely does this not speak of a "purpose" in the working of evolution? And what could the purpose be other than to create the world around us for our "use?" (See also the discussion of Faith, Ideology, Science, and Religion in Chapter 4).

The human world has for this reason, perhaps unthinkingly, viewed itself as the master of the natural world, with the latter made up not of coinhabitants but of "resources" which can therefore be exploited according to our conscious aims and plans.

And if, as we observe, human agents can think and act successfully in the world, then obviously perceptions, intelligence, and values allow decisionmaking to occur and social and economic systems to emerge and evolve over time. This train of thought leads us initially to a picture of great complexity, as the thousands of individuals with different perceptions, expectations, and skills attempt to adjust their behaviors and activities to the circumstances surrounding them.

Real Complexity and the Idea of Rational Policymaking and Decisionmaking Are Incompatible

In order to have a scientific basis on which to make policy, it is often thought that the requirements are those for "rational" decisions. The requirements are to:
•Have *goals* and values
•Be able to *predict* outcomes
•Use the former together with the latter to make a choice of action or policy

However, in the complex world of innumerable interacting, coevolving individuals, prediction seems impossible. How can the real complexity that surrounds us be made compatible with the idea of rational decisionmaking and policymaking? The answer is that it cannot be made compatible because it is not compatible.

The world is complex, and in reality we have a very poor understanding of the real consequences of our actions and policies. We are creatures of belief and emotional–social animals who enjoy crowds and the comfort of widely shared beliefs. Also, selfish motivations can sometimes color our views, so that convenient theories may be supported at the expense of inconvenient ones, especially because the truth of any is difficult to estimate in a complex world. Since it is difficult to decide what is a good idea and what is not, most of us would rather be wrong along with everyone else than stand alone, confident in the correctness of our position. As we shall see, evolutionary theory has a lot to say about these questions, showing us for example the power of the "imitative" strategy and destroying any lingering illusion we may have that "system optimization" results from evolution.

As a result of this difficulty that dealing with the complexities of the real world presents, we have witnessed the application of completely inappropriate "scientific" theories to the human domain. In this approach, all the complexity of the real world is evacuated by simply assuming that the system moves rapidly to equilibrium and that human beings are "rational decisionmakers."

On the Deficiencies of Economic "Science"

In economics, for example, people are assumed to know not only the present situation (of which they may possibly have some idea), but also the results of all possible experiences that have or have not taken place, so that they are in the position to choose from all these possibilities that action that best fits their particular goals. Thus, the curve of supply and demand, probably the most basic diagram in economics, is based on the assumption that at any given time there *is* a specific curve that expresses for any price how much of a good would be produced or bought.

Yet obviously, this situation is in reality very doubtful since it depends on the strategies of different producers' responding to their competitors' actions; and for consumers it depends on fashion, availability, and service, themselves dependent on the quantities sold. Nevertheless, instead of this real complexity, perfect competition and rational behavior are assumed to hold, so that markets clear miraculously with producers' optimizing "profits" and consumers' "utility" in an instantaneous onset of joyous "equilibrium." With such views it seems extremely inappropriate that economics should have been labeled the "dismal science." It is in fact quite the opposite! Not only is it naively optimistic in its premise that the neglect of a system leads to optimality, but also it isn't a science!

If these points are accepted, then of course a great deal is explained, if not excused. As noted earlier, there has been a totally unscientific borrowing of unsuitable concepts and ideas from the natural sciences; only the enormity of the economic edifice built thereon has permitted its continued existence. It is as if in physics the phlogiston theory of combustion had been consecrated by large institutions and a dedicated priesthood.

This view may seem a little harsh to some, but *Homo economicus* is neither an accidental nor a harmless monster. He is the product of a specific materialistic "cosmology" based incorrectly on the past successes of the natural sciences. In itself the transfer of equilibrium theory from the physical sciences to the much more complex situations of the human sciences represents a remarkable homage to mistaken inference--if physics is successful with these theories, then economics and the other human sciences can be too! This reasoning does not take into account the fact that the equilibrium state of a physical system can be *proved* to hold true under certain specific conditions which can exist in reality, whereas in economics such conditions do not hold true. Also, the fact that the improperly borrowed ideas outlaw much collective action and deny sociality means that they exclude many of the *real* paths that could otherwise be explored, thus diminishing choice for quite unsatisfactory reasons.

The Separation and Isolation of Expertise

The overriding fact is that individuals do *not* necessarily *understand* the world around them. How could they? Why should they? Evolution and survival require only that no fatal acts have been committed (yet). They imply in no way that survivors should have a perfect view of the world and its workings, or indeed that *such a vision exists*. Collective systems have multiple aspects: psychological, informational, social, cultural, economic, material, epidemiological, historical, etc. Despite the fact that all these different areas of interest are merely facets of the complex system that has arisen, the human mind has hitherto had no choice but to separate them out into a taxonomy of disciplines for further study. But the problem is even worse--they are usually studied *separately*. Economics, social science, psychology, etc. are considered academic subjects worthy of isolated study. What may have begun as a simple desire to specialize in a particular area, within the context of the others, has led instead to a disconnection and separation of "expertise."

If we accept the separate study of, for example, economics, then we must believe that the flows of goods, and the prices that are observed, can in themselves form some kind of closed, self-consistent system, and that explanation can be found from *within* this narrow structure. Inasmuch as, in reality, strong interactions traverse the boundaries of economics from the psychological, technological, environmental, and other domains, "explanation" could be found within economics only if there were some set of fixed relations among the variables which hold true over time. This is why a mechanical system is the only representation of change, and its rapid movement to a preexisting equilibrium is the bedrock of economic theory. These relationships are shown in Figure 2.1. Limitations of mainstream economic theory and practice are discussed further in Chapter 4.

The emergence and development of economic systems is, however, the story of real structural change, where innovations and new activities appear in the system and restructure it. The only "explanation" that a closed mechanical system can offer is therefore only a *description* of the system at a given time, with change coming unpredictably from outside. If we want really to understand some feature, then we have to consider the coevolution that system and context have undergone.

The Scientific Handling of Change, and Levels of Description

Understanding these complex systems not only requires an interdisciplinary (or transdisciplinary) approach, it also demands that we reexamine our most fundamental beliefs concerning science because some fundamental and unresolved issues are involved. If we and the rest of the living world are part of a complex system that is changing over time, then what is the *scientific basis* for understanding the path of events, an understanding that is necessary for making policy?

The answer must be: the scientific basis comes from the study of complex systems. And of course it is often said that the ultimate "scientific" expression of "understanding" must lie in physics. So we ask: can we explain what we see around us on the basis of universal physical laws? If we accept that we are the fruit of the evolutionary processes of biology, working upon underlying physical mechanisms; then a satisfactory science of our living world, and our place within it, can come only from an understanding of biological and sociocultural evolution based on the fundamental laws of physics--the reconciliation of Newton with Darwin.

However, the power of physics, particularly its ability to make *predictions* and hence to understand history as a "necessary" system trajectory, is founded upon fundamental laws of *conservation* and of *symmetry* which are independent of time. The only "law" in physics which clearly addresses the issue of change and the direction of time is the second law of thermodynamics. But this is not really a law like the others because in fact it merely *accepts* the existence of time's arrow, of irreversibility. It neither proves that irreversibility must result from more fundamental laws, nor does it show exactly how the phenomenon can be reconciled with the more fundamental laws. Evolution and change are concepts that science has traditionally been unable to deal with satisfactorily, and this situation is related to the use of mathematics as the preferred language of expression.

Equations merely express the fact that the left- and right-hand sides must be equal; therefore, they are merely two different ways of saying the same thing. When dealing with *change*, however, the "trick" is to suppose that the left- and right-hand sides of the equation might refer to later and earlier times, respectively. In this way an equation results from the assumption of some law of "conservation" over time since it obliges earlier and later values to be identical.

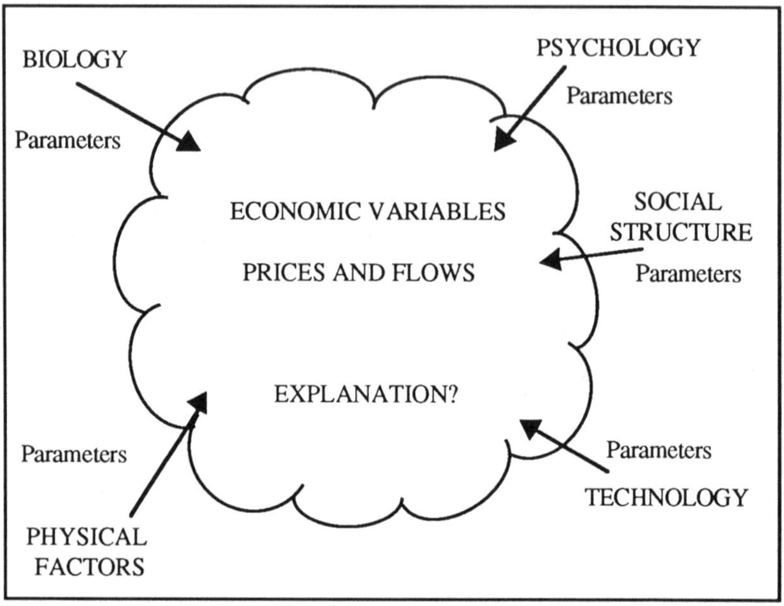

Figure 2.1. Economics is part of a complex system. Explanation within the system is false.

However, change and transformation are not about the "elementary particles" of the system which such equations must conserve. Instead, they are about the emergence and evolution of the state of organization of the individual elements of the system. They are about the emergence and evolution of complex organisms, objects, and societies, and *about the interactions among them.* We must recognize at least *two levels of system description:* that of its "elementary" components that are subject to an "accounting" principle, and that of the structure or organization within which they exist. Our science must aim toward understanding the interrelationship between these two levels if it is not to be trivial. It is the mutual coevolution of these two levels of description that is at the heart of the matter, and our focus of concern is the emergence of such new entities that have new powers and needs which open new dimensions of experience. The discussion of micro- and macrolevels of organization is pursued further in Chapter 4.

THE MECHANICAL AND ORGANIC OPEN-SYSTEMS PARADIGMS FURTHER CONTRASTED

As was noted above, the basis for scientific understanding has traditionally been the mechanical model (Allen, 1988; Prigogine and Stengers, 1984). In this view the behavior of a system can be understood and anticipated by classifying and identifying the system *components* and the *causal links* or mechanisms that act among them. In physical systems the fundamental laws of nature, such as the conservation of mass, momentum, and energy, govern these mechanisms and determine entirely what must happen.

By isolating or closing a system so that no new matter or energy could flow in to disturb it, such limits were placed on the behavior that made it possible for classical physics to predict quite generally the properties of the final state of almost any system, however complex. The final state was that of thermodynamic equilibrium, and the properties of matter that would characterize this equilibrium could be calculated in a very wide range of problems. This was such a triumph for classical science that it was believed (erroneously) that analogous ideas must apply in the domains of biology, ecology, the human sciences, and particularly of course economics (Arrow and Debreu, 1954; Debreu, 1959).

In such a view, however, the problem of change remains. If we study the system at a later time, then we find that its structure has changed. The mechanical system developed at a given time somehow *transforms itself* over time. In order to anticipate successfully the changes that will occur in the system, we must try to understand how this creative self-transformation can occur. Obviously, transformation is not contained in the set of mechanical equations that characterize the system at any one time. Transformation is beyond the capability of a *closed* system.

In fact, the systems of ecology and economics are always *open* to flows of matter and energy and attain thermodynamic equilibrium only with *death*. Ironically, the mathematics and the methodology borrowed from "all-conquering" physics and applied to the *living* world have been totally inappropriate. Living systems are in constant dialog (not equilibrium) with their environments, and even when not visibly evolving they maintain the *capacity*, related to their underlying diversity, to evolve and to change. Evolution and structural change are depicted in Figures 2.2 and 2.3.

In the classical scientific view, the future of a system is predicted by the simple expedient of evaluating the behavior of the equations that govern system motion. Explanation is reduced to an illumination of *how* the system functions without telling *why* the system is as it is. In order to write the equations of motion for a real system, it is always necessary to make approximations. It must be assumed that the elements making up the variables (e.g., individuals in a population or firms in a sector) are all identical to the *average* type. In such cases the model reduces to a "machine" which represents the system by a set of differential equations (perhaps nonlinear) that govern the variables. This is the *Newtonian vision* of the world as a vast and complex mechanism of clockworks.

Scientific prediction for such systems is concerned with both the system trajectory that must be followed, and the final state to which the dissipative forces of friction and viscosity must lead, just as a clockwork toy must run down eventually. The

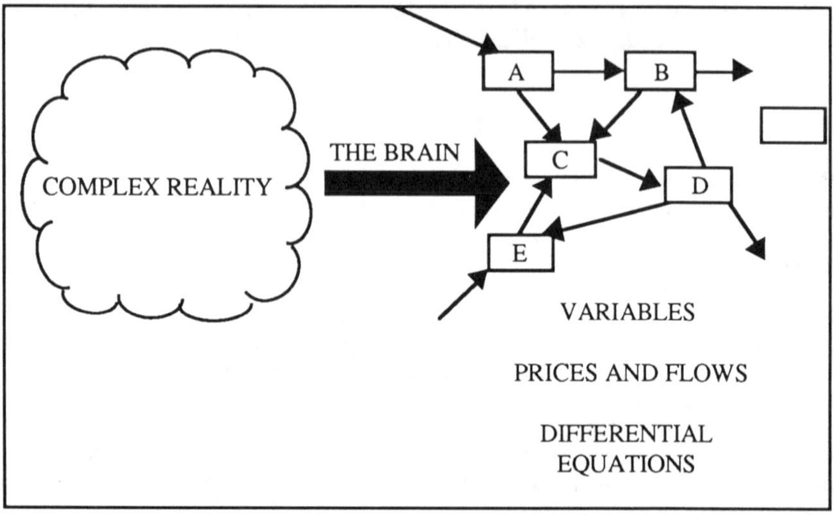

Figure 2.2. At any time the system has a particular structure.

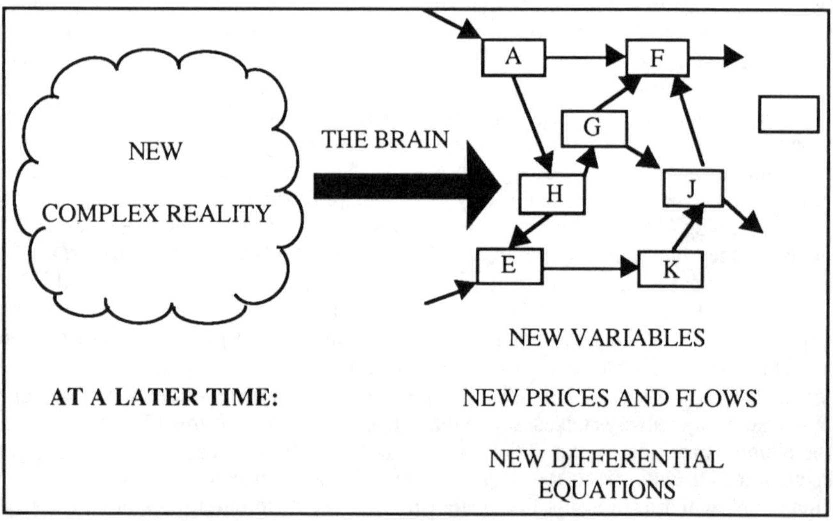

Figure 2.3. At a later time, the structure shown in Figure 2.2 may have changed.

idea of a predictable equilibrium comes from the simplistic assumption that there will be only a *single* solution to the equations--involving a point attractor--and that, therefore, whatever happens the system will end up there. This "point attractor" solution to the differential equations is viewed as expressing a maximum or minimum of some potential function, just as in physics the dissipative forces of friction and viscosity work to lead any mechanical system to a thermodynamic equilibrium expressing maximum entropy. But this approach is completely false for open systems!

Moreover, even in systems which we would all agree are purely physical, when they are open to the flows of energy and matter, there is no longer necessarily a unique final state expressing some optimal principle. We now know that systems of *nonlinear* differential equations can have a great multiplicity of possible solutions, from stationary homogeneous kinds through chaotic and heterogeneous kinds to solutions representing ordered spatial and temporal structures. The external conditions, therefore, no longer suffice to *determine* a unique future, as such systems can achieve structure in a variety of ways depending on the internal details of their components--details that cannot be controlled from outside the system.

In other words, there is a single, predictable outcome to experiments on isolated systems--thermodynamic equilibrium--but not for open systems where matter and energy can flow through the system (Nicolis and Prigogine, 1977). This situation changes profoundly both our notions of "explanation" and of scientific understanding.

When nonlinear mechanisms are present, the system may continue to change indefinitely, executing either a cyclic path of some kind or possibly even a chaotic movement around a "strange attractor." However, the system is still represented by a set of mechanical equations and therefore lacks the capability for self-organization.

Nevertheless, the capability for "self-transformation" and structural change is *not* contained in the dynamical equations, even if they are nonlinear. The systems represented by the equations are capable of *functioning* but not of *evolving*. Evolutionary change must result from what has been removed in the reduction to the deterministic description, that is, the *nonaverage*. The evolving system is therefore driven by two things: (1) deterministic average mechanisms operating among typical components, and (2) nonaverage local behavior that in nonlinear systems can be amplified and can lead to qualitative structural change in the average mechanisms. See also the discussions in Chapters 1 and 4.

Figure 2.4 summarizes many of these points.

EVOLUTIONARY DRIVE

Allen and McGlade (1987a, 1989) have mathematically simulated this evolutionary dialog between average (macro) processes and the nonaverage (micro) detail. This work has led to the new concept of *evolutionary drive*.

The interesting idea here has been to define a "possibility space," a space representing the characteristics and behaviors that could potentially arise for the different kinds of individuals present in the system. In practice, of course, the space is multidimensional, and we should be able to anticipate only a few of the principal dimensions. Nevertheless, it is extremely instructive to think about the evolutionary process in these terms. The central problem of change is that of understanding how, over time, the kinds of behavior present in a system can actually increase and become more complex. In terms of the possibility space, we may ask: if initially there is a single kind of individual present and occupying a single cell of this space, then how can new populations *appear* ? Figure 2.5 helps answer this question.

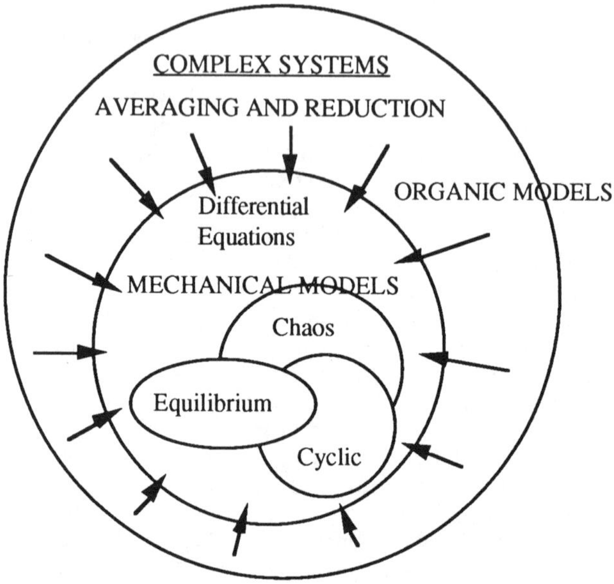

Figure 2.4. Models based on differential equations are an approximation to the broader class of complex systems.

Further, the possibility space will be explored by individuals if their behavior is plastic in some way. In biological evolution not only are there mutations, but more importantly sexual reproduction leads to offspring that are not exact copies of either of their parents. The genetic mechanism is precisely such that a large range of possibilities are explored, and offspring, offspring of offspring, and so on spread out over time from any initially pure condition. In human societies possibility space is explored by the variety of encounters and meetings that can occur, through individual differences, and as a result of imperfect information and learning.

Physical constraints automatically insure that some behaviors do better than do others, and so there is a differential rate of survival and of reproduction. If possibility space is viewed as a kind of "evolutionary landscape," with hills representing behaviors of high performance, then our simulations lead to the amplification of populations which are higher on the hill, and the suppression of those which are lower down.

Initial research showed how the imperfect reproduction of populations provided a capability to climb the hills of the adaptive landscape. By creating populations with different intensities of "error-making" or "exploration," it was possible to find the best amount of ascent for a given slope. However, because the "slope" is created by other populations, this finding means that evolution selects for populations that retain the *ability to learn* rather than for populations with *optimal behavior*.

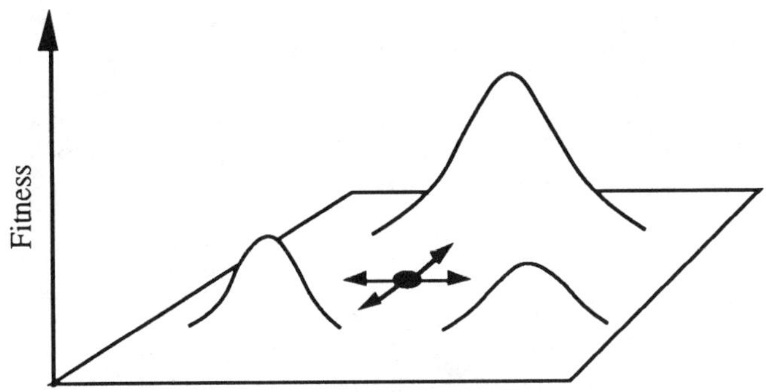

Figure 2.5. In possibility space an initially pure behavior will diffuse outward as a result of individual diversity, imperfect learning, and general entropic processes. Differential success makes average behavior climb hills.

This clear view of evolution allows us to understand how it was that Neo-Darwinism and neo-classical economics relied so heavily on the old ideas of equilibrium physics. In our simulations we found that if a population was on a hilltop, then it was advantageous to suppress exploration. Evolution was "over" and there was nowhere better to evolve to and nothing to learn. Obviously, in such cases, instead of requiring our complicated equations in possibility space, the final state of the system could have been predicted if the positions of the hilltops had been calculated initially. These positions would mark the inevitable end points of evolution and the equilibrium solution.

This view is completely false, however, because the shape of the hills in possibility space reflects the populations that happen to be present. In other words, the advantage to be gained from a particular behavior *depends on which other behaviors are present at the time*. The hills, the goals or objectives of a given moment, coevolve with the populations present. These relationships are shown in Figure 2.6.

These experiments show that this mixture of *exploratory diffusion* of individuals in some behavior space, and the differential *success* of the individuals, leads to the organic as opposed to the merely mechanical. As shown in Figure 2.6, there is a process of simultaneous "stretching" and "squeezing" of populations in the space of possible behaviors that is the core of our new understanding.

This interpretation, of course, bears a striking resemblance to the concepts of Yin and Yang and of dialectics; but in our case we not only have a vision of such a process, we also have mathematical equations that can represent the process.

Positive-Feedback Traps

In further computer experiments, it was found that if there was some characteristic or strategy that resulted in self-reinforcement, then once the characteristic or strategy emerged it could trap the population and block evolution, at least for some time. A biological example is the peacock's tail. A gene produces the beautiful tail of the male and makes the tail attractive to the female. In sexual reproduction anything that enhances the probability of mating produces a positive feedback on its own population dynamics and so fixes itself. However, such fixation may take place at the expense of functionality with respect to the external environment. Peacocks' tails are not an aid to finding food better nor to escaping predators. They simply mark a positive-feedback trap.

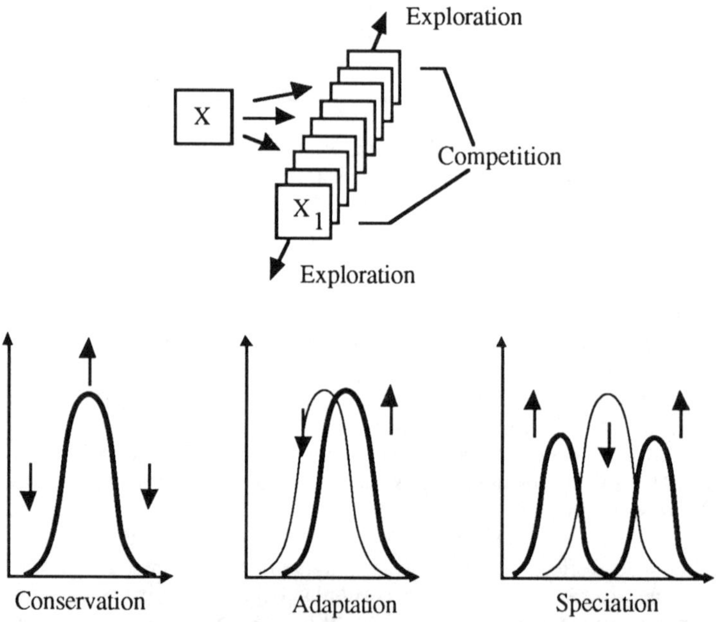

Figure 2.6. Evolution corresponds to the simultaneous "stretching" and "squeezing" shown here.

In human systems there is an abundance of such positive-feedback systems (see also Chapter 7). Much of culture may well consist of behavior that has become fixed. In most situations imitative strategies cannot be eliminated by the evolutionary process. Thus, fashions, styles, and indeed whole cultures rise and decline without necessarily expressing any clear functional advantages. Indeed, culture can be viewed as being not so much the "best" way of doing things somewhere but rather as perhaps the result of ignorance of other ways of doing things. Human activities in general, from economic "science" to Patagonian folk dance, exhibit these properties of autocatalytic self-organization; whereby ritual and shared ideology emerge and serve as the identity and focus of a social group, irrespective of the precise merits or truth of the ideology itself. A great deal of human attention is focused on playing roles in groups, where values are generated internally and the physical world outside is largely irrelevant.

Competition of Closely Similar Activities

The work discussed above has been further extended to show how adaptive landscapes are generated by the *mutual interaction* of behaviors. In the space of possibilities, closely similar activities are most in competition with one another, inasmuch as they feed off the same resources and suffer from the same problems. However, there is some distance in character space; that is, some level of dissimilarity at which the two types do not compete with each other.

Initially, a behavior grows until it reaches the limits of underlying resources set by the competition. At this point there is a positive payoff for explorers or entrepreneurs who escape somewhat from competition. It could be said that, although initially there was no hill to climb, the behavior effectively digs a valley for itself until there is a hill to climb on either side of the present behavior "centroid." However, over some distance in this space, the growth is restricted because of the "competitive shadow" of the original behavior, and the participants diffuse in small numbers up the slope away from the original type. Further, after a certain time, small initiatives arise which are sufficiently different from the original type, such that they can grow and multiply based on some other resource.

In its turn, this new behavior or strategy increases until it too is limited by internal competition for the given resource. Once again there is a payoff for deviants, particularly for those on the outside of the distribution, as the participants climb another self-made hill toward unpopulated regions of possibility space. In this way, well defined clusters of activities appear successively; and colonists diffuse out from each of them as they reach a competitive limit, gradually filling character space with a set of behaviors separated approximately by a distance that is characteristic of the resource diversity that can be tapped.

From a single activity our model generates a simple and dynamic ecology, inasmuch as the identity of each behavior is maintained by the balance between a continual diffusion of deviants outwards into character space and the competitive field that surrounds it. Random events that occur during the "filling" process will affect *which* populations arise. Hence, it is not true that the evolution represents the discovery of preexisting niches.

Such systems are *beyond* the mechanical paradigm because their responses to external interventions can involve changes in both structure and the nature of the activities or technologies in the system. Suppressing particular activities in these systems will provoke a complex response from the system as its other activities adjust. This situation may, for example, result from some environmental change.

The particular pattern of clusters that emerges depends on the accidents of system history. Clearly, the precise paths and directions of "exploration" are very random, and the pattern may be essentially unpredictable (Allen, 1990). Although the "inventiveness" of the population is constantly present, as there is diffusion into the possibility space, it is fascinating to see that our experiment shows that only at certain times does this inventiveness lead to structural change. In other words, the system evolves in phases of apparent stability, separated by periods of instability and fairly rapid reorganization, even though the disturbing pressure of exploration and creativity is relatively constant. Figure 2.7 depicts an evolution of structure.

APPLICATIONS FOR POLICY EXPLORATION: FROM SETTLEMENT TO FINANCE

Let us now turn from these theoretical developments to some practical applications of these new ideas to problems of policy exploration and long-term planning. The point emphasized here is that the kind of evolutionary model that we have developed can capture the response of the actors in a system and in the natural environment to their changing circumstances. Purely mechanical systems do not allow for learning or adaptation of the system affected. Hence, most plans end up where they were not intended to go or merely replace one problem with another.

Spatial Economics and Settlement Patterns

In the first applications of these ideas to human systems (Allen and Sanglier, 1979, 1981), a nonlinear dynamical system of equations expressing the supply and demand of different products was made to evolve by the random occurrence of entrepreneurs at different places and times in the system. The supply side was characterized by a non-convex production function for different economic activities, and consumer demand was assumed to reflect relative prices. The random parachuting of entrepreneurs onto the plain of potential demand resulted in the gradual emergence of a stable market structure and pattern of settlement.

As discussed in the previous section, we find "positive-feedback traps," where the system can get itself locked into a somewhat unsatisfactory market structure, as a result of a particular history. Basically, a very large number of possible stable structures could result from the experiment, involving different numbers of centers in different locations, and necessarily not offering the same level of efficiency.

Even in this very preliminary form, the models already showed many important principles: many final states are possible, precise prediction in the early stages is impossible, and approximate rules apply (e.g., to center separations) but always with considerable deviation and local individuality present. The results are affected by the particularities of the transportation system, as well as by the information flows associated with the mental maps of the consumers (Gould and White, 1978). The evolution of structure resulting from changing technology, transportation, resource availability, etc. can also be explored because the changing patterns of supply and demand affect each other in a complex, dynamic, spatial process.

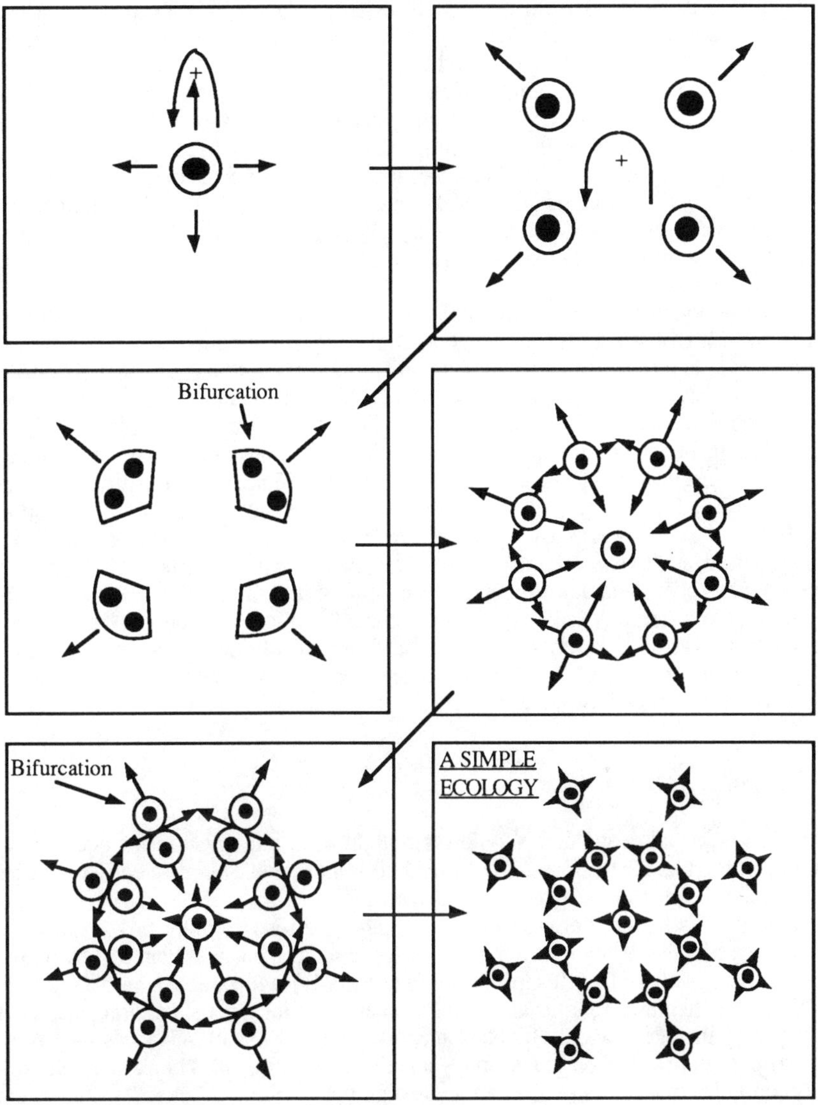

Figure 2.7. The evolutionary process leads to the emergence of "ecological structure." The particular clusters of behavior that emerge are interdependent, and the response to perturbations is therefore more than mechanical.

The fundamental basis for the models lies in the decisions of the different kinds of individual actors, and these decisions reflect the actors' values and functional requirements. These may be represented by very simple rules. However, the spatial dynamics give rise to quite complex patterns of structure and flow and to a structural emergence at the collective level. In such systems the microscopic and macroscopic levels are not related in a simple fashion. It is not true that the large structure is simply the small "writ large." This is because macroscopic structure emerges, and the result affects the circumstances of the microscopic parts as they find themselves playing a role in the larger, collective entity. Each actor is *coevolving* with the others. Micro-macro interrelationships provide one of the key topics in this book and receive further treatment in Chapter 4.

These models have been continually developed from the early simulations in order to provide realistic tools for understanding urban and regional evolution; here the patterns of structure and flow result from an ongoing evolutionary process of self-organization (Allen, 1985; Sanglier and Allen, 1989). Applications of these ideas have been made in the USA, Belgium, Holland, and France; the applications include those that led to the understanding of intraurban evolution in cities in Belgium and France.

The latest version of the models is designed to help in the formulation of development policy for Senegal. It allows the user (a constant) to change the transport network, to estimate energy costs, to develop new agricultural or industrial infrastructure, and to explore the consequences over time as the different links among demography, economic activity, and the natural environment affect each other. A representative interactive display for the Senegal application, that can be used for policy exploration, presents material both pictorially and alphanumerically. Displayed are a map of Senegal or a subarea and separately boxed data on populations, prices, salaries, etc.

Evolutionary Economics

The idea that economics should be thought of in biological terms occurred to authors at least as far back as Alfred Marshall (1890), and of course earlier Charles Darwin had been inspired partially by Adam Smith. However, for various reasons (amounting to a kind of "physics envy"), which are discussed more fully elsewhere, economics adopted the equilibrium perspective of a mechanical system that has come to rest. Economists like Joseph Schumpeter (1934) and later Richard Goodwin (1951), who attempted to introduce dynamical and nonlinear considerations, were not followed by the mainstream. But Nelson and Winter (1982) have advanced the evolutionary perspective in economics, and Kenneth Boulding's (1981) book on evolutionary economics makes many of the points being made in this chapter. However, the mathematical models presented here provide a new practical path of which Boulding was apparently unaware.

We have developed a description of the self-organization of a "market system," where the possibility space represents the possible products that could be supplied by firms for a particular function or end use. The possibility space can be viewed as a "quality space," and by using a simulation model strategies can be explored by competing firms. Also, evolution will create a market "ecology," resulting from the processes of research, development, and economic competition. Figure 2.8 depicts this evolutionary situation.

Another set of models has been used to study the behavior of financial markets. In these models the speculative game of investment is simulated by assuming that the differences among actors' perceptions, requirements, and beliefs allow transactions to take place. A given asset is considered to be over- or underpriced depending on the beliefs that a given actor has about the alternatives. The parameters that characterize an actor are, therefore, those which capture the manner in which he forms his beliefs and updates them in the light of events.

The model can include "fundamentalists" who try to assess future values, speculators who try to anticipate events, and "technical analysts" who try to ignore the real complexity and to extract buy-and-sell signals from some time-series data (e.g., on changing prices). The models are capable of producing quite useful and realistic behavior as well as market structure. Once again, it is the diversity of different actors that allows the system to function at all, and what we see emerging is a "financial ecology."

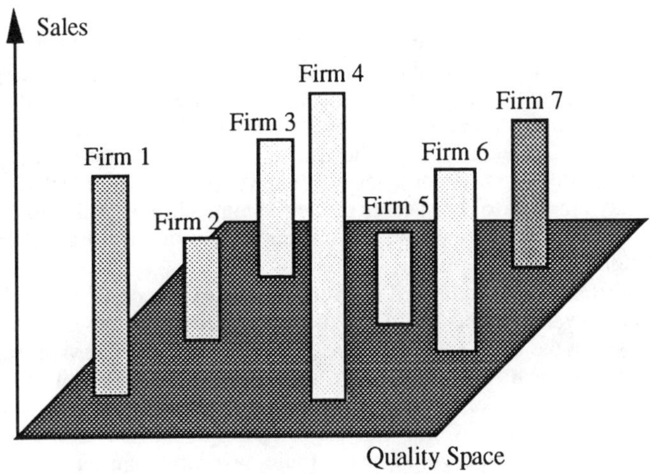

Figure 2.8. A self-organizing model of a market system leads to an ecology of firms.

Management of Natural Resources

This evolutionary approach has also been used to develop dynamical models for the management of natural resources (Allen and McGlade, 1986; McGlade and Allen, 1985). Following is a brief summary.

The essential point revealed was that success in fishing, as in life as a whole, requires two almost contradictory facets of behavior. First is the ability to organize one's behavior so as to exploit the information available concerning "net benefits" (that is, to be rational). We have called this kind of behavior "Cartesian." A second, more surprising, behavior is the ability to ignore present information and to explore the domain beyond present knowledge. We have called fishermen showing this kind of behavior "Stochasts." The first kind of behavior makes good use of information, but the second generates it! At the root of creativity is always the second kind. (The concepts of "chance and necessity," of stochasticity and determinism, of the limits to rationality, and of the generation, growth, and exhaustion of information and of ideas are of immense importance in this book. They are discussed in detail in Chapters 1 and 4.)

In the short term the more "rational" actor must outperform the less rational. In this sense, taking steps to maximize present profits is better than not doing so. Nevertheless, over a longer period the best performance will not come from the most rational behavior but instead from some complex compromise between the two kinds of behavior. For example, a fleet of Cartesians, which goes where available information indicates the highest profits, will in fact lock into zones for much too long, remaining in ignorance of the existence of other, more profitable zones simply because there is no information available concerning these other zones.

You don't know what it is you don't know!

New information can come only from boats whose skippers have chosen not to fish in the "best" zones, from boats which lack a necessary technology, or from boats whose skippers do not share the consensus values--that is, from boats that *generate* information. These skippers behave like risk takers, but they may or may not perceive themselves as such. They may act as they do through ignorance or through a belief in some myth or legend. Whatever the reason, or lack of it, for their behavior, they are vital to the success of the fishing endeavor as a whole. It is their exploration that probes the value of the existing pattern of the fishing effort and lays the foundation for a new effort.

As information is generated concerning the existence of new, rich fishing grounds, so the value of this information starts to fall as the news spreads and exploitation rates at the new grounds increase. We see a cyclic pattern in the discovery of value in a zone that involves the spread of information and with this spread the saturation or exhaustion of the discovery, calling for fresh explorations (Allen and McGlade, 1987b, 1987c). Complexity develops in the system as fishermen try to discover and conceal information, using radio eavesdropping, encoded messages, scramblers, night sorties, monitoring of cafe gossip, cooperatives, alliances, and structured fleets.

The model can be used as a simulator for the overall management of the fishery, or for the benefit of any particular fleet wishing to improve its performance. The parameter values that appear in the mechanisms governing the decisionmaking for fishing boats are calibrated so as to give realistic behavior (Allen and McGlade, 1987b). It is also possible to construct a model that will run competing strategies, and that will discover for itself the robust and successful sets of strategies–generating an "ecology" of fishing strategies.

A pictorial and alphanumeric display can be generated, as in the Senegal application, so as to allow policy to be explored. For example, a map of Nova Scotia and surrounding land and sea areas can be presented. The location of and data pertaining to various fleets are then indicated.

DISCUSSION

Reflecting on the nature of the new scientific paradigm of complex, self-organizing systems, we find that it suggests why the world is facing major crises like those summarized in other chapters. For instance, instead of finding that the world economy is moving swiftly to the theoretical "optimal" situation, the El Dorado of equilibrium, we find that our very existence is under potential threat from climatic change, global warming, and sea-level rise. Similarly, even in the developed countries, water resources, coastal resources, soils, and marine ecosystems are all under severe stress. In developing countries pressures are even greater as population grows and economic growth is demanded. There are enough examples so far of failed development policies to awaken us to the need for "sustainable development."

Sustainable development, however, can be a new catch phrase, and it must be understood in terms of the ideas that have been developed here. It should not be interpreted as the search for the perfect equilibrium. The world will never stop changing, so what sustainability is really about is the capability to respond, to adapt, and to invent new activities. The power to do this lies not in extreme efficiency, nor can it be had by simply distributing cash. The power lies in *creativity*. And in turn creativity is rooted in diversity, cultural richness, openness, and the ability and will to experiment and to take risks.

The evolutionary models discussed here can be of help in examining the long-term, broad aspects of policies in order to see what the response of the system might be and where the response might take the system. But these models are not a panacea (nor are any other models). There is no easy solution to the human condition. Looking at the easy rhetoric of most political ideology, and at the empty theories that serve as support for the different kinds of ideology, it seems extraordinarily presumptuous to dare to think that complex systems can be so easily dealt with! To someone from the natural sciences, the thought of basing large parts of national (and international) policy, and the associated political ideology, on ideas with such flimsy underpinnings seems the height of folly. But then, who said that human beings are wise?

Instead of viewing the changes that occur in a complex system as necessarily reflecting progress up some preexisting, if complex, landscape, we have shown that the landscape of possible advantage itself is produced by the actors in interaction, and that the detailed history of the exploration process itself affects the outcome. Paradoxically, *uncertainty is therefore inevitable*, and we must face up to this constraint. Long-term success is not just about improving performance with respect to the external environment of resources and technology; it is also affected by the "internal game" of a complex society. The payoff of any action for the individual cannot be stated in absolute terms because it depends on what other individuals are doing. Strategies are *interdependent*.

Ecological organization is what results from evolution, and it is composed of self-consistent sets of activities or strategies that both pose and solve the problems and opportunities of mutual existence. Innovation and change occur because of *diversity*, the nonaverage individuals with their bizarre initiatives. Whenever these initiatives lead to exploration into an area where positive feedback outweighs negative feedback, growth will occur. (See also Chapters 1 and 7.) Value is assigned afterwards. It is through this process of post hoc explanation that we *rationalize* events by pretending that there was some preexisting "niche" which was *revealed* by these events, although in reality there may have been a million possible niches and one particular one arose.

The future, then, is not contained in the present, in the sense that *the landscape is fashioned by the explorations of the climbers* and adaptability will always be required. This does not mean that total *individual* liberty is necessarily always best. Our models show also that adaptability is a *group or population* property. It is the shared experiences of others that can offer much information. Indeed, it pays everyone to help to facilitate exploration by sharing the risks in some cooperative way, which takes some of the "sting" out of failure. Performance is generated by *mutual* interactions, and total individual freedom may not be consistent with good social interactions; hence, such freedom will make some kinds of strategy impossible. Once again, we must differentiate between an "external game," where total freedom allows wide-ranging responses to outside changes; and an "internal game," where the division of labor, internal relations, and shared experiences play a role in the survival of the system.

It is thus naive to assume that there is any simple "answer" to the world's difficulties. The world is just not made for simple, extreme explanations. Shades of gray, subjective judgments, post hoc rationalizations, multiple understandings, and complex motivations are what characterize the real world. Neither total individual freedom nor its opposite provide solutions since there is no "problem" to be solved. Such freedom and its opposite are possible choices among all the others, and each choice gives rise to different strengths and weaknesses, a different spectrum of possible consequences, and different successes and failures. Much of this situation cannot really be known beforehand. We can only do our best to imagine possible futures, and to try to carry on modifying our views about reality and about what we want.

Mismatches between expectation and actual outcome may either cause us to modify our (mis)understanding of the world or, alternatively, simply leave us perplexed. Evolution in human systems is therefore a continual, imperfect learning process, spurred by the difference between expectation and experience, but rarely providing enough information for a complete understanding.

Finally, instead of the classical view of science's eliminating uncertainty, the new scientific paradigm accepts uncertainty as inevitable. Indeed, if this were not the case, then it would mean that things are preordained, a situation that would be much harder to live with. Evolution is not necessarily progress, and neither the future nor the past can be considered preordained. Creativity really exists. It is the motor of change, and the hidden dynamic that underlies the rise and fall of peoples, regions, and civilizations (see the discussion in Chapter 1). Evolution both encourages and feeds on invention. Recognizing this, the first step towards wisdom is the development and use of mathematical models which capture this truth. The topic of modeling is pursued further in detail in Chapter 7. A comparison of theoretical bases for modeling is made in Chapter 4.

REFERENCES

Allen, Peter M., 1985. "Towards a New Science of Complex Systems." In Shuhei Aida (Ed.), *The Science and Praxis of Complexity*. Contributions to the Symposium Held at Montpellier, France, 9-11 May 1984. Tokyo: United Nations University Press.

Allen, Peter M., 1988. "Evolution: Why the Whole is Greater than the Sum of Its Parts." In W. Wolff (Ed.), *Ecodynamics*.. Berlin: Springer-Verlag.

Allen, Peter M., 1990. "Why the Future is Not What It Was." *Futures*, (July/August), 555-570.

Allen, Peter M., and Michèle Sanglier, 1979. "Dynamic Model of Growth in a Central Place System." *Geographical Analysis*, 11(3), 256-272.

Allen, Peter M., and Michèle Sanglier, 1981. "Urban Evolution, Self-Organization and Decision Making." *Environment and Planning A*, 13, 167-183.

Allen, Peter M., and Jacqueline M. McGlade, 1986. "Dynamics of Discovery and Exploitation: The Case of the Scotian Shelf Groundfish Fisheries." *Canadian Journal of Fisheries and Aquatic Science*, 43(6), 1187-1200.

Allen, Peter M., and Jacqueline M. McGlade, 1987a. "Evolutionary Drive: The Effect of Microscopic Diversity, Error Making and Noise." *Foundations of Physics*, 17(7)(July), 723-739.

Allen, Peter M., and Jacqueline M. McGlade, 1987b. "Modelling Complex Human Systems: a Fisheries Example." *European Journal of Operational Research*, 30(June), 147-167.

Allen, Peter M., and Jacqueline M. McGlade, 1987c. "Managing Complexity: a Fisheries Example." Report to the United Nations University, Tokyo.

Allen, Peter M., and Jacqueline M. McGlade, 1989. "Optimality, Adequacy and the Evolution of Complexity." In Peter L. Christiansen and Robert D. Parmentier (Eds.), *Structure, Coherence and Chaos in Dynamical Systems*. Manchester, England: Manchester University Press.

Arrow, Kenneth J., and Gerard Debreu, 1954. "Existence of an Equilibrium for a Competitive Economy." *Econometrica*, 22, 265-290.

Boulding, Kenneth E., 1981. *Evolutionary Economics*. Beverly Hills, CA: Sage.

Debreu, Gerard, 1959. *Theory of Value*. New York: Wiley.

Goodwin, Richard M., 1951. "The Nonlinear Accelerator and the Persistence of Business Cycles." *Econometrica*, 19, 1-17.

Gould, Peter, and Rodney White, 1974. *Mental Maps*. New York: Penguin.

Marshall, Alfred, 1890/1927. *Principles of Economics*. Eighth edition. London: Macmillan.

McGlade, Jacqueline M., and Peter M. Allen, 1985. "The Fishery Industry as a Complex System." In R. Mahon (Ed.), *Towards the Inclusion of Fishery Interactions in Management Advice*. Canadian Techniques of Fisheries and Aquatic Science Number 1347, pp. 209-216. Ottawa, Ontario, Canada: Fisheries and Oceans.

Nelson, Richard R., and Sidney G. Winter, 1982. *An Evolutionary Theory of Economic Change*. Harvard University, Cambridge, MA: Belknap Press.

Nicolis, Grégoire, and Ilya Prigogine, 1977. *Self-Organization in Nonequilibrium Systems: From Dissipative Structures to Order through Fluctuations.* New York: Wiley.

Prigogine, Ilya, and Isabelle Stengers, 1984. *Order out of Chaos: Man's New Dialogue with Nature.* New York: Bantam.

Sanglier, Michèle, and Peter M. Allen, 1989. "Evolutionary Models of Urban Systems: an Application to the Belgian Provinces." *Environment and Planning A*, 21, 477-498.

Schumpeter, Joseph A., 1934. *The Theory of Economic Development.* Cambridge, MA: Harvard University Press.

CHAPTER THREE

ADAPTIVE CONTROL AND CONFLICT RESOLUTION FOR INTERNATIONAL STABILITY

by Harold Chestnut

INTRODUCTION

In Chapter 1 the evolution and structural change of the world society and environment were emphasized, and within this context conditions that may lead to war or peace were discussed. In Chapter 2 some formal models of evolutionary processes were presented. In this chapter formal models of a cybernetic nature are utilized.

In thinking about policymaking in the 1990s, it will be assumed that at least some of the decisionmakers are desirous of providing a common and sustainable future for their nation and its people. This future is one that can provide a fair measure, or better than fair, of international stability for the nations of the world, and one that strives to meet the long-range needs of the planet and of the people who live on it. The report of The World Commission on Environment and Development (1987) on "Our Common Future" provides a vision of what policymakers might have for their goals for the 1990s and some ideas on how these goals can be achieved. The Stockholm Initiative is another fine description of objectives for the future (Prime Minister's Office, 1991).

International stability should be built around the concept that changes of people and nations with time are necessary and inevitable, but that those changes which occur in an orderly and foreseeable fashion are to be preferred from those which are sudden and unexpected. Achieving international stability should include consideration of the people's essential needs for such goods and services as food, clothing, housing, health, education, transportation, communication, finance, and entertainment using world laws rather than military force as the primary means for resolving conflicts.

Important elements of the systemic basis for policymaking in the 1990s are the principles of systems engineering which are aimed at consideration of the overall needs of the nations and their peoples for the present and the future. Inherent in system design, manufacture, and operation are the principles of closed-loop control, in which stable system operation is an essential and important factor (Chestnut, 1965).

Also of particular importance in the systemic basis of policymaking is the recognition that the planet is made up of subsystems such as food, population, industry, environment, energy, health, etc., many of which are nonlinear and may change with time. Thus, the control that policymakers may incorporate in their decisionmaking procedures should be capable of adapting to the various conditions of the system of nations and should have built into them means and procedures for resolving the conflicts that are more than likely to occur.

As a basis for considering the many different conditions which may exist between and among nations, it is important to take into account the several different parts of the Peace/War/Peace cycle which policymakers will have to include in their decisionmaking processes for providing international stability. Depending on where nations stand in the Peace/War/Peace cycle, conditions may be considered to be normal, alert, or emergency, and the policymakers must adapt their decisions and control to the appropriate conditions for that part of the cycle which are thought to exist.

Our common future requires that there be a more peaceful and stable world. By reducing national expenditures for armaments, it should be possible for the leaders and decisionmakers of the world to obtain the funds and resources for a satisfactory environment and development. In order to increase the likelihood that these funds and resources be made available, it will be necessary for world leaders and policymakers to develop ways and means of resolving conflicts with the aid of laws, treaties, and other decisionmaking techniques, many of which are already established and have been used successfully in different fields of endeavor and in various parts of the world (Fisher and Ury, 1981).

Fortunately, the field of peace research/peace studies has developed an appreciable literature, from which the following are selected references: Beer (1981), Boulding (1978), Galtung (1975), Singer (1980), SIPRI (1979), Wilkinson (1980), and Wright (1964).

Getting from Here(1992) to There(2000)

Today's world is a large and complicated system. To work well, it must be stable and able to change in an orderly way. The world system itself is made up of the many interdependent subsystems that are the world's nations, about 160 of them. In addition, our planet Earth is part of a solar system that impacts on the weather and environment of the nations. The emphasis of the member nations of the world has been on the security, well-being, and safety of each nation, with insufficient attention being given to the overall or *international* point of view. Now that technology has made it possible for new understanding, new accomplishments, and new opportunities for the world's people, these people are now able to ask new questions of their leadership. Many of these questions relate to what can be done so that the civilization of the whole world can look forward to a more attractive future in which to live and develop with a greater measure of certainty concerning its future existence (Gorbachev, 1988).

At present many people in many countries are spending their time and money to provide new and improved weapons for wars should they develop. However, for the most part, these wars are not wanted by most of the people of the world's countries. There appear to be too few people devoting their efforts and money to using the present time, when the military forces of the world's major powers are not at war, to strengthen and build up the understanding and confidence among countries that will lead to continued peaceful existence.

"If you don't know where you are going, any path will take you there," is a saying that seems to have particular significance at present. On the other hand, suppose that each of the major powers, collectively if they so desire, or separately if they choose not to act together, establish a "steering committee" to direct their country, and collectively the world, to a set of goals and objectives to get the people of the world from the present, 1992, to the year 2000 without a major war among the industrial national powers. One of the important findings of the technology of systems and cybernetics is the steering or guidance principle which states that one important way by which to achieve a particular desired result is to have firmly in mind the result desired and to understand the principles by which it is possible to achieve the result desired (Chestnut, 1965).

An organized effort on the part of the physical scientists, engineers, technologists, social scientists, political scientists, and others it needed to understand better the present processes of the world's nations; to identify desirable objectives for the future for them separately and collectively; and to develop some mutually acceptable plan for getting them from here (1992) to there (2000). These efforts at understanding the major world processes as well as the behavior and attitudes of the nations of the world; of identifying desirable and acceptable objectives for the future; and of helping to guide the necessary people and resources for getting from here to there in an orderly fashion are ones which challenge the best skills of the practitioners of the world's intellectual and political communities.

The overview which follows sets forth some of the premises that may be helpful in providing an adaptive control system as well as the conflict resolution means for a more systemic basis of policymaking in the 1990s.

OVERVIEW OF INTERNATIONAL STABILITY AND CONFLICT RESOLUTION ISSUES

Present Period Is One of Rapid World Change

The world is in a period of very rapid change in terms of the numbers of people on it and their way of life. The nature of the future for the increasing number of nations in the world is likewise subject to change. In particular, the nature of the emphasis on science and technology has changed our idea of life and its expectations. However, although the world has become more and more interdependent in its economic, technical, social, and cultural life, its political and military approaches continue to emphasize *national* values rather than international well-being.

We as a world civilization have become less fearful of natural destructive forces and more concerned with the possibility of man-made destruction from military armaments. Especially have the nuclear armaments developed in the last 40 years, now present in greatly increased numbers measured in the tens of thousands, caused millions of people in most of the developed countries of the world to be concerned about their personal and collective future. Will civilization survive to the year 2000?

Changes from the Past are Possible in the Future

In the world of centuries past, in which ideas changed relatively more slowly, it was possible to anticipate the future with more assurance, especially since the life expectancy was relatively short--perhaps 30 to 45 years. Now with the life expectancy increasing to 55 to 75 years in many countries, and with global weapon delivery systems capable of destroying millions of people in less than an hour, there is an increasing concern about what the future of life on earth will be.

However, there is a feeling on the part of many people that the future is not foreordained and that more people should have a greater influence on what will the future be for them. They want to live and have a greater chance for the life for their children. They feel that there should not be a predetermined future which is inevitable. If there is to be a desirable and attractive future for the majority of mankind, it is not likely that any *one* of the present nations can achieve this desirable future all by itself. Cooperation between nations at some level of activity will be required.

Methodology Needed for Constructive Change

It would appear that there must be developed a methodology for looking at nations in a systemic fashion in terms of their several sectors of activities, as well as in terms of the several constituencies of their members. Sectors should involve significant aspects such as demography, households, production means, utilities, labor, government, and finances, each of which may have different elements or facets in their make up. National constituencies as mentioned above refer to groups of people such as those that constitute the beneficiaries of the goods and services supplied, the operators of the processes, the suppliers of the equipments and infrastructures, the designers and builders of the processes, and the providers of the resources such as energy, materials, and information. The methodologies to be developed should enable various alternative scenarios or possible futures to be evaluated in terms of their effects on national, regional, international, as well as local, groups under medium and longer time-frames of reference.

Clearer Understanding of International Stability Needed

In today's highly interdependent world, activities in one country can have an important impact on the stability and events of other countries as well. A clearer understanding of the factors affecting international stability must be obtained so that national governments and other national decisionmakers can be aware of the influence of their decisions on world peace. Rather than using the presence or absence of war as being the major criterion of whether or not a nation is in a state of peace, it is desirable to use some quantitative or, if necessary, some qualitative indication of the state of the nation's stability. In this way it should be possible to establish whether there is a likelihood of there being a continued state of peace or a change to war in the foreseeable future.

Peace/War/Peace Process Understanding Needed

Improved understanding of the many facets of the Peace/War/Peace process can do much to enhance the likelihood of international stability and peace. The

Peace/War/Peace cycle is an important framework within which one can anticipate movements in the direction of peace or war. The geographical location or core-periphery phenomenon is another such important framework within which one can anticipate movements toward war or peace. The concepts of peacekeeping, peacemaking, and peacebuilding are also useful frameworks for influencing activities that are more or less conducive to peace. The influence of human expectations and perceptions represents another facet of the Peace/War/Peace process that needs better exposure as a means of improving international stability and peace. By no means to be overlooked is the importance of the improved understanding of the influence of peace on the many factors which have been considered to be the major causes for war. Through the study of the many wars and warlike situations of the past and the present, it may be possible to learn enough of the Peace/War/Peace process that specialists in the various skills involved in war and peace will be able to become more skilled in improving the ability of nations to live in a more stable and peaceful fashion.

Different Value Systems and Bases for Judging Nations Exist

In studying the influences of various scenarios and alternative possible actions of nations on one another, it is desirable to understand the value system by which the various decisionmakers in different countries may judge the importance of different events or activities (Kile, 1987). The needs and values of people as individuals in different countries differ. Further the factors that make up the Quality of Life (QOL) in different countries are not all the same, let alone the weighting of them. It is not essential that there be agreement on what are the factors or what their weighting should be; it is desirable that there be an awareness of what factors are important to the different countries involved. See also Chapter 8.

Although the economic approach to judging the performance of a nation would not be the only basis for judging a country's performance, it is worthwhile to understand that there are certain quantitative economic data that are quite available and can be useful during some initial analyses. Since there are these economic data, it is worthwhile to be able to use them.

Role of Closed-Loop Concepts in Steering Nations to the Future

An important principle developed in connection with the control of physical processes is that of cybernetics or closed-loop control. The closed-loop control concept involves the establishment of one or more reference signals or desired conditions which are compared to measured values of the actual conditions of the process or processes. The difference between the desired and actual conditions is used in some suitable way to initiate actuating signals, which in turn direct the actual process conditions into correspondence with the signals representing the desired conditions for the process. This closed-loop principle is used extensively and successfully in many servomechanisms and other weapon-control systems. It would appear likely that this closed-loop control can provide some useful insights into satisfactory means for steering nations and their peoples to a more stable and peaceful future. In the case of directing nations to a more stable and peaceful future, the nature and significance of the desired reference conditions are of particular importance since they greatly influence the outputs that are produced by the national and international processes.

Adaptive and Predictive Aspects of Guiding Nations

Adaptive control is a specialized and advanced form of control which can have special applicability in trying to achieve a desirable form of interaction between nation states (Chestnut, 1965). Further, the predictive control form of adaptive control appears to be highly attractive as a means of establishing effective means for identifying suitable ways of moving nations and people from the present to a desirable future. Fundamental to the adaptive control principle are the ideas that there may be a variety of different states or conditions of system operation. The predictive control principle emphasizes computation and communication, as well as a range of time conditions, for which different control requirements may be employed.

It should be understood that there is an important and critical dependence on establishing political, social, and military objectives as well as economic and technical ones.

Roles of References, Constraints, Perceptions, and Ideology

An understanding of the principles of closed-loop control performance indicates the critical importance of the references and constraints of a system. In a psychological and social science sense, frequently the perceptions people have of themselves and others are also very significant in the effects they have in terms of what people think is possible.

Based upon such information sources as the national Constitution, various historical declarations of purposes, the annual statements of the national leaders, the financial budget and statement of accounts of the government and the various other national sectors, one is able to establish to a reasonable extent what are the ideology and traditions which serve as references to the government, industry, and the people of the nation.

In addition to the ideology and traditions of each nation, it is important to recognize the many real differences which exist between any two countries and which serve to form the basis for the perceptions that each country may have of itself. Nations differ in such important factors as the size and climate of their territory, the number of inhabitants, the availability of natural resources, the degree of industrialization, the education and skills of their population, their social and religious beliefs, their military and political attitudes, and the nature of the leadership processes. These various conditions and others tend to produce the situations, whereby each country must be looked at as a special case in terms of its references; that is, its desired performances, as well as its perceptions of itself and other countries. In addition to trying to understand itself and its perceptions, each country must make a concerted effort to understand the comparable characteristics of the other countries with which it deals.

Conflict Resolution for Settling Disputes

Over recent decades there has been an increasing awareness of the inevitability of there being a conflict of interests on various levels--industrial or sectoral, national or labor-management, as well as international along ideological or religious lines. Likewise, there has been developing an increasing number of skills in facilitating, negotiating, arbitrating, and in general in resolving such conflicts. Although there is increasing competition in many fields, the rules are becoming better defined, so that conflicts can be terminated successfully by mutually acceptable agreements instead of by killing (Kahn, 1985).

In many cases there has not yet been developed a common awareness of the language, methods, and structure involved in these conflict resolution means that are used in different countries and cultures. It would be advantageous for there to be a clarification of the understanding of the methods used in different countries for these purposes. Likewise there should be a further clarification of the methodology of conflict identification and structure as well as of the overall conflict resolution process itself.

A need exists for developing impartial and trustworthy people and organizations that can be used as disinterested third parties, fact-finders, arbitrators, etc. in the resolution of the various conflicts that will inevitably arise. People with such skills should be identified, trained, and recognized for the skills that they possess.

Education to Facilitate Change

Work in education for a more peaceful world is taking place internationally and on a national basis in many countries. More effort is needed and more adequate funding is required to make progress in coming up with ideas concerning peace education that has more of an international peace emphasis--that is, peace for several countries, not only our own. In particular is a comparison desired which coordinates educational ideas of the different countries, so that an awareness is developed of the influences of different cultures and perceptions as well as the ideological differences that are to be found in the different countries of the world. The influence of such means as advertising, radio and TV, and sports and cultural exchanges should be included in thinking about education for peace.

A Cooperative Security System

The advent of technological advances in the past few decades has made possible a more interdependent set of relationships among nations that can emphasize cooperative rather than individual security. Mutual economic and technical organizations, as well as multi-national corporations, represent examples of institutions that foster cooperative rather than individual results. Rather than counting primarily on military means for bringing about agreement among nations, use is made of cooperative, informational means of working out mutually acceptable bases for the solution of differences in which commonly agreed-upon criteria for judgment are employed. The possibility of establishing a Cooperative Security System (CSS) should be explored, and its requirements should be defined. The characteristics of a CSS should be looked at from such viewpoints as geography, hardware, software, categories for potential conflict, and system states for decisionmaking, as well as the states of the various nations involved, and from the several time regimes of interest.

It must be recognized that different countries have different needs and different ways of operating. The possibility of developing adaptive control and decisionmaking means for the nations involved should be explored and clarified as part of the investigation of the CSS.

Work Breakdown Structure of Activities

Various aspects of the problem of improving the understanding of international stability have been presented, but there has not yet been an identification of *which* of the tasks or activities are to be done in what order and to what degree of detail. Truly there is much to be done, and certainly it is not possible to do everything at once. To provide a list of options of what tasks could and should be done, a Work Breakdown Structure (WBS) should be developed and presented which itemizes a basic list of activities for improving international stability and indicates the various detailed activities which should be done as part of each.

From a review of this WBS it should be possible to have a basis for discussion of which major activities appear to be most important or most critical, which should be done first, and which might be deferred initially and performed at a later time. Some of the details of the WBS activities might show that they should have more or less emphasis for various reasons.

It is essential that one view the process of improving international stability as an evolutionary, learning experience in which some progress is made initially which serves as a foundation upon which more progress can later be made. The Wright Brothers showed that man could fly a heavier than air machine; they didn't build a Boeing 747.

Obstacles to Peace

The preceding material has tended to imply the thought that, if people were to put their minds to work on the matter, then international stability and peace could be readily achieved. In reality, of course, the situation is somewhat reversed. The system of wars that we have endured over the years seems for some people to have been the best one that civilization could develop. In other words, alternative systems have been available and the one that we have is better in some sense than the others which we do not have.

The following lists some important generic obstacles to be overcome in the process of bringing about improved international stability and increasing the likelihood of achieving a more peaceful world:

- The perception of the process of attaining a more peaceful world may differ from country to country.

- People must be educated to accept the concept and the operational necessity for there not to be a war.
- Many people are not convinced that peace is possible.
- The role of technology in providing change must be better understood.
- War and the preparation for war have certain desirable features for some people.
- Cultural differences, prior conditioning, and the nature of the decisionmaking process are different in different countries.

Some day, somewhere, some people are going to find a way of enabling the people throughout the world to live without war between and among the nations. We who are alive today have a better chance than any people on earth so far to make peace and stability possible.

We have an unparalleled opportunity if we would just take advantage of this chance. Let us address ourselves to the challenge of trying to find better ways to enable the major nations of the world to live together without wars.

THE PEACE/WAR/PEACE PROCESS AND SOME OF ITS ELEMENTS

In the preceding sections the subject of international stability and peace have tended to be used somewhat interchangeably. It appears to be reasonably self-evident that the chances for there to be improved international stability are considerably enhanced by the presence of peace. Under the conditions of war, the likelihood of there being satisfactory stability among nations is highly unlikely since rapid and costly changes are apt to take place as the military destruction occurs (Chestnut, 1984).

Frameworks

In this section an effort is made to identify the Peace/War/Peace process so as to develop frameworks in which it will be possible to gather data and analyze them from the point of view of whether the present situation is one that is likely to develop into one where war is more probable than heretofore. These frameworks represent a number of different perspectives from which it should be possible to anticipate whether the likelihood of instability to occur will increase or decrease.

Block Diagram of Nations. The first of these frameworks is that of the block diagrams which look at the nation and the activities that are taking place within the country as well as the impacts that activities taking place in other countries are having on the first country, and vice versa. The primary emphasis of any country is meeting the needs of its people. The information available to the policymakers and decisionmakers at all levels of that country, as well as the national processes--technical, social, economic, and political--influence what can be done in that country. In addition, activities in other countries may influence the activities of the first country as well as others. It is essential that an effort be made to understand the nature of the interrelationships of the sort that can be shown by means of block diagram.

Peace/War/Peace Cycle. A second framework of significance is the

Peace/War/Peace cycle in time. In this case there tend to be times when relations among countries are peaceful, perhaps after some time of war. At a later time the countries have marshaled their strength again and they prepare themselves to fight. A war breaks out again, and the nations fight until one side or the other is victorious. Peace may again occur after which the Peace/War/Peace cycle is renewed.

Human Expectations and Perceptions. Human expectations and perceptions represent a framework element that can have an important influence on peace and war. Expectations or anticipations are the result of education, conditioning, current events, or propaganda. Expectation serves as a reference in terms of what people believe is happening or is about to happen. Perceptions are like feedback in a closed-loop control. When expectations and perceptions are essentially the same, then the likely behavior tends to be well defined. When expectations and perceptions are not the same, the possibility may exist for some unexpected results to occur and the likelihood of war is greater.

Internal and External Communications. There are many considerations which hinder effective communication and therefore are of importance as impediments to peace and stability. These include preoccupation with internal elements in a country that may obscure external happenings; emotional blocks in which certain words have an inflammatory effect, ideological hostility, charisma of a leader without regard to ethics or morals; past history with one or another countries; and hidden agenda of local objectives that override the needs and best interests of the country as a whole. Since much of what is involved in the matter of stability involves communication, those elements which contribute to noise or unpredictability in the system represent a framework of concerns that should be included.

Conditions of Peace. The words "peacekeeping," "peacemaking," and "peacebuilding" described by Johan Galtung (1975) represent a third framework of concerns that are important to the Peace/War/Peace process. These terms are ones that are part of the vocabulary of peace research and peace science. They may in fact be more closely related to the Peace/War/Peace time/activity cycle than has been heretofore recognized. Peacekeeping is a dissociative process of keeping people and armies apart. Peacemaking is a conflict resolution process and includes such activities as negotiation, mediation, and in general using reason rather than force. Peacebuilding refers to an associative process of creating a synergistic situation in which all parties involved can benefit in various ways. It may be that in these terms the control logic of the process of improving stability can be found.

Through the study of many wars and warlike situations it may be possible to learn enough of the Peace/War/Peace process that technical and other experts will be able to become more skilled in improving the ability of nations to live in a more stable and peaceful fashion.

INTERNATIONAL STABILITY AND PEACE

In describing the subject of international stability, as noted above, the term "peace" has often been used as an interchangeable expression to international stability. It would seem to be reasonably evident that the chances for there to be improved international stability are considerably enhanced by the presence of peace. As a part of the educational process of improving the likelihood of peace, there needs to be an expanded effort in educating people in the meaning of peace and ways for increasing the likelihood that peace will be achieved.

A significant contribution to international stability can be made by bringing together people from various countries to develop educational material concerning peace and its interpretation as seen from different countries. Although it is not likely that initially there would be full agreement on the meaning and implications of peace, nevertheless, it is likely that a better understanding of the issues involved in improving international stability would be obtained and the possibility for peace would be enhanced. The discussion which follows endeavors to describe "what is peace" and presents a number of descriptions of what peace researchers and educators have in mind when they speak of peace. These ideas draw heavily on work of Dennis F. Carey and Maire A. Dogan, titled "Peace Studies: An Introduction," and done while they were at Kent State University.

What Is Peace?

The question "what is peace?" is straightforward. It would be quite natural for one to expect that there would be a neat one-sentence answer to that question. However, peace is more complicated than that and any such short definition of the concept of peace would be too simplistic. In fact, there was until fairly recently a great deal of disagreement among peace researchers and educators on the meaning of the word "peace." This lack of agreement and understanding of the meaning of peace may be a contributing factor to the lack of achievement of peace among nations.

Peace scholars have attempted to reduce the variety of the approaches to peace, and the associated confusion with its meaning, by developing analytical definitions. As a result, the definitional debate on the nature of peace has been focused on two key concepts: negative peace and positive peace. Although there is still some variety among the scholarly definitions, almost all can be subsumed under one of these two terms.

Negative Peace

Negative peace can be defined by the simple phrase, the absence of war. However, the problem with this definition is that the word "war" is somewhat ambiguous since some wars are not declared, and other wars such as the cold wars are not fought in any classical sense. A more precise way of defining negative peace is "the absence of large-scale physical violence." Physical violence is "the infliction of actual bodily harm and/or the destruction of material which is immediately and directly necessary for human survival." It is large-scale when it is undertaken by large social units such as nation-states, cultural/ethnic groups, or alliances of such units.

To be more concrete, an example of an act of large-scale physical violence is the bombing of a city because this results in the killing and injury of human beings and the destruction of basic goods on which they depend. If the individuals are not injured, they still face imminent bodily harm because sources of food, water, and shelter have been destroyed.

Some people, however, have been dissatisfied with the definition just offered even though it can be made quite concrete. This is because there are situations in which there really is peace between nations and other situations which are merely brief intermissions between wars, and it would be misleading to call both of these situations peaceful. Yet, according to the above definition, no distinction would be made between these situations and they would both be classified as peaceful. In an even more extreme case, a "peaceful" period could be merely a case of regrouping each side's military strength in order to repursue the war. To prevent this distortion and to refine the definition of peace, a distinction has been introduced between "stable peace" and "unstable peace."

Stable Peace

Stable peace is a "situation in which the probability of war is so small that it does not really enter into the calculations of any of the people involved." The concept of stable peace goes beyond that of negative peace in that it includes not only the absence of war but also the expectation that this condition will continue. An example would be the relationship between Canada and the United States. Although the history of the relationship includes actual and threatened physical violence, these events are not within living memory of citizens of either country. Residents of Northern New York or Montana do not fear that terrorists from across the border will attack their towns and farms in the night, nor do Canadian border residents fear an invasion from the south.

Unstable Peace

Unstable peace, on the other hand, is a situation in which no large-scale physical violence is going on at the time, but the threat of violence is real. Relations between the former Soviet Union and the United States constituted a case of unstable peace. Even though these relations had not involved physical violence between the two governments for well over 60 years, people did not expect this situation to continue. In fact, a major bulwark of the foreign policies and calculations of both countries was the fear that the other side would attack and the perception, true or false, that the only thing that kept the other side from attacking was the fear of retaliation.

Although the typical reader may never have consciously distinguished between stable and unstable peace in this precise way, the definition of stable peace more accurately describes situations which we tend to call peaceful. Considering the two examples discussed above, we are much more likely to describe Canadian-US relations as peaceful than to apply that term to former US-Soviet relations. The reason for this cannot be explained simply by referring to peace as the absence of war or large-scale physical force because that condition applies in both cases. The difference is that in the first, we expect the peaceful condition to continue while in the second one we feared it would not.

Positive Peace

The very term "negative peace" implies the existence of another way of viewing peace which is positive. Evolving as a concept more recently, "positive peace" is the second, and increasingly more predominant way of defining peace within the peace studies field. Whereas negative peace is the absence of physical violence, positive peace means the absence of both physical and structural violence. Structural violence is a term with which most readers may not be familiar.

Unlike physical violence, which as noted above is typically overt and direct, structural violence tends to be less visible and indirect. It exists where groups of people are systematically prevented from meeting their basic needs and/or developing their full potential. It can also be referred to as social injustice. Slavery and apartheid are good, if extreme, examples of structural violence. In those cases, it is not only that human beings have limited freedom and that their everyday lives are almost completely controlled by external influences, but that their actual physical lives are limited by a shorter life expectancy and by higher infant mortality rates.

In the above discussion of peace, both negative and positive peace have been defined as the absence of something; what it is not, when it does not exist. This is somewhat ironic, since we would expect "positive peace" to be a "positive" concept; in fact, "positive peace" can be defined in positive terms.

Adam Curle (1981), a proponent of the use of the positive peace concept, defines peace as "a condition from which the individuals and groups concerned gain more advantages than disadvantages." This is in opposition to unpeacefulness which is "a situation in which human beings are impeded from achieving full development either because of their own internal relations or because of the type of relations that exist between themselves (as individual or group members) and other persons or groups." "By contrast with the absence of overt strife, a peaceful relationship would, on a personal scale, mean friendship and understanding sufficiently strong to overcome differences that might occur. On a larger scale, peaceful relationships would imply active associates, planned cooperation, and intelligent efforts to forestall potential conflicts."

The preceding definitions have served to indicate that considerable effort has been given to the definition of peace and that in all probability much more is needed. Nevertheless, definitions and concepts of peace do exist in sufficient form for work to go ahead on developing ways for more stable and peaceful relationships. An important on-going effort in connection with improving international stability must be a clarification of just what constitutes peace for each of the specific cases involved, as well as a description of the efforts and activities necessary to achieve peace for each case. It seems quite evident that the identification of what constitutes activities that promote peace among nations differs from the identification of activities directed toward how a country can win a war with another country should war occur. More efforts at identifying what activities promote positive peace, and bringing these into being, are required.

NATION BLOCK DIAGRAMS

Figure 3.1 shows a block diagram for a *single nation* in which the accomplishments of the nation are the output of the national processes that are measured and have data gathered from them to provide perceptions of what is taking place in the country. These perceptions are compared to the expectations of the country that have been developed through understanding and comprehension. The discrepancies between the national expectations and perceptions represent inputs to a decisionmaking and control process which produces outputs of actions that should modify the national processes to bring them into correspondence with the national expectations (Chestnut, 1982).

It will be noted that the national processes have natural and man-made external disturbances that may upset them from their anticipated values. Likewise, there may be perceived external disturbances which may influence the decisionmaking and control process and cause it to function in other than the best fashion. Thus, it is apparent that although the expectations-accomplishments loop appears to be quite controllable, it is evident that there is enough uncertainty and delays in some of the national processes that it is by no means certain that control of the national processes is a straightforward possibility.

With reference to Figure 3.1 it should be noted that the expectations of the nation are themselves an input from an understanding and comprehension process. This process itself has inputs from information derived from basic knowledge and understanding, from data gathering and measurements from this and other nations, as well as from disturbances. From this outer loop of information it is possible to provide still other influences on the national processes, which can cause them to behave in ways different than would be the case if they were operated by themselves alone.

This *interrelationship between and among countries* is brought out in Figure 3.2 that has simplified the representation of the Nations 1 and 2 and has emphasized the coupling between them. In this representation, information as well as basic knowledge and understanding are shown to be the overall driving force or reference to both nations. The feedback of information from data gathering and measuring of the two nations constitutes the additional inputs. Although the national comprehension for Nations 1 and 2 is shown to be separate, it is quite conceivable that some joint decisionmaking might well be beneficial. From Figure 3.2 it is evident that national behavior is influenced by the actions of more than a single country. Simulation and study of various alternative strategies and tactics for two or more nations can and should be undertaken. It should be noted that the accomplishments of both nations also serve to interact on one another as is shown in a simplistic fashion in Figure 3.2.

Perceptions

People tend to be motivated by those events or activities which significantly affect their well-being, either favorably or unfavorably. However, since many events or activities have influences on different aspects of people's well-being--such as the health, financial, and emotional aspects, and religious feelings--it is not always possible to readily determine in advance how people or nations will react to a particular event or incident. Often one's reaction to an event will be strongly influenced by both one's perception of the event and one's expectation for the occurrence.

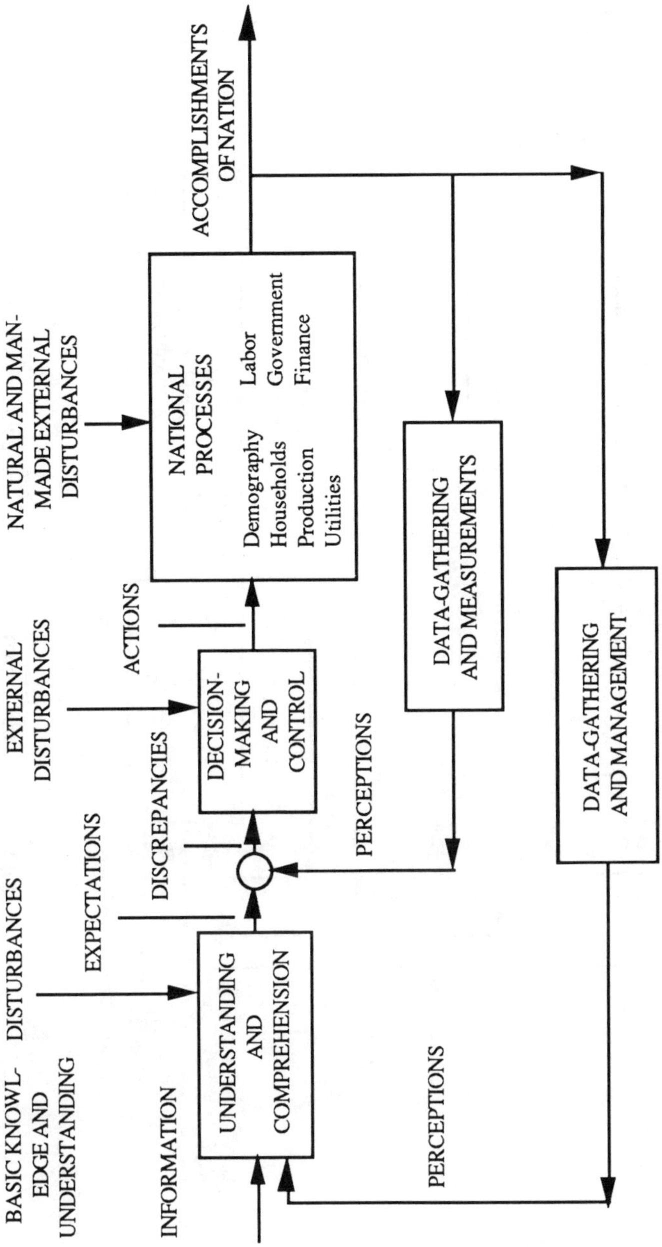

Figure 3.1. Block diagram of a single nation.

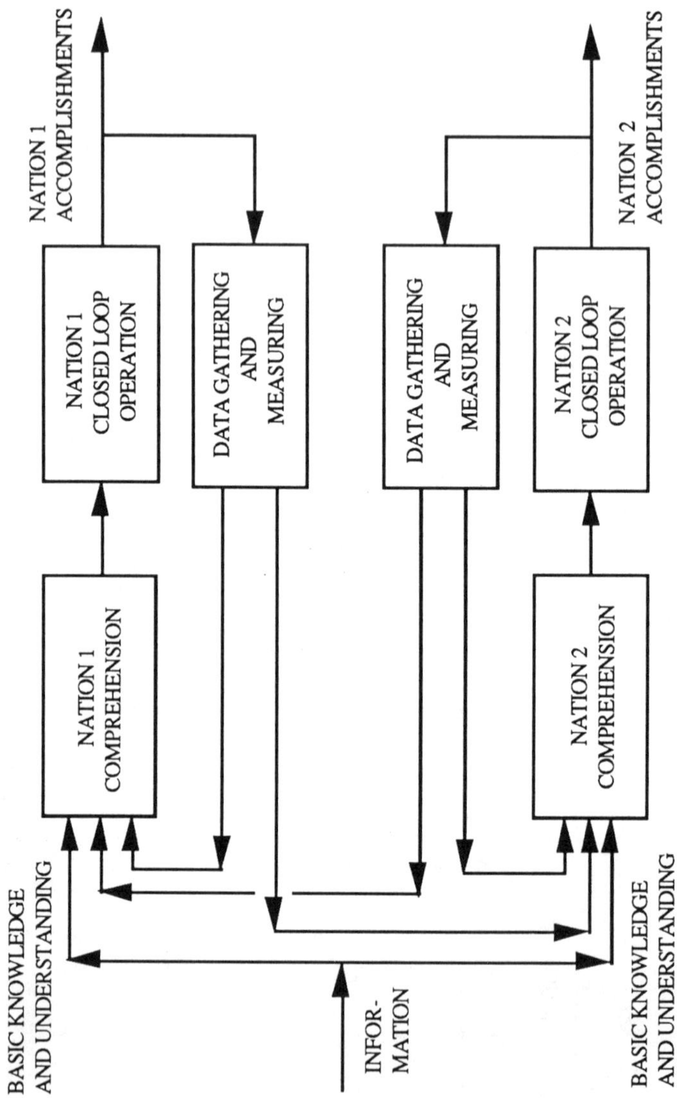

Figure 3.2. Block Diagram showing interaction of two nations

For example, suppose relative to the subject of peace that one expects a strong military position of a country to bring peace. In this case, as the country is building up its armaments, one might perceive that the nation is on its way to achieving peace by being sufficiently well armed, so that no single nation nor likely group of nations can successfully threaten the country building up its armaments. This approach can be described as the confrontation method of achieving peace.

An alternative expectation of the means for achieving peace could be that peace can be achieved through trade and cooperation which could be mutually advantageous to the two or more nations involved. In this case, the perception of a nation not going about building up armaments might create an impression of peaceful relations with other countries. This approach could be described as a more cooperative one. The nations of the European Economic Community, now in the 1990s, tend to operate toward one another somewhat on this cooperative expectation and perception of achieving peace after many centuries of a more confrontational mode of expectations.

Presumably, there are people in most countries who can be found to believe that confrontation is to be preferred, and there are still other people in these same countries who believe in cooperation. And no doubt there are still other people who have intermediate positions. Rather than thinking that all the people in a country perceive international relations from a single viewpoint, it is worthwhile to seek out the expectations of more than one group within the various countries involved (Fisher and Ury, 1981).

PEACE/WAR/PEACE TIME (ACTIVITY) CYCLE

Peace and war represent activities which, though they may be interrelated in time, are characterized by quite different phenomena. It is important that ways be developed to identify the events which are characteristic of these different phenomena, so that it will be possible to recognize from measurements at any time where in the Peace/War/Peace time cycle a country, in relation to other countries, appears to be located.

Figure 3.3 illustrates the four major stages in the Peace/War/Peace Time Cycle. The four stages shown are illustrative of the principles of the Peace/War/Peace cycle which might well be described in more or fewer stages while embracing the same concepts. These stages can be described as follows:

1. Peace and Industrial Development,
2. Preparedness for Military Activity,
3. Exercises and Precipitating Incidents,
4. War, Armed Conflict, Destruction, and Cessation.

It is not essential for a country to experience all of the four stages, and in fact it would be most desirable that all countries not go beyond the first stage of peace and industrial development. If the country is just recuperating from a recent war, some cleaning up and rebuilding will take place during this first stage.

The second stage is described as a preparedness for military activity and can be a rather unclear situation. In most cases this period of military preparation is characterized in terms of strengthening the defensive capability of the country. As such, this stage is well within the normal responsibilities of the country's leadership to carry out. In many cases this stage enables the country to put into place many of the longer lead-time items necessary to conduct a military war. If peace is to be maintained over the long run, certainly Stage 2 is a critical period in terms of the Peace/War/Peace cycle.

The third stage is described as consisting of exercises and precipitating incidents. These exercises may be conducted in the open as a way of "showing the flag" and impressing the potential enemy that they should yield to the implied force of the country(ies) carrying out the exercises. In other cases the exercises and precipitating incidents may be covert and be represented by sneak attacks or military surprise.

The fourth stage, represented by war, armed conflict, destruction, and cessation of the war may take years or may, in the future, last a relatively short time measured in hours or days. It is during this stage that the operational, tactical, or strategic weapons may be used. This can reasonably well be identified as the war part of the Peace/War/Peace process cycle. During this fourth stage of war the national decisionmaking by force takes place; this decisionmaking process is being used in lieu of negotiation, arbitration, or conflict resolution. In recent times even the declaration and conduct of war may be a somewhat ambiguous stage.

It should be noted that it is not necessary that each stage of the four shown occur, and ideally only the first one of peace and industrial development should be used.

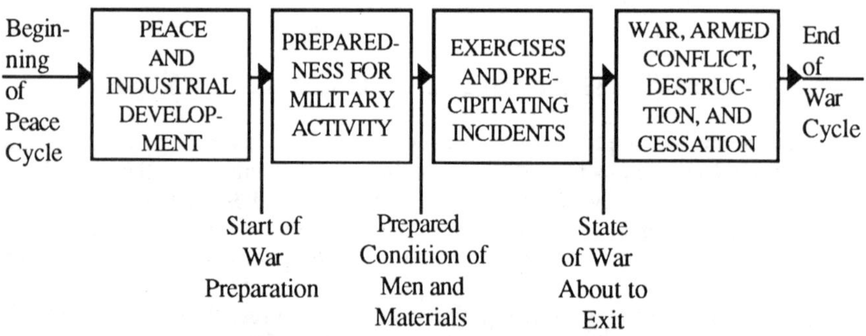

Figure 3.3. The four major stages in the Peace/War/Peace Time Cycle.

This thought that it is not necessary to go to war in the Peace/War/Peace process is illustrated in Figure 3.4 where various alternative options for a nation's leadership are shown. At the end of each major stage there is indicated a feedback path which shows that a choice can be made by the nation's leadership to decide not to proceed to the next major stage of the cycle.

Thus, after the Peace and Industrial Development stage the national leadership, as presently in the case of Japan, decides to continue with a condition of peace and industrial development without a significant commitment to preparedness for military activity. Although this peace without war preparedness is frequently cited as being a highly desirable goal, in reality, there are relatively few countries which actually practice this stage of operation.

The alternatives after the second stage of Preparedness for Military Activity include a peace with preparedness option. Basically this is a defense-oriented option such as practiced by Sweden and Switzerland in which there are an army, air force, and perhaps a navy which are available for defensive purposes. However, offensive military weapons are almost non-existent, and the emphasis is primarily on preventing invasion as opposed to any offensive military capability. This condition of national policy is the one more frequently used.

After Stage 3 of Exercising and Precipitating Incidents, an alternative exists that can be called peace through confrontation. In this case, there are examples of "showing the flag," there are exercises, etc.; but the activities stop short of war and the war part of the cycle is not carried out. Perhaps this is the more common condition for many countries in these times, when the industrial and military stages are in place, and the "standing military forces" are kept in a condition of combat readiness by means of exercises that may be indistinguishable from precipitating incidents.

Once Stage 4, or war, breaks out, there are still alternatives such as beginning the peace cycle again, having a short war, or electing to have a prolonged war in which there may be a major change in the whole social and economic structure. In the case of a prolonged war, the initial stage proceeds to build up the industrial capacity for production of military equipment and other required supplies, the military preparedness is greatly expanded, and military exercises are replaced by continued warfare.

The representations of the various alternatives shown in Figure 3.4 provide a basis for supervisory monitoring of nations by skilled observers to try to establish the stage of each nation at this time. Does the situation appear to be tending to go to the next stage of military escalation, or is it tending to reduce its stage to a more peaceful set of conditions?

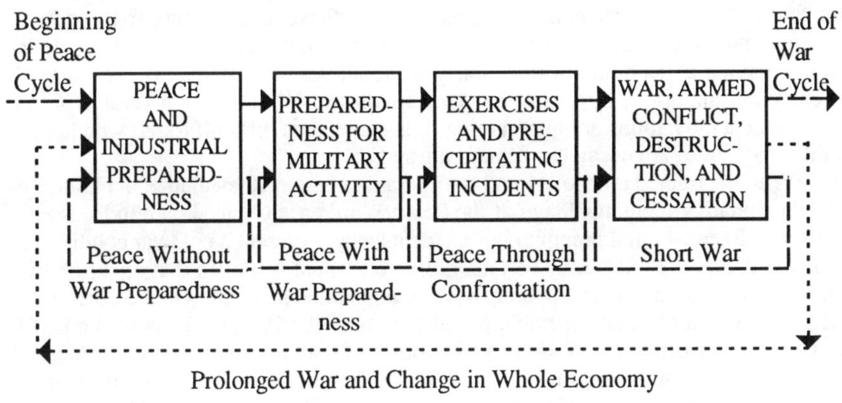

Figure 3.4. Various alternatives for the Peace/War/Peace Cycle.

To provide an indication of the next level of detail of each of the major items of activity shown in Figure 3.4, a brief description is depicted in Figures 3.5 to 3.8 of activities that might make up some possible subactivities. The Peace and Industrial Development stage of the Peace/War/Peace cycle is described in Figure 3.5. It might start with the rebuilding of the destruction of a previous war. Next an improved infrastructure might be put into place. Based on this improved capability, economic expansion might occur and foreign competition might be encountered either domestically or abroad. Industrial and military strength might then be gradually acquired. At this time the start of a war preparation stage can begin. In these days high levels of industrial support must be available before much of any military activity can be successful.

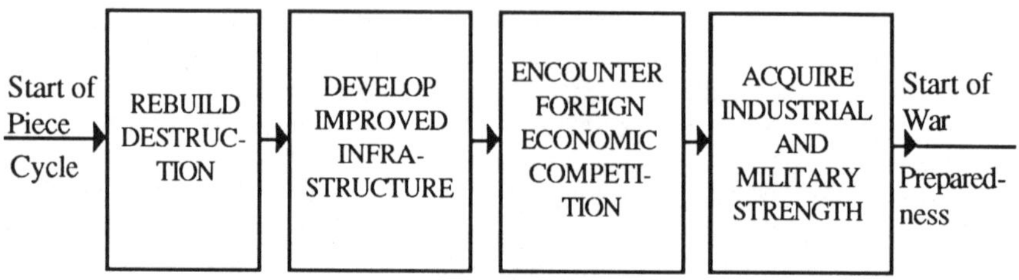

Figure 3.5. Elements of Peace and Industrial Development Stage of the Peace/War/Peace Cycle.

The preparedness portion of the war cycle is indicated in Figure 3.6. To be successful Preparedness for Military Activity requires considerable planning in a logistics, budgeting, design, fabrication, recruiting, and training sense. Budgets must be approved and actual appropriations must be made. The allocations of budgetary expenditures generally precede the actual appropriations for the design of hardware and facilities as well as for the recruitment efforts for obtaining people. Training of men and fabrication of equipment all must precede the achievement of a prepared condition for men and materials. The preparedness portion of the Peace/War/Peace cycle involves a major expenditure of time, manpower, and resources which is very difficult to carry out unnoticed by people in other countries. With the high level of armaments that some nations are maintaining, it is sometimes difficult to tell whether the military activities are normal or beyond normal.

With the trained manpower and the operational material available, the Exercises and Precipitating Incidents Stage of the Peace/War/Peace Cycle shown in Figure 3.7 can start. Exercises and precipitating incidents may represent a coldwar condition in which men and materials may be undergoing continual activities that could be triggered at any time into a state of war by some unplanned incident. On the other hand, there might be a planned timetable like the cases of the 1941 invasion of the USSR by Germany or of the attack on Pearl Harbor by Japan against the US. Here again the fact that a country may engage in exercises does not preclude its deciding not to go to war; but in the sense of being alerted to the possibility of war, such maneuvers represent exposures that can lead to war.

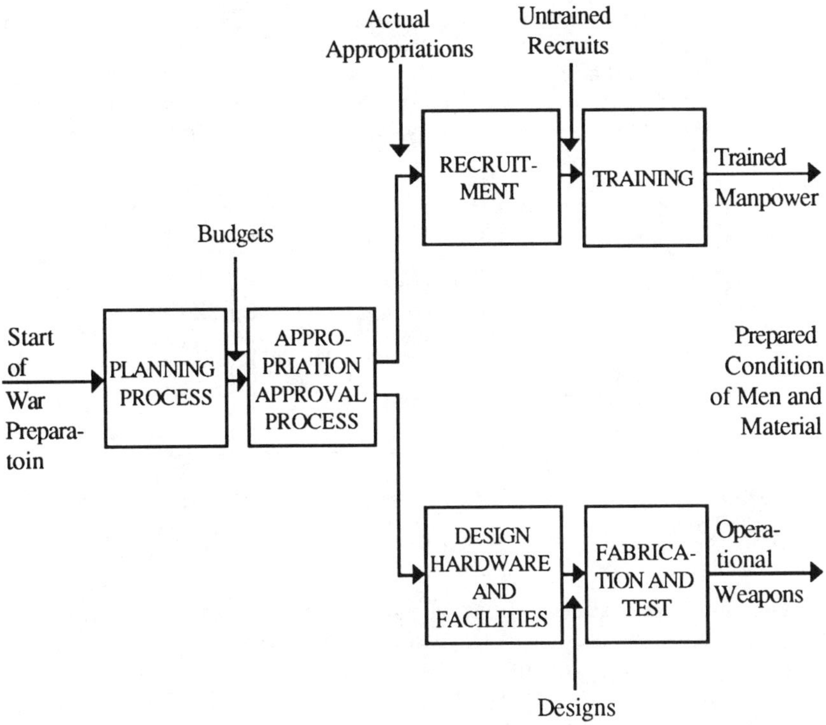

Figure 3.6. Preparedness for Military Activity Stage of the Peace/War/Peace Cycle.

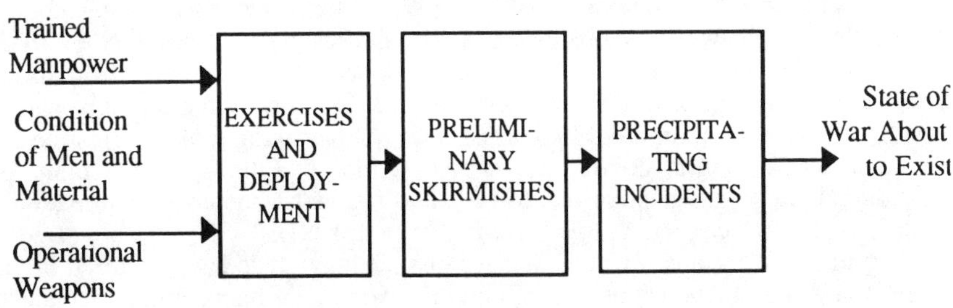

Figure 3.7. Exercises and Precipitating Incidents Stage of the Peace/War/Peace Cycle.

The last stage of the Peace/War/Peace cycle is shown in Figure 3.8 and is represented by such activities as war, armed conflict, destruction and attrition, and weakening of public morale in the conduct of the war. Cease fire and agreement on terms represent other activities which can be time consuming and important. The start of the process of rebuilding the destruction of war may in some instances take place while the war processes and armed conflicts continue. In the future some wars may be of shorter duration but higher intensity, while others may drag on for long periods.

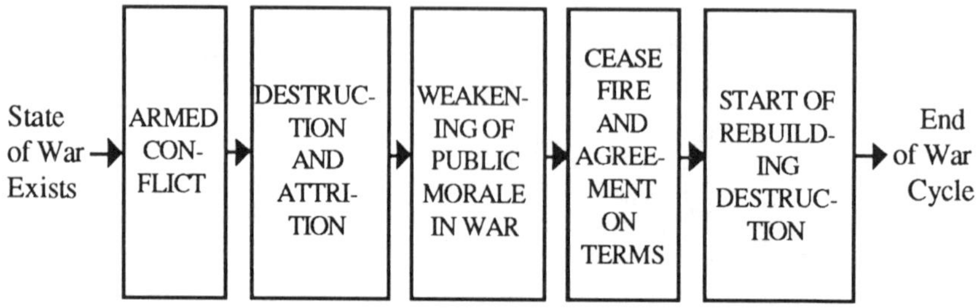

Figure 3.8. War, Armed Conflict, Destruction, and Cessation stage of the Peace/War/Peace Cycle.

Use of a Peace/War/Peace cycle diagram could be made by a joint review board to help nations to establish the perception of other nations as to where in the cycle conditions seem to be at any time. Each of the parties involved could provide the other with a factual appraisal of their impression of where things stand.

NORMAL, ALERT, OR EMERGENCY CONDITIONS OF INTERNATIONAL POLICYMAKING AND DECISIONMAKING

The preceding descriptions of the Peace/War/Peace cycle serve to indicate that the relationships between nations may change significantly with time. Not only is it possible for the degree of the friend (cooperation) or foe (confrontation) relationship to change with time, but also the state of cooperation or confrontation may be altered in sign. In order for the basis for policymaking to be appropriate for the current inter-nation conditions, it is essential that the policy content of decisionmaking be an adaptive one to take into account whether the international conditions are *normal, alert,* or *emergency* (Chestnut, 1984).

To illustrate instances where changes of such magnitude have occurred in the Peace/War/Peace relationships among countries which are major powers, consider what has occurred between the USA, the USSR, Germany, and Japan, before World War (WW) II, during WW II, and after WW II (as of the present). Whereas the USA, Japan, and Germany were considered to be anti-Communist and opposed to the USSR before WW II, during WW II Germany and Japan were enemies to the USA (Capitalists) and still opposed to the USSR (Communists). Currently, after WW II, the Japanese, Germans, and the USA are considered to be military allies, but they operate as economic rivals. And such relationships continue to change.

Clearly, there is an ongoing need for the major nations to maintain information contact with one another in order to be better aware of ways in which their international relationships are dynamically changing with time. Summit meetings of heads of state and other influential decisionmakers are taking place at annual or more frequent intervals to help meet the need for carrying out these information exchange functions. Presently there are some exchanges of military information on major troop movements between the USA and the former USSR. Although progress is being made in some cooperative security efforts between potential rival nations, there is a need for a more comprehensive and effective information system.

In the light of the changing international conditions that are likely to exist, it is desirable to establish a more complete information system which is able on a current basis to provide better awareness of the many variables and conditions between the nations involved. The system should be able to structure itself and adapt to these changes between nations during times of peace and stability, and to maintain and continue the desired conditions of peace while maintaining intact the essential national interests of the several nations involved. The international information system being suggested is one that has adaptive control capability and includes *joint review board* features that stress stability and peace as well as common security for the several nations involved (Chestnut, 1987).

ADAPTIVE CONTROL FEATURES FOR INTERNATIONAL RELATIONSHIPS

In considering the decisionmaking and control involved in guiding a country, it has been implied that the state of the nations involved has been that of the normal conditions for the countries being considered. Experience with a number of large-scale systems such as electric power networks, air transport, and large-scale industrial processes has shown that there may be at least three cybernetically different states for a system for which different decisionmaking and control logic must be considered: (1) normal, (2) alert, and (3) emergency (Chestnut, 1987).

The *normal* state corresponds to the conditions of the system for which the disturbances are within normal limits, the equipment is operating satisfactorily, and in general there are "no surprises" from the expected satisfactory operation. The *alert* state refers to conditions in which such things are not normal, but they are not sufficiently abnormal that it is unreasonable to expect that they can be brought back to normal without an emergency occurring. Under the alert state the disturbances to the system may exceed the normal limits for these disturbances, some equipment may not be operable or available for use, or some other uncertainties may exist such that the system can be considered abnormal but not sufficiently abnormal as to be in an emergency. The *emergency* state refers to abnormal conditions for the system and/or its environment such that the system is in real danger of not being able to survive. It is quite apparent that, *depending on the state that the system is judged to be in, the sort of control action to be employed may be quite different* (De Greene, 1985). See also De Greene (1990).

The basis for decisionmaking and control under each of these three states may be quite different and should be clearly understood. Conditions, either internal to the system or external from other systems than the given one, will influence whether the normal, alert, or emergency states will exist for the country at each time period. Thus, the decisionmaking and control process can be thought of as being an adaptive control which is made up of two parts. The first part senses or measures the internal and external conditions to determine the state in which the nation finds itself. The second part determines the appropriate actions or activities for the system based on the particular present conditions of the system and the appropriate state of the system as determined by the first part. Figure 3.9 illustrates this adaptive control application.

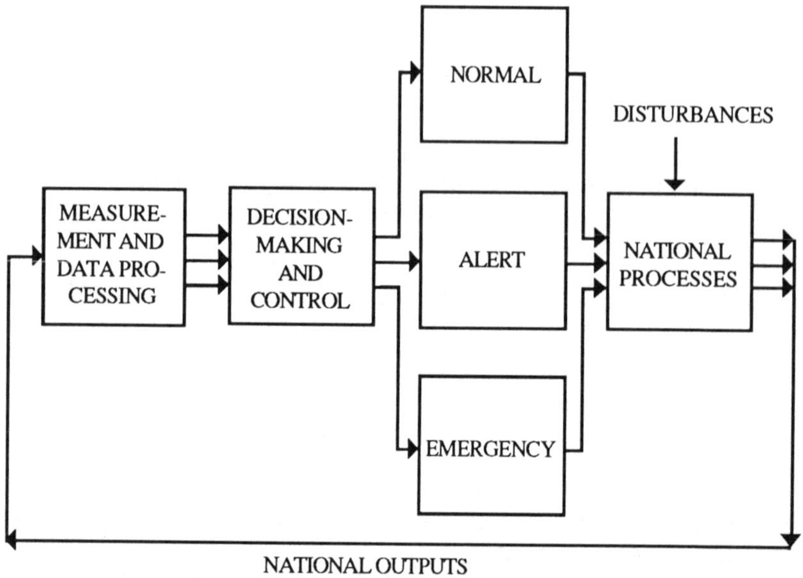

Figure 3.9. Adaptive control of National Processes using Normal, Alert, and Emergency State Information.

The National Processes shown on the upper right have their actual national conditions or activities, as well as the external system influences, measured and the data processed. The Decisionmaking and Control block analyzes the system data and decides whether the system state is normal, alert, or emergency. Depending on the system state which has been established and the measured conditions as determined from the data, the normal, alert, or emergency control block logic is used to provide inputs to the National Processes which are influenced by this closed-loop operation.

In considering the adaptive control aspects shown in Figure 3.9--and in particular the second part in which the appropriate control action is determined for the state that has been estimated--it is highly important to include the added dimension of the attitude or perception aspect of what the control action should be. As noted earlier in the matter of international activities, there may be different attitudes which some groups or people in various countries have in their approach to international relations. These attitudes may be the result of the basic approach to life that people have because of their religion, education, or family upbringing; suffice it to say differences in attitudes often can be found (Kile, 1987).

BASIC ASSUMPTIONS ABOUT THE INTERNATIONAL SYSTEM AND CONFLICT

In defining a system such as an information system for helping to resolve international conflicts, it is necessary to describe the basic underlying assumptions that are to be included in the system that is to be developed. Depending on the nature of these assumptions, the subsequent international conflict-resolution system that results may be different. Although different designers may start with different assumptions, for the purposes of this chapter, a set of *typical* assumptions will be set forth to provide an indication of how different assumptions may exist, and how the presence of such basic assumptions can influence the international conflict resolution process.

International System Is Made Up of Sovereign States

Each nation is assumed to be looking after its own interests as far as national security is concerned, although in some instances a country is willing to be associated with one or more other countries in some sort of protective alliance. For the present most countries are willing to join the United Nations, but they are not willing to assign the responsibility for their national security to some international police force. Thus, each country is looking after its own interests as perceived by its governing bodies or leaders. The nature of the relationship between any pair of countries can range through a number of conditions from stable peace through several degrees of less peaceful existence, to a condition of unstable peace, to a less peaceful condition of unstable war, that in turn may lead to a condition of sustained and continuous conflict that can be called stable war. Depending on the current peace/war conflict condition that exists between two nations, the procedure for resolving conflicts between them may differ. In other than military matters, the sovereign states may be willing to enter into trade or banking agreements, in which a fair measure of trust may be demonstrated despite the existence of national sovereignty interests (Galtung, 1975).

International Conflicts are Inevitable

The advent of greater interdependence within nations as well as between nations has heightened the likelihood of conflicts since people with different needs and interests share common facilities and activities. Although efforts in the past have often tried to encourage people to live in such a way that there would be no conflict between or among people or nations, the assumption that is made here is that conflicts between countries will exist and peaceful means for resolving them must be sought and developed. In particular, groups of experts in the particular fields in which the conflicts arise will be asked to seek out solutions acceptable to all the nations involved.

There are Organized Procedures for Resolving Conflicts

Within each nation there are one or more ways for resolving conflicts that are used to settle disputes. Although there may be drawbacks to each of the methods involved, in general all of the methods employed try not to use military force as the basis for conflict resolution within a nation (Fisher and Ury, 1981; Kahn, 1985).

In the system for resolving international conflicts, it is also desirable to establish international conflict resolution means that are based on ways that do not require military force. It is important in these methods of conflict resolution that there be an awareness of the perceived needs and values not only of the nation of the decision-maker, but also the needs and values of the other nations involved (Kile, 1987). Further, a knowledge of the conflict resolution procedures of the nations involved is highly desirable and important (Fisher and Ury, 1981).

As an example of steps useful in international conflict resolution, consider the following: (1) Determine Parties and Objectives; (2) Find the Facts; (3) Identify Alternative Options; (4) Evaluate Alternatives as Perceived by Each Party; (5) Negotiate the Agreements; (6) Carry Out and Monitor the Agreements; (7) Terminate the Conflict; and (8) Review the Methodological Aspects. These steps are discussed more fully below.

Steps such as these, or some procedures similar to them, should be established and agreed upon in times of normal relations when the procedures are not needed, so that they will be available when they are needed.

Potential Areas for Conflict Needing Resolution Can Be Identified

In order to make progress in efforts toward successful conflict resolution, it is desirable to identify potential areas where conflicts might arise, to establish groups from the two or more nations involved, and to assign to these groups the responsibility for reducing the likelihood of conflict in their assigned categories. Thus, the basis for conflict will be considered in the light of what is happening in several categories, not simply on what is happening in any single category (Chestnut, 1987; Taylor, 1978; Wright, 1964).

Each category identified should be capable of being monitored, explored, and understood in such a way that successful solutions may be arrived at. These categories can be identified in terms of national sectors as used by each nation, or in terms of specific topics such as food, housing, water, energy, boundaries, etc. Even though the initial choice of these categories may not prove particularly useful, these choices should form the basis for a subsequent improved selection.

Measurement Means and Criteria for Monitoring Can Be Found

Of particular significance in the matter of international conflict resolution is the matter of whether the parties involved consider the conditions to be normal, alert, or emergency vis-a-vis the other country. Depending on this perception by each of the parties of the actions of the other country, the other party will see what should be its appropriate action. If necessary, it will be required to look at the problem in more detail and to establish a better base of information.

Various Possible Alternative Actions Can be Explored

Through the use of models, simulations, discussions with knowledgeable experts from both the nations involved, as well as with third-party experts, it should be possible to get various impressions of what may be the outcome of possible alternative actions. It may not be necessary to get precise information for the future, but it is often sufficient to learn what are thought to be the likely possibilities. Much work has been done on national and global models, and there seems to be an increasing confidence that such models can be useful in general, even though they may not be precise in detail. See Chapter 7.

Incentives and Motivations for Peace Can Be Developed

In each of the countries involved the people responsible for the policymaking and decisionmaking that cause a country to go to war can be provided with financial incentives associated with keeping a country out of war, with honors or prestige of a patriotic nature, or with political rewards that may have particular appeal to the persons responsible for the peacemaking efforts. The Nobel Peace Prize is an example of one form of incentive. Recognition at various levels of duty could also be designed and awarded for appropriate accomplishments. With the huge amounts of money being spent for arming for war, there are potential savings to be made that could loom large to persons who contribute to lowering somewhat the expenditures for the military. Motivation for accomplishment of progress for peace must be possible (Gorbachev, 1988).

ELEMENTS OF CONFLICT RESOLUTION

In applying adaptive control principles to international conflict resolution, it is desirable that there be an understanding of the elements that can make up the conflict resolution process. As a start in this effort it is worthwhile to have a better mutual understanding of the different approaches that are used in the different cultures and countries involved. Some of these methods are highly humanistic and involve emphasis on social and personal relations. Others are analytical and mathematical and emphasize quantitative and logical results. It is useful for these several different approaches to be understood and the ones appropriate for the particular situation to be employed (Fisher and Ury, 1981; Kahn, 1985). Contrast in this book, for example, the more formal modeling approaches of Allen and of Forrester with the more values-oriented approach of Harman.

Listed below are eight elements of conflict resolution based more or less in order of their use in the solution process. These elements are present in the resolution of a particular conflict and emphasize that each conflict is probably a part of an ongoing problem which may be resolved for the present but will appear again later in perhaps a modified form. Keeping in mind the learning-by-doing principle, one no doubt will move on to developing new methods and elements that may be superior to the existing ones. The following is a brief description of the elements of conflict resolution built around the methods of the book *Getting to Yes* by Roger Fisher and William Ury (1981).

The Eight Elements in Order of Use

1. Determine Parties and Objectives. In any nation there may be different parties or ethnic groups involved which may have different viewpoints and objectives. Furthermore, depending on the duration and severity of the conflict, the basis for resolution may be different. It is important to identify the various parties involved and their associated interests.

2. Find the Facts. This fact-finding work element is an important one where technology can be of help in international conflict resolution. Sensing, data-gathering, information-handling, simulation, etc. are representative of the uses where technology can be of assistance to the policymakers and decisionmakers. Use of multiple and redundant sources of data is helpful in fact-finding.

3. Identify Alternative Options. The resolution of conflicts will probably involve consideration of a finite number of alternative options which must be looked at in a broad manner initially. As the poorer alternatives or options are identified or put aside, those options which appear to be attractive can be looked at in greater detail and thoroughness. Again systems analysis, simulation, operations research, and decisionmaking and control technology can be very useful for this work element (Chestnut, 1965).

4. Evaluate Alternatives as Perceived by Each Party. In considering the alternative options, these should be evaluated in terms of the benefits and costs to each of the parties involved. Hopefully, alternatives which provide win-win solutions--at least to some degree--for each of the parties involved can be found. This activity may involve some of the same technologies as the preceding elements, but the social, political, and other skills required may be of greater significance than the purely technical ones (Kile, 1987).

5. Negotiate the Agreements. Although the alternative options considered initially may have yielded results that appeared to be satisfactory, the actual conditions that may be acceptable for each of the parties involved may require some changes in form, substance, timing, or degree from the originally acceptable conditions. Statements about the proper degree of formality should be drawn up and agreed upon.

6. Carry Out and Monitor the Agreement. Once the agreements have been reached and the resolution of conflict processes begin to occur, an effort should be made to see that the agreed-upon procedures are carried out according to plan. The technologies involved here are of the project management sort for which the required hardware and software are generally available.

7. Terminate the Conflict. Hopefully, the preceding steps can be carried out in such a way that the conflict can be considered to be terminated. Actually, it may be necessary to start the process over again with a *repeat* of the methods of the conflict resolution process described above.

8. Review Methodological Aspects. If the conflict cannot be considered as having been successfully resolved, then it may be necessary to use new methods, new people, or--in general--a fresh approach to bringing about a successful resolution of the conflict.

Comment. As one reviews the eight elements of the conflict resolution process, it appears that emphasis is on the management of conflict, and the use of technology tends to be in the fashion of a management information system. Nevertheless, the structure, organization, and effectiveness of such technological contribution to management can be very helpful in arriving at useful results in a more satisfactory manner. Further, such a well-organized procedure may be helpful in establishing an efficient data base upon which future solutions may be better than those of the past.

It should be emphasized that the international conflict resolution process described above is one which is a *continual* effort involving regular meetings of the people and organizational representatives. At such meetings there is a chance to identify any misunderstandings, and there are opportunities to modify the procedures if this is necessary to continue to achieve satisfactory solutions to the problem being resolved.

POTENTIAL AREAS FOR INTERNATIONAL CONFLICT

Resulting from earlier studies as well as current studies of potential areas for international conflict requiring resolution, many causes for conflict have been identified (Taylor, 1978; Wright, 1964). Table 3.1 lists some of the more likely areas that might well serve as the starting point for some sort of *joint review committees*--such as the Standing Consultative Commission (SCC) of the SALT I Treaty--to be established with bilateral participation to identify current concerns for handling existing or potential conflicts. These joint review committees could also be compared in function with a Federal Aviation Agency (FAA), in each of the areas in which they have responsibility, for providing advice and recommendations similar to that necessary for insuring that safety in air transportation is maintained--carrying through from design, to manufacture, to operation.

It should be recognized that each pair of countries at any particular time will not necessarily confront critically all the major possible causes of conflict, so there is a need to identify those causes of conflict that are worthy of special attention at the present. However, for each of the causes for conflict requiring resolution, questions like those listed below should be asked in the sense of a failure modes and effects analysis (FMEA) for a systems analysis reliability study.

1. What could be the likely effects produced by this *cause* for conflict?
2. How can this cause for conflict be detected so that one can verify what seems to be happening?
3. What can be done by whom to avoid having this become a cause for conflict?
4. What reasonable actions can be taken by whom to reduce greatly the likelihood of conflict?
5. What course of action is required to implement the above-stated actions to reduce the likelihood of conflict? What are the costs to whom? What are the benefits to whom?

The establishment of a high-level organization in each involved nation to work out a Peace Plan to carry out preventative conflict resolution is worthy of further study. Such a study could be set in motion through such organizations as the US Institute of Peace or the State Department with the assistance of such organizations as the National Academy of Science and the National Academy of Engineering. In other countries there are similar organizations for which the role of establishing the joint-review committee function described above would be appropriate and could be initiated.

Table 3.1
Major Possible Causes for War among Nations

1. Geographic	5. Raw Materials
* Boundary Disputes	* Pricing
* Invasions	* Natural Resources
2. Religious	* Monopoly Practices
* Persecution	6. Political
* Value Clashes	* Human Rights
* Discrimination	* Oppression
3. Economic	* Ideology
* Dumping of Goods	* Minorities
* Monopoly Practices	* Personal Gain
* Price Barriers	7. Military
* Trade Embargoes	* Aggression
4. Environmental	* Blackmail
* Acid Rain	* Puppet Governments
* Standards	* Covert Activities
* River Pollution	8. Other
	* Power
	* Personalities

PROGRESS TOWARD INTERNATIONAL STABILITY AND PEACE

For more than 30 years now, scientists and engineers around the world have been working to understand and to apply control systems principles to help improve the ability of such control systems to meet more effectively the needs of people in both the more developed and the less developed countries of the world. About 10 years ago the Working Group on International Stability of The International Federation for Automatic Control (IFAC) Systems Engineering Committee started to explore ways by which systems engineering could be used in conjunction with people and techniques in other disciplines such as political science, international relations, economics, social sciences, global modeling, and conflict resolution. Other professional organizations, such as The Institute of Electrical and Electronics Engineers (IEEE), and The International Society for the Systems Sciences (ISSS) have shown similar interests.

It is appropriate that we recognize that considerable progress has been made toward improving stability and peace. Also we should take advantage of the fact that it is a fortunate time in world history when information systems can be employed to emphasize the likelihood of there being significant improvements in peace among the major industrialized nations in the foreseeable future. It is worthwhile to note, however, that at the same time that the use of information systems is able to increase the likelihood of peace, automatic control and systems have also increased the lethality of war. It is all the more important that increased emphasis be placed on exploring how the same control principles can be used to increase the likelihood of continued peace (Gorbachev, 1988).

COMMON SECURITY OBJECTIVES

A particularly significant step forward toward stability and peace between and among the global systems of nations is the recognition of the importance of common security of the many nations as contrasted with national security for one or more of a limited number of nations. Common security options can be grouped in three categories:

1. Restructuring the military with an emphasis on defense and on reducing the reliance on increased numbers of conventional and nuclear deterrents.
2. Deepening the security agenda to include international environmental, economic, social, and health development categories.
3. Advocating political and other agreements to resolve conflicts, eliminating intervention, and increasing cooperation and confidence in international organization (Chestnut, 1990).

Increasing international emphasis is being placed on preventing wars through conflict resolution rather than having wars start and then setting out to win them. There are an increasing number of persons and organizations worldwide working on such global systems. Not only is the United Nations making efforts at improving international stability, so also are other international functional organizations working on specific areas such as food, health, energy, transportation, communication, and other human needs (Prime Minister's Office, 1991; World Commission on Environment and Development, 1987).

For the civilizations on which Earth's inhabitants depend to continue to develop and prosper, an increased effort must be made to understand better the natural and man-made dynamic systems and how they interact. The following points are emphasized:

1. For the international system at the global level to work in a more stable and effective manner, a *cooperative security system* is needed to assist national and international policymakers and decisionmakers.
2. There is a need for international and non-governmental organizational (NGO) efforts to understand better the role and logic of national policymakers and decisionmakers, especially as studies are made of the influence of peace and war on the National Processes and the means for resolving international conflicts.

Parallel efforts on the nature of possible cooperative security systems should be undertaken by interdisciplinary and transnational teams of participants. Both governmental and NGO opinions can be valuable. Multiple teams should be used for making possible the obtaining of majority logic decisions (e.g., two out of three or better) to be obtained. Simulations of and by man-machine systems using interconnected national models can be helpful in understanding the national processes involved (Prime Minister's Office, 1991; World Commission on Environment and Development, 1987)

INFLUENCE OF THE PAST, PRESENT, AND FUTURE ON NATIONAL ATTITUDES

One of the principles associated with systems is that systems may change with time. Such evolutionary change is a major theme of this book. This idea should help highlight the fact that present inadequacies or limitations in the system may be able to be overcome with the use of time, energy, materials, information, and/or other resources to make the desired changes possible. Thus, the opportunities for change exist, and ways and means should be sought for converting opportunities to accomplished changes (Chestnut, 1965).

From this perspective the emphasis can and should shift from a concern, focused on who or what is responsible for the present situation, to a concern for placing attention on what can be done during the future to make a present set of conditions change to more satisfactory and acceptable future conditions.

The managers of technology should be encouraged to seek ways to achieve more of people's needs rather than to continue to be blamed for the inadequacies of the present. Whereas things could always be better than they are, by the same token, dwelling with increasing emphasis on the inadequacies of the present may cause them to indeed become worse. In the Cooperative Security System that is being considered favorably for improving international stability, the emphasis is on helping the nations involved to improve their relations in the future (Gorbachev, 1988).

When people consider the relationships between and among nations, there are generally some built-in assumptions present in their thinking, and these assumptions may influence people's attitudes toward the other nation(s). Although there may be different conditions for the inter-nation relationships than those shown in Table 3.2, the differing conditions from war to peace shown on that table serve to indicate the nature of the range of conditions that may exist.

At present these conditions of inter-nation relationships are not generally quantified or publicly described. Necessary are ways of exchanging information between well-trained persons on the nature of current inter-nation conditions in order to identify when there are marked differences between the perceptions of people on these inter-nation relationships. More progress in defining and monitoring such conditions would be helpful (Chestnut, 1990).

Table 3.2
Range of War/Peace Conditions Existing Between Nations

1.	Stable War	5.	Peacemaking
2.	Unstable War	6.	Peacebuilding
3.	Unstable Peace	7.	Stable Peace
4.	Peacekeeping		

DECREASE THE LIKELIHOOD OF WAR, INCREASE THE LIKELIHOOD OF PEACE

The process of resolving international conflicts can be approached from two or more views. One way is to wait until an international conflict exists and then to apply methods to resolve the conflict. A second way is to monitor continually the relationship between the two or more nations and take suitable actions that will prevent minor potential conflicts from developing into a major international conflict of war.

Since there is still the possibility in international conflicts today that escalation of the conflict may develop into a nuclear war, it is important to find ways of preventing international conflicts at all times. In the discussion which follows, attention will be placed on the second method mentioned above--namely, seeking to monitor conditions between nations to prevent conflicts from getting out of control

In many instances peace is defined as the absence of war, in which case it is not clear what should be measured that will indicate whether there is more or less peace than heretofore. One way of developing the needed peace indicator is to look continually at the various causes for war between countries as provided from a list like that in Table 3.1. Depending on a continuing review of these causes of war in a suitably organized fashion by representatives of each of the countries involved, and perhaps some outside experts, it should be possible to provide a useful indication of when the relationships between the countries are at the normal, alert, or emergency stages.

When in the alert or emergency conditions, steps should be taken, such as peacekeeping or peacemaking to bring the abnormal stage conditions back to the normal stage. When at the normal stage, peacebuilding methods should be employed to strengthen the condition of normalcy so that the likelihood of the relationship becoming one of alert or emergency is reduced.

The activities of peacekeeping, peacemaking, and peacebuilding have been described at length (Galtung, 1975) and noted above, but short forms for the description of these terms are now reviewed as follows.

Peacekeeping is a dissociative process of keeping warring forces apart. This may be the least constructive in a long-range sense, but it does provide a way of "stopping the fighting now." Peacekeeping should be used when needed, but this should not be the principal approach to peace.

Peacemaking is a conflict resolution process. Peacemaking emphasizes ways to reduce conflicts, but it may provide neither the basic means for avoiding conflict in the first place nor the means of preventing the resumption of conflict after a short period of time. Peacemaking should be used when needed, but, here again, it should not be the principal approach to peace.

Peacebuilding is an associative approach which stresses cooperative actions. Peacebuilding endeavors to find out what a healthy set of relations consists of and to work to accomplish these activities. It endeavors to build new structures and organizations or other means for making peace more possible in the future. In many cases the technology is already present and is being used to a certain extent, but there are certain barriers which seem to inhibit a fuller utilization of these techniques. In other cases, what is needed is the application of management and organizational skills to make possible the distribution of what is possible to where it is needed. More emphasis must be placed on peacebuilding.

From the above descriptions, it appears that a continuing effort must be made to evaluate the conditions which exist in the relationships between countries and to determine whether peacekeeping, peacemaking, or peacebuilding efforts are most appropriate. Further, within each of these three possibilities, there may be many alternatives, some of which are more favorable than others. There exists a need for means of acquiring current information and utilizing decisionmaking and control skills to put in place a form of supervisory control somewhat akin to that for a complicated and dangerous technical system. In this way one should be able to decrease the likelihood of war between nations by increasing the likelihood of their being at peace.

When one considers that a country is likely to be at peace more than 80 to 90 per cent of the time, much of the time it can be seeking ways for getting along without conflict with its national neighbors. More time, money, and effort must be placed on seeking ways of maintaining and strengthening its peaceful relations, rather than on counting so heavily on winning a war that may in fact be unwinnable.

WHAT IS TO BE DONE NOW?

The cybernetic principles of establishing goals or references of desired conditions or trajectories, measuring actual accomplishments, performing decisionmaking and control, and carrying out the actions appropriate to the requirements have been indicated as a worthwhile course of action for developing systems. These principles highly successful in the development of large-scale sociotechnical systems, have been described in terms of how they might be applied to a system for carrying out the task of increasing the likelihood of there being peace in the future and thereby decreasing the likelihood of there being war.

From these considerations, it appears that the following policy-related activities should be undertaken at the present time:

1. Recognize that a stepped up level of activities directed toward increased US and United Nations efforts for peace, now and in the 10-year period ahead, is needed.
2. Select one or more persons or agencies to head up a more active US peace effort. Perhaps the United States Institute of Peace, currently underway, should be used for the initial effort.
3. Identify people who are willing to work on the problems of how to strengthen the United Nations as a means for improving cooperative efforts; to determine the benefits and costs of alternative courses of action; to determine how to identify, measure, and quantify normalcy and non-normal states; and to determine how to identify what appear to be the major causes of conflict in the 10-year period ahead.
4. Select three to five suitable groups of the identified people, say of 10 to 15 persons in each group, and charge each group to prepare a plan of action of lessening the causes for conflict. The action plans might be built around cybernetic principles and use modeling and simulation to facilitate understanding of the plan. However, each group should be free to choose whatever methods it prefers for drawing up its own plan.
5. Identify a few promising alternatives to the proposed plan indicated by each group.
6. Obtain support for one or more of the plans. There should be technical support for the ideas and then financial support for the level of work to be done. Information on the approximate schedule, financial level of effort, and the personnel and equipment needs should be included in the proposed plan.
7. Establish teams to implement the plan and proceed with the work to be done. Monitor the accomplishments against the proposed schedules of the work to be done.

This plan of activities *provides a novel approach* embodying cybernetics principles that have provided successful results in many previous instances. It offers solutions at far lower costs and in shorter times than ones currently being funded. As such this plan warrants serious consideration as a parallel effort to the existing military programs presently underway--it provides *supplemental ways for increasing international stability*.

CONCLUSIONS

During the time period over which the ideas contained in this chapter were acquired, there have been two very significant reports to the United Nations which are directed at the solution of the many problems of the world--natural and man-made. One of these reports on "Our Common Future" by The World Commission on Environment and Development (1987) has been useful as a basis for the United Nations Conference on Environment and Development (UNCED) in June 1992 in Rio de Janeiro. It sets forth to the UN "The Challenge for Institutional and Legal Change." It also recommends "A Call For Action" to bring about the changes identified.

The second of these reports is titled "Common Responsibility in the 1990s" and is referred to as "The Stockholm Initiatives on Global Security and Governance" (Prime Minister's Office, 1991).

The six categories of proposals that are set forth follow:

1. Peace and Security
2. Development
3. Environment
4. Population
5. Democracy and Human Rights
6. Global Governance

These two sets of reports contain many ideas in the way of recommendations and proposals. These proposals should be considered as starting points for the policymakers in the 1990s. Depending on the support available in manpower and money, a more specific plan of action should be started and carried forward. United Nations, NGOs, as well as interested institutes and individuals should be encouraged to participate in these endeavors to help implement ways for proceeding toward a stable and peaceful common future for the people of the world.

REFERENCES

Beer, Francis A., 1981. *Peace Against War: The Ecology of International Violence.* San Francisco: Freeman.

Boulding, Kenneth E., 1978. *Stable Peace.* Austin, TX: University of Texas Press.

Chestnut, Harold, 1965. *Systems Engineering Tools.* New York: Wiley.

Chestnut, Harold, 1982. "Nations As Large Scale Systems." In Yacov Y. Haimes (Ed.), *Large Scale Systems,* pp. 155-183. Amsterdam: North-Holland.

Chestnut, Harold, 1984. "Use of Systems Techniques to Improve International Stability." In Harold Chestnut, Robert Genser, Peter Kopacek, and Andrzej P. Wierzbicki (Eds), *Supplemental Ways for Improving International Stability* (Proceedings of the IFAC Workshop held 13-15 September 1983 at Laxenburg, Austria), pp. 87-92. Oxford, UK: Pergamon.

Chestnut, Harold, 1987. "Applying Adaptive Control Principles to Resolving International Conflict." In Rolf Isermann (Ed.), *Automatic Control World Congress* (held at Munich), Vol VII. Oxford, UK: Pergamon.

Chestnut, Harold, 1990. "Progress Toward International Stability and Peace." In U. Jaaksoo and V. I. Utkin (Eds.), *Automatic Control World Congress* (held at Tallinn, Estonia), Vol. VI. Oxford, UK: Pergamon.

Curle, Adam, 1981. *True Justice: Quaker Peace Makers and Peacemaking.* London: Quaker Home Service.

De Greene, Kenyon B., 1985. "International Stability and Policy-Related Thinking About Equilibrium, Complexity, and Structural Change." In Bela H. Banathy (Ed.-in-Chief), *Proceedings of the Society for General Systems Research,* Vol. II, pp. 1051-1060. Seaside, CA: Intersystems.

De Greene, Kenyon B., 1990. "Supplementary Systems Paradigms for Different Stages of Societal Evolution with Special Reference to War and Peace." *Systems Research,* 7(2), 77-79.

Fisher, Roger, and William Ury, 1981. *Getting To Yes.* Boston: Houghton Mifflin.

Galtung, Johan, 1975. *Essays on Peace Research,* Vol. II. Copenhagen: Christian Ejlers.

Gorbachev, Mikhail S., 1988. *Perestroika.* New York: Harper & Row.

Kahn, Lynn S., 1985. *Peacemaking.* Lanham, MD: University Press of America.

Kile, Frederick O., 1987. "Social Values and Active Peace." In Harold Chestnut and Yacov Y. Haimes (Eds.), *Contributions of Technology to International Conflict Resolution* (Proceedings of the IFAC Workshop held 3-5 June 1986 at Cleveland, Ohio). Oxford, UK: Pergamon.

Prime Minister's Office, 1991. *Common Responsibilities of the 1990s: The Stockholm Initiative on Global Security and Governance.* Stockholm: Prime Minister's Office.

Singer, J. David (Ed.), 1980. *Correlates of War-II.* New York: Free Press.

SIPRI, 1979. *Armaments or Disarmament? The Crucial Choice.* Stockholm: Stockholm International Peace Research Institute (SIPRI).

Taylor, A.J., 1978. *How Wars Begin.* New York: Antheneum.

The World Commission on Environment and Development, 1987. *Our Common Future.* Oxford, UK: Oxford University Press.

Wilkinson, David, 1980. *Deadly Quarrels: Lewis F. Richardson and the Statistical Study of War.* Berkeley, CA: University of California Press.

Wright, Quincy, 1964. *A Study of War.* Chicago: The University of Chicago Press.

CHAPTER FOUR

FIELD-THEORETIC FRAMEWORK FOR SYSTEMS THEORY AND SYSTEMS THINKING

by Kenyon B. De Greene

Several recent and profound theoretical developments are subsumed under the rubric "field theory." These developments and their relevance to policymaking are discussed in this chapter which also extends a number of concepts first introduced in Chapter 1. The systems theories originated mainly in the physical sciences but are finding increasing application in the biological, behavioral, and social sciences. This presents a problem because we must avoid the long condemned reductionism and physicalism. It is emphasized that field theory as presented in this chapter is much more a framework or highly generalized theory than a particular or specific theory. Field theory, like all theories and models, can best be thought of as a heuristic or aid to thinking and problemsolving. No theory or model or set of beliefs is or can be absolute and everlasting truth.

The material in this chapter may seem to be highly esoteric and different from that to which policymakers have been traditionally exposed. But—once again—changing times and increasing complexity require new ideas.

THE NATURE AND HISTORY OF FIELD THEORY APPLIED TO NONLIVING AND LIVING SYSTEMS

Terms like "fields" and "forces" are widely used in the sciences, and these uses may be rather casual or they may receive the exact definitions given in the literature of this chapter. Field theory, like scientific theory in general, has evolved from earlier theories, observations, and calculations.

Nonliving Systems

Consider two major challenges that confronted physicists in the mid-to-late 19th Century. The first was called "action at/from a distance." The second dealt with how interactions at a more microlevel generate a field with emergent new properties at a more macrolevel.

Physicists had for several centuries wondered whether space was just a geometric abstraction or whether it possessed physical properties. Most believed the latter, and Newton had coined the term "ether" to describe a space, the properties of which provided much of the impetus for the study of gravity and electromagnetism. The ether, an entity filling all space, was observed to be modified in the vicinity of matter. For example, a gravitational potential, following the familiar inverse square law, was seen to cause matter to move from regions of higher to lower potential (energy). Expressed differently, a force was driving the bodies of matter together. The ether was also seen to be necessary for the transmission of light, electricity, and magnetism across space.

Michael Faraday discovered the important field property of electromagnetic induction: if electricity can produce magnetism, then magnetism can produce electricity. Faraday appears to have been the first person to use widely the terms "field" and "electromagnetic field." James Clerk Maxwell developed a more rigorous theory of the electromagnetic field that dealt with the space in the vicinity of electrical and magnetic bodies.

The new idea of the intermediary field explained much of electromagnetic behavior. The field appeared to "carry" electromagnetic interactions between material bodies, and field behavior was completely determined by contiguous processes. But the question arose: is there an ether that carries the field, or should the field have an independent, irreducible existence of its own? Parsimony called for an answer favoring the latter point. Albert Einstein (1950) developed the idea of a consistent field theory embodying all the elements of the theory in both space and time, as opposed to the argument for a continuous field punctuated with discontinuous material points.

Einstein formulated a law of gravitation without referring to force or action at a distance; emphasized rather were the properties of space itself which produce a tendency of material bodies to approach one another. Again, however, the presence of matter modifies its whole vicinity, producing a gravitational potential. According to special relativity, space detached from any physical content does not exist. This physical reality can be represented by a field, the components of which are continuous functions of four independent variables, namely, the coordinates of space and time. Gravity is not a force, per se, that acts between objects but is rather a property of space itself. Gravity is imbedded in the field itself, and masses can be represented as singularities in the field.

This arcane material has no direct transfer to societal systems and ecosystems, the main objects of policymaking. Rather, it serves to stimulate thinking about such directly relevant areas as system-environment interaction and figure-ground interrelationships.

Let us now move into the present. Physicists recognize four basic forces in Nature: (1) electromagnetism (electricity, magnetism, and light); (2) the weak nuclear force yielding radioactive decay; (3) the strong nuclear force associated with quark binding in atomic nuclei; and (4) gravity. The first three forces operate probabilistically at microscopic levels according to quantum theory. Gravity operates at a macroscopic level and, according to Einstein, is deterministic. The first three forces are carried by short-lived subatomic particles called bosons that bind protons, neutrons, and electrons together or push them apart. Bridging these gaps has led physicists to pursue a *unified field theory.*

Easy-to-understand presentations of some of these ideas are given by Boslough (1989) and by Contemporary Physics Education Project (1990). A history of physics from a mid-century perspective, that therefore avoids subsequent filtering and reinterpretation, is provided by Richtmyer and Kennard (1942). For a treatment of electromagnetic field theory see Schwab (1988). For highly technical treatments of field theory see Amit (1978) and Arnowitt and Nath (1976). The last two sources include discussions of phase transitions, critical phenomena, and order parameter, all very important to the underlying theory of this chapter. Wilson (1979) provides a relatively easy-to-understand summary of the last three areas.

The concept of critical phenomena arose out of observations that (1) superficially very different substances (e.g., metal alloys, magnets, and superfluids) behave under certain conditions in almost identical manners; and (2) these substances undergo phase transitions or qualitative changes (e.g., nonmagnetized to magnetized) at critical points. The simple behavior of the elements of the system under the appropriate conditions leads to the emergence of collective or coherent behavior of the system as a whole. Critical phenomena are due to strongly nonlinear interactions among subatomic particles or atoms. The system may show, as a function of fluctuations in state near the critical point, pockets or order or structure embedded in regions of disorder embedded in still larger regions of order, and so on. Well above a critical point of, say, temperature, the system elements are arranged essentially randomly; the system possesses only short-range order. As the critical point is reached from above, larger-scale order begins to emerge. At the critical value itself, disconnected patches of order expand to infinite size, but fluctuations of lesser scales still persist within the larger pattern. A critical point, therefore, represents a discontinuity above and below which the system is qualitatively different.

If a single element in the system is perturbed, the disturbance can propagate neighbor-to-neighbor over the entire field. Distant elements, now structured in the same way, are said to be *correlated. Correlation length* represents the maximum distance over which a correlation can be detected. Near the critical point (or threshold) the correlation length grows rapidly. The basic reactions remain the same and connect only neighbors, but long-range order has emerged out of short-range forces. At the critical point itself, the correlation length becomes infinite. Phase transition has not yet occurred, but the system is now hypersensitive to very small perturbations, the effects of which can nearly instantaneously engulf the entire system.

This systems approach attempts to predict the macroscopic emergent properties of the system from the known microscopic properties. *This approach has been of fundamental importance in all science!* Consider economics or attempts to understand the mind. This is the micro-macro problem to which we shall return. In the simple physical systems being discussed here, qualitative structural change emerges out of the known laws of mechanics, quantum mechanics, and thermodynamics. It is not necessary to know the precise form of the interactions. The macroscopic collective behavior occurs at the statistical level, and only probabilistic predictions are made as to the behavior of any individual element of the system.

The fact that very different physical substances behave identically in the neighborhood of the critical points is called the *universality hypothesis*. The critical behavior of these systems is determined by only two values: the dimensionality of space, and (2) the dimensionality of the order parameter. Systems that share these dimensions have identical "critical exponents." The *order parameter* represents a macroscopic property or field emergent from interactions at a more microscopic level. Examples in physical systems are spontaneous magnetization, correlation length, density differences between the liquid and gaseous phases of a fluid, and concentration differences. The strong fluctuations that characterize these phase transitions take place in the order parameter field. The order parameter concept plays an important role in my theory of the world system-field of forces (e.g., De Greene, 1982a, 1989). Even so, do not generalize too far from *equilibrium* phase transitions.

Finally, there are many uses of the term "field theory" in physics, astronomy, and chemistry today. Examples of terminology include "unified field theory," "quantum field theory," and "mean field theory." In a galaxy like the Milky Way, a gravitational *force field* is generated by 100 billion or more stars and by dark matter whose makeup is unknown. The field can be determined from the measured motion of stars and gases. In physical chemistry, spectral analysis can yield a measure of the stiffness of various parts of a molecule; that is, the force field can be established. Also, the electric "field strength" is the negative gradient of the electric potential (Atkins, 1986).

What all this means is that there are many very valuable concepts and constructs that appear again and again across the sciences. The search for and utilization of these concepts provides a key structural underpinning of this book. Thus, this book is *transdisciplinary*, rather than interdisciplinary or multidisciplinary. This book also represents advances in *general systems theory*. Consider the mean field theory mentioned above. This can be used to describe the average behavior of a system or environment, an average behavior that damps desirable fluctuations in organizational, social, and ecological systems. Chapters 2, 3, and 7 provide other general-purpose theories and models that are applicable to a wide variety of systems. And in a related sense Nicolis and Prigogine (1989, p. 217 ff.) emphasize, as a first step in model building, the extensive use of analogies and archetypes: physicochemical systems yielding transition phenomena, long-range order, and symmetry breaking far from equilibrium can serve as an archetype for understanding other complex systems.

Living Systems

Whereas the development, evaluation, and application of field theory in physical science have been essentially continual, corresponding efforts in the biological, behavioral, and social sciences have been rather sporadic and limited. Two areas will be discussed next: embryology and psychology.

Embryology. A long-standing challenge in developmental biology is to understand how a fertilized egg "knows how" to differentiate into the numerous complex tissues and organs of the adult. Embryologists have postulated, and in many ways experimentally verified, the existence of chemical substances called morphogens. Morphogens are distributed in the embryo in concentration gradients, and cells are differentiated as a function of how much morphogen they are exposed to by their surroundings, the *morphogenetic field.* The area with the most morphogen has been variously called the organizer, the organizing tissue, and the zone of polarizing activity. When the organizer is transplanted to host organisms, it can induce the generation of structures that otherwise would not have developed in the given area. The organizer is the only group of cells that can initiate a whole embryo.

The term "field" was introduced into embryology in the early 1920s by Hans Spemann. The field character of organization was expanded by Weiss; see Weiss (1939) for a discussion of the early development of the concept. Basically, embryological development is a function of the interaction between the properties of the field and the properties of the given cells or tissues. The field does not have unlimited capabilities to induce structural development.

Interestingly, problems of morphogenesis led René Thom (1975) to develop his catastrophe theory, a topological means of passing from the local to the global. Living phenomena are manifestations of the geometry of the life field.

Psychology. An elaborate field theory in psychology and social science was developed by Lewin (1938, 1951). Lewin stated the basic aspects of his field theory to be: (1) human behavior is derived from a totality of coexisting facts; and (2) the facts have the characteristics of a dynamic field, and the state of any part of the field depends on every other part. A field possesses general characteristics, such as atmosphere, as well as more specific features like goals, stimuli, needs, and social relations. Like the physicists whose work is summarized above, Lewin was concerned about the nature of space. (A simple definition of a field is that of a scalar or vector function of space.) Like many of the physicists, Lewin believed that Euclidean geometry was inappropriate.

Further, to Lewin a force in psychology has magnitude, direction, and point of application and can be represented by a vector (Lewin's theory was often called "topological and vector psychology"). In psychology as in physics, force is related to motion (locomotion). But locomotion is related more to a resultant of forces than to a single force. However, the laws that determine the magnitude and direction of even two forces are unknown. Examples of psychological forces are instinct, motivation, goal, and libido. The strength of a force is a function of the attraction/repulsion or valence of a goal and the distance between the person and the goal. Valences affect the structure and the spread of structure in the field, and qualitatively different force fields may emerge. One person can generate a force field that induces behavior in the overlapping force field of another person. For example, a power field indicates the influence one person has on another.

Forces can thus lead to the restructuring of the field, as in learning, conflict, and decisionmaking. (Previously the Gestalt psychologists had defined certain perceptual changes and insight learning to be sudden reconfigurations of the psychological field.) For example, decisionmaking can be interpreted as a restructuring of the field, so that one of two or more overlapping force fields becomes dominant. The outcomes are resultants stemming from opposite forces in the overlapping fields.

Lewin avoided too literal adoptions from physics. Direction of a psychological force is not the same as physical direction. Lewin invented a non-Euclidean geometry that he felt was more appropriate to psychology. He called this space hodological space. It deals with regions much more than with points and is concerned with the total psychological field and with its cognitive structure. The space often shows abrupt steps or nongradual paths.

ALTERNATIVE OR COMPLEMENTARY PERSPECTIVES ON FIELD THEORY AND ON WORLD SYSTEMS

This section summarizes some perspectives on field theory and on world systems that expand this approach to social systems.

The Aftermath of Lewin's Field Theory

In the case of many pioneering scientists, when the master dies the majority of the theoretical work ceases. It is sometimes suggested that a measure of scientific productivity should be the ratio of new knowledge resulting from a given theory to the amount of knowledge derived from the increasing focus on methodology and on fine-tuning the theory. Typically, this ratio falls to and even below unity. We meet this problem often in this book. Consider the many examples of exhausted paradigms and our discussion in Chapter 1.

Following Lewin's death, his students developed concepts, such as the behavior setting, and performed many empirical studies, especially using the statistical technique of factor analysis. The organizational development practices of sensitivity training and of unfreezing, changing, and refreezing, were partially descended from Lewin's late work on group dynamics. These practices found their heyday in the 1960s and early 1970s. In sociology workers often referred to social forces. Some field-theoretic concepts became embedded in the body of psychology and social science. Field theory provided one input to the development of general systems theory. Nevertheless, Lewinian field theory in psychology and social science is now neither a frontier area nor part of the mainstream.

Fortunately, a legacy of Lewin's field theory operates in social systems science. Consider the work of Rummel (1987), who has studied cooperation and conflict as distinct catastrophe-theoretic equilibria of forces in a social field. The system jumps between these equilibria as a function of the gap between a structure of social expectations and a balance of power, and as triggered by some minor event. Expectations and power tend to become more incongruent over time. Conflict and cooperation, disorder and order, and war and peace are viewed as representing differential stability situations. As the gap widens, tensions develop in the field, making the outmoded structure more susceptible to a triggering event. A somewhat similar model is provided by De Greene (1978, 1982a, 1982b). See also the cybernetic interpretations in Chapter 3.

The founding fathers of sociotechnical systems theory and practice, Trist and Emery, were also influenced by Lewinian field theory in several ways. Most important to our discussion here is the classifications of the external environments of organizations; and of the four main categories (a fifth is the vortical), the most complex environment, the *turbulent-field* environment, is of the greatest interest. The concept of the turbulent-field environment (Emery and Trist, 1973) has greatly influenced my own thinking (e.g., De Greene, 1982a, 1987, 1990a).

Briefly, the dynamic processes in the turbulent field arise not merely out of the interactions among systems but also from activities triggered in the environment itself. New structures in the environment assume autonomous lives of their own. Emery (1977), reviewing its evolution, proposed that five major trends led to the emergence of the turbulent-field environment: (1) the immense growth in organizational size and power, leading to points of diminishing returns; (2) the growing interdependence among all parts of society, particularly abetted by advances in technology; (3) harnessing of R&D to augment organizational power; (4) the revolution in information, communications, and transportation which, however, have not reduced uncertainty but rather have generated a false sense of knowing; and (5) the increasing bureaucratization of all walks of life. Interestingly, and apart from Kondratiev theory (cf. Chapter 1), Emery placed the emergence of the turbulent-field environment as the years 1967-1969. For Emery's most recent thoughts on policymaking, see Chapter 6.

World Systems

Immanuel Wallerstein and his colleagues started a world-system area of study in the 1970s. They deal with the world economy, historical systems, and civilizations. Like me this school stresses a holism: Social actions and economic processes take place at the level of the world-system, a spatiotemporal whole. That is, a supranation-state organization emerges as a unique entity with unique properties, beyond the sum of the participating nation-states. System elements cannot be identified apart from their being embedded in the whole. Structures, even that of the world-system, at different points of time, are not repeatable, independent replicates of a given phenomenon.

The world-system approach is complementary to my own world system-field of forces approach (notice the difference in hyphenization). As stated above, their approach appears to *imply* field theory. The main differences between the two approaches appear to be: (1) a much stronger reliance on my part on field theory and a much greater attempt to advance field theory as a result of recent advances in systems theory and in mathematics; and (2) a much greater participation by sociologists and related social scientists in the world-system effort. For an example of overlapping interests, see Goldstein (1988) and Schaeffer (1989) for studies of Kondratiev cycles, hegemonic cycles, and war and peace, themes of particular concern in this book.

Finally: field theory is an important *organizing principle* for policymaking.

SYNOPSIS AND INTERFACING OF CURRENT SYSTEMS THEORIES

This section contains brief, mostly qualitative overviews of the given theories, overviews that are most relevant to policy analysis and policymaking. Note that the classification by chapter subsections that follows is arbitrary. For example, most theories are concerned with stability and several with bifurcations. For mathematical and other detail, see the referenced literature.

Stability Theories

Stability and instability are key concerns throughout this book. Stability is often closely associated with equilibrium, especially in classical stability theory. In Lyapounov stability the system does not return to equilibrium after a sufficiently long time following a small initial perturbation (perturbing force). The system is simply displaced from some reference state and is said to "remember" the effects of the perturbation. In state space, system behavior may resemble undamped oscillations whose amplitudes and periods have been changed. In phase space (an abstract space), the "phase portraits" resemble sets of ellipses. In asymptotic stability, in state space behavior of the perturbed system may resemble damped oscillations, and in phase space it may resemble spiraling in to an equilibrium point. See Nicolis and Prigogine (1989).

More generally, stability is the system property of reacting to perturbations and fluctuations in the environment while maintaining the same dynamical behavior over time (Casti, 1979). In classical stability theory the system itself does not change; as just noted it is merely displaced. This kind of stability situation appears to apply to at least some *thinking* about policymaking (cf. Chapter 1). See also Ingrao and Israel (1990) for a discussion of the history of equilibrium thinking in economics.

Much more important to the complex living systems of concern in this book is *structural stability* which involves the qualitative changes in the behavior of the system when it is perturbed. A structurally stable system displays a correspondingly small change in its behavior following a small perturbation. Bifurcation theory, catastrophe theory, and dissipative-structure theory variously deal with this kind of situation. Structurally stable systems are rare in more than three dimensions (May, 1974).

In living systems stability and instability are often associated with the degree of population fluctuations in stochastic environments. Equilibrium itself is an average around which the system fluctuates. Stability tolerates relatively small fluctuations.

Stability is also closely associated with the existence of *attractors*, regions in phase space that, like a magnet, attract system trajectories. See again Figure 1.6 and the related discussion. Different stability/instability conditions are generated by the three generally recognized kinds of attractors: simple, static equilibrium points; limit cycles; and strange or Lorenz attractors. Catastrophe-theory models, predator-prey models (e.g., Lotka-Volterra), and models of deterministic chaos, respectively, exemplify these conditions.

The most important aspect of stability, for our purposes, is its interrelationship with complexity. See Chapter 1. This interrelationship has been extensively studied in ecology by May (1974, 1981). Both in mathematical models and in reality, *increasing complexity tends to yield decreasing stability*. A stable environment may allow a complex system to evolve, but a complex system is especially vulnerable to instability in the environment. The greater the amount of connectance in an ecological web (cf. our earlier discussion of correlation in fields), the greater the number of modes of oscillation. A given mode is as likely to be unstable as stable, so additional modes increase the likelihood that the entire web is unstable. Mutations, symbiosis, and competitive interactions are destabilizing in society as well as in ecosystems. Systems characterized by lack of variety (e.g., monocultures in agriculture, forestry, and human organizations) are inherently unstable.

The domain of stability of complex systems may be small, and the system will therefore be dynamically fragile. The system can survive only under conditions of tightly limited values of environmental parameters. The system will collapse following significant perturbations to *either* population characteristics *or* environmental characteristics. Complexity, identified by great numbers of system elements, great interdependence among elements, and increasing strength of interaction, generates dynamic fragility. Models show that such complex systems are never more, and are usually less, stable than are simple systems (May, 1981; see also De Greene, 1991a).

Realworld systems have of course, over evolutionary time, developed a great deal of persistent hierarchical structure of subsystems; and of resilience, the capability to maintain internal structural interrelationships despite repeated subjection to the forces of perturbation and fluctuation. Indeed system resistance can decrease and the boundary of stability implode in the absence of continual testing by the environment. One can note many instances of system degradation in the absence of environmental stimuli. One is the loss of muscular and skeletal capability by astronauts and cosmonauts following periods spent in the absence of gravity. Another is the stagnation and decline of informationally closed human organizations. Overall, however, it is a mistake to count on unlimited stability, resilience, and adaptation of complex societal systems and ecosystems.

Attempts to validate scientific theories and hypotheses are often made by perturbing the system and seeing what happens. Many of the systems theories presented in this book involve perturbing the model of the system with an external force or input and then testing the stability features of the model. Testing is often made for the stability of an equilibrium point and for the existence of a bifurcation.

Synergetics

Synergetics is a very active school of systems thinking, better known in continental Western Europe than in the US and UK. It was developed by German theoretical physicist Hermann Haken and his colleagues (Haken, 1983). The school has now made theoretical and experimental contributions to a number of areas of interest to this book, including cognitive psychology (Haken and Stadler, 1990) and sociology and economics (Weidlich and Haag, 1983). In many ways synergetics can be interpreted as being a direct outgrowth of modern field theory in physics. Both areas are concerned with such theoretical constructs as nonequilibrium, order parameter, critical points, fluctuations including critical fluctuations, critical slowing down, self-organization, equilibrium and nonequilibrium phase transitions, attractors, system stability and instability, molecular and deterministic chaos, and control parameters. Synergetics also overlaps dissipative-structure theory.

Synergetics is defined as the science of collective static or dynamic phenomena in closed or open multicomponent systems having cooperative interactions among the system units. The living systems of interest to this book are of course dynamic and open; but we must emphasize again that much human thinking about the world is static and closed-system. Synergetics deals with both deterministic and stochastic processes. A particular strength of synergetics is its emphasis on micro-macro interrelationships. Interactions at a more microlevel induce or generate a macrolevel collective field or *order parameter* that, expressing circular causality, governs the behavior at the microlevel. See also De Greene (1989).

Some of the main features of synergetics, as somewhat modified by me, are summarized as follows:

1. Macroscopic spatial, temporal, or functional structures emerge spontaneously in a self-organized manner when control parameters are changed.
2. The control parameters do not "control" the emergent structure; rather they lead the system through instabilities.
3. New structure can be likened to a nonequilibrium phase transition.

4. The macrolevel structure or collective field generated by microlevel interactions can be viewed as an order parameter.
5. The nonlinear order parameter dynamics may show a variety of qualitatively different behaviors including multiequilibrium, multistability, and deterministic chaos.
6. Order parameters spontaneously appear or disappear at critical points.
7. Order parameters may compete or cooperate.
8. As a control parameter approaches the vicinity of a critical point, value, or threshold for a phase transition (a structural change), the system displays *critical fluctuations*, a sort of early warning of incipient change in the order parameters.
9. Because of *critical slowing down*, the structure or order parameter may take longer and longer to recover from perturbing forces. It is as if a form of Lyapounov stability has been introduced.
10. Branching at bifurcation points leads to alternative structures dominated by previously unknown or irrelevant order parameters.

Consider several examples. In a laser, a simple physical nonequilibrium system, the control parameter is nonspecific. The control parameter is the amount of externally applied light necessary to excite the atoms of the substance. But, depending on the magnitude of the control parameter, the system may show qualitatively different kinds of behavior. Below the critical point, when the degree of excitation is low, the system shows independent behavior of the emitted light. Above the critical value, coherent light is emitted. At still higher values of the control parameter, the system shows ultrashort laser pulses and then deterministic chaos. Note that in this example the "control parameter" can be clearly related to some controlling action by a person or a machine. In complex systems such clear relationships seldom obtain.

Many kinds of systems, including perception, cognition, and the economy, exhibit bistable or multistable phenomena. The equations of synergetics can be applied, for example, to the face-profile/vase, wife/mother-in-law, and man's face/reclining-girl illusions long studied by psychologists. If the stimuli are presented as a sequence from left to right and then from right to left, the perceptual phase shift does not occur at the same stimulus location. Rather, there is a lag, called *hysteresis*, as in catastrophe theory which has also been used to study this phenomenon. At the level of the economy, Mensch [referenced in Chapter 1; see also Weidlich and Haag (1983)] has developed a bistable model of the post-World War II situation.

Synergetic interpretations can be made of the age-old mind/body and brain/mind problems. These have implications for the design of neural networks and other artificially intelligent systems. A variety of cognitive behaviors can also be noted. Behaviors include mental ruts or blocks, *idées fixes*, rigid mindsets, insight learning, and creative breakthroughs. Again, such behaviors can be observed at both the individual and collective human levels.

Many kinds of macropsychological, social, economic, and political behaviors can be better understood by using these theoretical concepts. Once again: interactions among individual human beings and between people and machines generate the macroscopic field or order parameter, which then helps define the realm of the possible at the microscopic level. A great many values, beliefs, attitudes, opinions, and individual behaviors in social contexts, and transitions between forms, can be studied in this way. De Greene (1989) interfaced the approach of Weidlich and Haag (1983) with studies of consumer attitudes and spending by George Katona.

Several key points are summarized here:

1. Changes in values, beliefs, attitudes, opinions, and expectations exemplify the individual psychological fluctuations in society, and they are capable of driving societal structural change; that is, they can induce new order parameters.
2. These fluctuations do not cancel one another out to produce some unmodifiable average value.
3. The order parameter of collective attitudes, etc. possesses some stability and may take months or years (or even centuries) to change; but change can be hastened by surprise or by major perturbations, such as shocking news, that generate widespread anxiety.
4. Far from bifurcation points changes in attitudes, etc. may be slow, and change may not be greatly hastened by the electronic media; as is the case of a correlated physical field, reinforcement of selected new information requires neighbor-to-neighbor confirmation by word of mouth.
5. Attitudes, etc. have perceptual, affective, and cognitive components; they are not the ingredients of mainstream economic models, nor of rational decision-making.
6. Social learning, an order parameter, occurs much more slowly than does individual learning; but once social learning has occurred, it is not easily reversed or otherwise changed.
7. As discussed in Chapter 1, people fit new information into a preexisting stereotype or paradigm; social learning has filtered out (or "averaged out") much of the variety of individual learning.
8. Social learning is much slower than are technological and social change, making adaptation more difficult.
9. The world societal-field of forces is characterized by a number of cognitive/emotional order parameters that differ in life-cycle length and stability. Consider again our discussion of the various cycles and life histories in Chapter 1. At the other extreme of permanency and stability are various fads, panics, and hysterias.
10. The evolution of the order parameters or collective fields can be expressed in terms of equations of motion. There are several forms of these equations. They may represent general probabilistic laws applicable to the statistical dynamics of multicomponent systems. For example, individual transitions in attitudes, etc. lead to changes in macroscopic dynamics describable by

appropriate equations of motion. Transition probabilities estimate a change from a given individual behavior to another and provide a basis for estimating the probability of transition from one collective system state or order parameter to another.

Finally, let us close this section with a further example and with a further definition. The important concept of order parameter is not always easy to understand. Haken (1983, p. 353) states that languages are order parameters that slave the subsystems consisting of individual human beings. A language changes very little over the lifetime of the individual person. A person learns a language (i.e., he is slaved by it), and over his lifetime the person contributes to that language.

Macropsychology, which overlaps order parameter, is hereby defined. Macropsychology is the science of collective psychological phenomena, that is, collective perception, collective consciousness, collective intelligence, collective anxiety, etc. See Table 1.3 and associated discussion. Macropsychology operates at the same hierarchical level as do economics and sociology.

Dissipative-Structure Theory

Dissipative-structure theory has been developed by Nobel Prize-winning chemist Ilya Prigogine and his colleagues (Nicolis and Prigogine, 1977, 1989; Prigogine,1980; Prigogine and Sanglier, 1987; Prigogine and Stengers, 1984; Schieve and Allen, 1982). In the physical science literature, dissipation (of energy) and dissipative force have long been contrasted with conservation (of energy) and conservative force. Prigogine emphasized the continuous flow of energy through open systems far from equilibrium (i.e., under nonequilibrium conditions). These systems dissipate free energy and generate entropy which is transferred to the external environment. Dissipative structures can be spatially, temporally, or spatiotemporally organized. Dissipative-structure theory provides a theory of evolution. With increasing complexity internal fluctuations (deviations from mean values or qualitatively new events) may arise that the system can no longer absorb or damp. The system is particularly susceptible to fluctuations at critical or bifurcation points, beyond which lie new solutions to differential equations and new (i.e., dissipative) structures. Bifurcation points are reached as control parameters are changed. Evolution can be represented irreversibly by cascades of successive instabilities and self-organizations. Evolution is not preordained, and there are potentially many different branches in evolution. *Of great importance to policymaking, the evolutionary outcome may be quite unpleasant and quite irreverent to any particular ideology or set of human goals.* See also Chapter 2.

Dissipative-structure theory stresses the close interplay between stochastic (probabilistic) and deterministic factors and forces; that is, between chance and necessity or innovation and confirmation. This concept played a strong role in our arguments in Chapter 1 regarding the eventual exhaustion of ideas, designs, and paradigms. Fluctuations or mutations, which may be genetic, personalistic ("great persons"), social, or technological, are the stochastic or chance factors. Under certain conditions the fluctuations may grow into nucleations. Growth may be driven by self-amplification or autocatalysis via positive-feedback processes, or by cross-catalytic activity involving two or more fluctuations. Far from critical or bifurcation points, any growing fluctuation or nucleation tends to be damped in conflict with the larger system or environment. At bifurcation points the system is particularly sensitive to fluctuations, and the structural stability of the system often breaks down permitting amplification of the fluctuations. Mathematically, the stability of the original system can be studied before and after the occurrence of a fluctuation. The growth of a new system, if it takes place, is deterministic and can be expressed by an equation(s) for macroscopic variables that show average behavior. Deterministic processes include those of interaction, reaction, and diffusion which require different kinds of equations. One may also infer a potential function.

Dissipative structures are strongly contrasted with equilibrium structures and can also be viewed as nonequilibrium phase transitions. The laser, described above, is a simple physical example. Further, structural changes can be driven, not only by internal fluctuations, but by external fluctuations (environmental noise) and by external perturbing forces. The various forces can act multiplicatively, with even more complicated and unpredictable results.

Structural stability here refers to the susceptibility of the system to fluctuations or mutations. The system is unstable if mutations like innovations can multiply and take over and modify system processes.

Symmetry breaking is associated with the process of bifurcation. When the control parameter(s) cross a critical point, the symmetric solutions to the rate equation(s) become unstable. Symmetry breaking reflects the intrinsic differentiation between parts of the system or between the system and its environment. A homogeneous system may spontaneously become inhomogeneous. Processes become possible that were impossible in the undifferentiated state.

Societies and ecosystems, and their constituent living elements, are thermodynamically open systems operating far from equilibrium and are dissipative structures. The complex systems are separated from equilibrium by instabilities. Dissipative-structure theory has advanced understanding in sociobiology; for, example, in explaining mound-building behavior in termites and food finding behavior in ants.

Figure 4.1 shows a simple pitchfork bifurcation diagram of the kind used by dissipative-structure theorists. The figure illustrates the constructs: far-from-equilibrium, critical bifurcation point, instability of the equilibrium branch, and bifurcating solutions to equations or evolutionary pathways.

Figure 4.2 shows the evolution of the Kondratiev cycles/structures discussed in Chapter 1 (De Greene, 1988a) as a cascade of dissipative structures, each separated by an instability. Note that the system is stable during the phases of prosperity and recession and is most susceptible to change during the phases of depression and recovery.

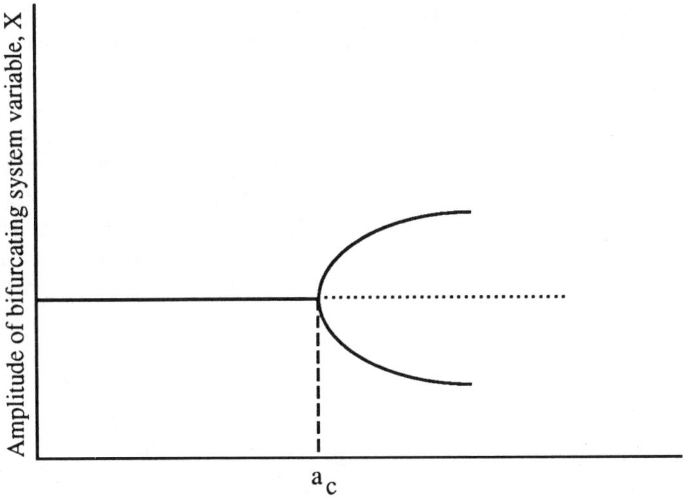

Figure 4.1. Simple pitchfork bifurcation. The control parameter, a, leads the system farther and farther toward the instability at the critical point. At a_c, the critical or bifurcation point, two new stable structures or solutions to the model equations emerge, and the stability of the old equilibrium structure breaks down. Solid lines indicate stability, and the dotted line indicate instability. Near the bifurcation point the system is hypersensitive to fluctuations.

Catastrophe Theory

Catastrophe theory is the invention of French mathematician René Thom (1975), who had in mind biological development like the morphogenesis discussed earlier; but it has been made much more understandable by British mathematician E. Christopher Zeeman (1977). See also Gilmore (1981). Catastrophe theory is particularly valuable in interpreting multistability and multiequilibrium behavior. The catastrophe itself is a sudden jump or drop from one equilibrium level to another. Elementary catastrophe theory is defined as the study of how the equilibria of the gradient of a potential function change as the control parameter(s) change. The catastrophe takes place when the values of the control parameters cause the system to flip from one attractor region to another. The attractors are static, fixed equilibrium points. There can be no limit-cycle attractors nor strange attractors (generators of deterministic chaos) in the system.

Catastrophe theory has many things in common with the other theories and models discussed in this section. For example, catastrophe theory is concerned with structural stability, control parameters, state variables, critical points or sets, nonlinearity, bifurcation, and hysteresis. However, there are also differences. Not every complex system is a gradient dynamical system, nor can the dynamics always be derived from a potential function. The attractors may not be just static equilibrium points. Catastrophe theory is descended from differential topology and is itself based on a number of limiting assumptions. For example, catastrophe theory assumes smooth differentiability in all orders; none of the derivatives has a discontinuity. This means that some "obvious catastrophes" are technically not catastrophes. Such matters are beyond the scope of this book. Catastrophe is an important heuristic.

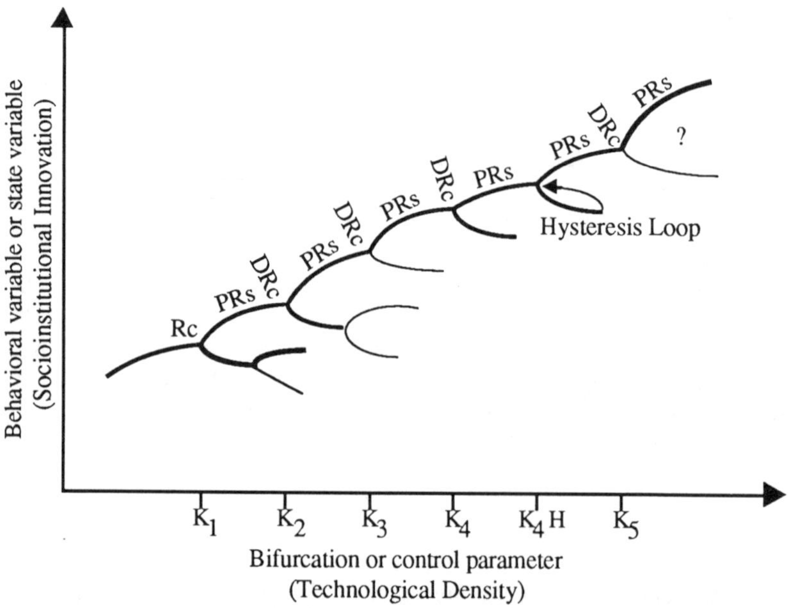

Figure 4.2. Evolution of the Kondratiev cycles/structures shown as a cascade of bifurcations. The K_i s represent bifurcation points at the onset of each cycle/structure. H represents a bifurcation of the catastrophe-theory kind, with hysteresis, within the fourth Kondratiev. Heavier lines indicate (potentially) stable evolutionary pathways. PRs indicates the prosperity-recession phases, DRc the depression-recovery phases. Cycle length is not shown exactly.

In catastrophe theory a system is structurally stable if its qualitative properties are unaffected by a perturbation of arbitrary strength.

Catastrophe theorists, dissipative-structure theorists, and others emphasize the *qualitative* nature of mathematics. This is an important point. Many of the equations of the models discussed in this book can not be solved directly, and reference must be made to other equations or to numerical solutions or computer simulations. As discussed in Chapter 1, for policymaking it is much better to develop an intuitive understanding or insight about the dynamics of complex systems than to be able to calculate everything in a limited context but understand nothing of the whole. For this reason much policy advice is limited or wrong.

Elementary catastrophe theory is limited to systems that can be described by one to five control parameters and usually one or two state variables. The control parameters can be viewed as slowly changing causal forces whose source is exogenous to the system. Also a force can be defined as the negative gradient of the potential. The model tries to minimize (or maximize) the potential. The control parameters change incrementally (some say this is comparative statics more than true dynamics), and the critical point or singularity beyond which lies the sudden catastrophic jump is almost always unknown. The state variables are sometimes called behavior or fast variables. The reader should refer to the literature for the many graphical depictions of the fold, cusp, butterfly, and other catastrophes. Here, as Figure 4.3, we show only the simplest catastrophe, the fold, which is generated by one control parameter and one behavior/state variable. Even so, the fold illustrates many of the constructs of catastrophe theory.

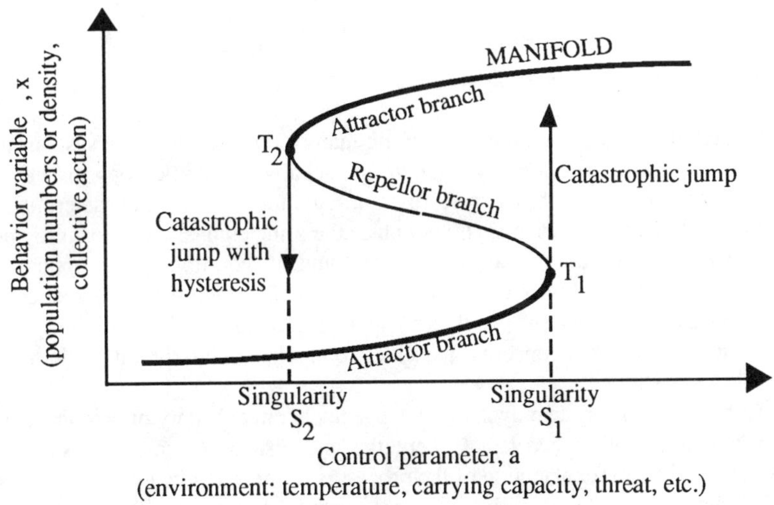

Figure 4.3. Fold catastrophe. The repellor branch is unstable.

Catastrophe theory has received a large number of applications including attitude and opinion shift, stock market crashes, aggressive behavior, population explosions of forest insect pests, and population collapses of fisheries. There are many examples of multiequilibrium systems wherein one population level may hover near extinction, another level may represent greater numbers under control, and still another level may reveal itself as an explosive outbreak.

Consider this use of catastrophe theory as a heuristic in policymaking. The Los Angeles area in southern California periodically experiences findings of a form of Mediterranean fruit fly ("Med fly") whose usual habitat is Hawaii. When a few flies are detected, the state government responds with a program of massive aerial spraying with malathion. The government argues that the flies could spread to the San Joaquin Valley and cause extensive crop damage there. However, large numbers of people resent the spraying. The spray leaves a dirty film on houses and automobiles. The spray is harmful to beneficial insects, some of which prey on the Med fly. The pesticide has unknown long-term effects on birds, pets, and people and it may be carcinogenic. The government insists that each new find of Med flies indicates a new, imported infestation. The alternative view is that the fly is now endemic in very low numbers. Considering the many unknown side effects, the costs of the spraying, and the quick evolutionary adaptation of insects to poisons, government policy should be rethought. Continued spraying could act as a control parameter, flipping the fly population into another basin of attraction, that of higher population levels. But of course policies often yield vested-interest bureaucracies with momenta and lives of their own.

Cybernetics and System Dynamics

Cybernetics and system dynamics share many constructs with the systems theories just discussed— nonlinearity, positive and negative feedback loops, growth and decline, stability and instability, perturbation, regulation and control, determinism. Cybernetics and system dynamics do not place the same emphasis on stochastic processes, activities in the external environment, and nonequilibrium (as opposed to equilibrium and disequilibrium) situations as do the other theories. Cybernetics and system dynamics (like catastrophe theory) apply excellently to the functioning of existing systems but do not concern themselves so much, as I interpret it, with the origin of new systems.

Cybernetics and system dynamics have a much longer history of practical application than do the other theories. Cybernetics and system dynamics far exceed the other theories in their explanation of the processes of information and control in systems. Cybernetics and system dynamics are concerned with actual system behavior as compared with desired system behavior, as made evident by available information, and the control (policy) steps that can be taken to reduce the disparity. Cybernetics and system dynamics apply to informationally closed-loop systems; they are not primarily concerned with flows of matter and energy as in dissipative-structure theory.

Both cybernetics and system dynamics have a common origin in the theory of information/communications and control. Both can be related to the area of control systems engineering. However, cybernetics may, for the most part, be more closely associated with the design of hardware devices like servomechanisms and manual and automatic control systems (the work of Stafford Beer on organizations is one exception). System dynamics has developed its own extensive conceptual basis and methodology. System dynamics, as a computer-simulation methodology, is especially accessible to users, including relatively naive users. Recently, system dynamicists have expanded the conceptual framework to include things like bifurcations and deterministic chaos.

Cybernetic (adaptive-control) theory is applied extensively to the international stability or war/peace problem by Harold Chestnut in Chapter 3. The main ideas, historical experiences, and recent applications to teaching of system dynamics are presented by its inventor, Jay W. Forrester, in Chapter 7. See also Forrester (1987, 1990).

Deterministic Chaos

There are at least two kinds of chaos treated of in systems science. Molecular chaos describes a completely entropic system, as when the universe reaches a final state of heat death. This is not our concern here. Deterministic chaos may arise in systems with no probabilistic properties yielding behavior that looks stochastic. Such chaotic behavior shows up in many physical, biological, and social systems, ranging from the simplest like a periodically forced, damped pendulum to the very complex like the human brain. There are at least two or three different kinds of deterministic chaos, one form of which was mentioned above in our discussion of the laser. Some understanding of chaos is important to policymaking and decisionmaking for three main reasons: (1) chaos appears in many iterated, recursive, or simulation computer models and therefore may place limitations on the fidelity of these models; (2) in real systems (e.g., the mammalian cardiovascular and nervous systems) chaos *provides information* regarding system state—there is evidence that the normal heart shows chaotic oscillations whereas the pathological heart does not; and (3) as in other nonlinear, dynamic situations, one may never be able to predict when a decision-action sequence will lead to the crossing of a critical threshold, with the unleashing of surprising behavior.

The chaotic dynamics of nonlinear deterministic systems have been recognized for several decades, and the systems context that generates chaos is described in an increasing literature [e.g., Grebogi, Ott, and Yorke (1987); May (1974); and Mosekilde, Aracil, and Allen (1988)]. Gleick (1987) provides a popular treatment. Grossmann and Mayer-Kress (1989) and Saperstein and Mayer-Kress (1988) apply chaos theory to interpretations of the US-USSR arms race.

As already indicated deterministic chaos is associated with phase space, attractors, basins or domains of attraction or repellence, and fractals. A phase space can best be visualized when a variable is plotted against its first derivative on a graph; or better, when two interacting variables (like a predator population and a prey population), with time having been eliminated, are so plotted. A figure generated by a phase space diagram is called a phase portrait. A domain of attraction is an open region in phase space that surrounds an attractor; in some situations trajectories may be repelled rather than attracted. The dimension of point attractors is zero, that of limit-cycle attractors one. The dimension of chaotic attractors may not be an integer and is said to be fractal. In dissipative systems attractors can change their form and size, and phase space itself can expand and contract.

There are two kinds of nonlinear equations that in particular, when iterated, exemplify deterministic chaos. One is the single difference equation, like the "logistic difference equation," that can represent the growth of insect populations whose generations are separated in time. See Chapter 9. The second mathematical model consists of three or more coupled differential equations. In the early 1960s E. N. Lorenz developed a model of this kind for weather prediction. The model failed to predict the weather but instead generated deterministic chaos, and Lorenz's name has been given to the attractors involved.

The kinds of chaos differ according to the gradualness or suddenness of onset, the continuous or discrete nature of the system, the nature and mixture of periodic and aperiodic wave forms, and the range and limit of the chaotic regime. Chaos reflects strong feedback loops.

An important feature of one kind of deterministic chaos is the extreme sensitivity of the system to initial conditions. All systems possess some degree of uncertainty as to their state, and the very study of systems invariably introduces differences in initial conditions and errors in measurement and in the rounding off of calculations. Such an uncertainty or error in knowledge can be amplified under conditions of system instability and expand exponentially over time, so as to make long-term prediction impossible. Among other things, scientific experiments and models can never be exactly replicated. It has been estimated that an error or random fluctuation as minute as one part in a trillion can grow exponentially, so as to dominate a computation or behavior of the system after only 20-25 steps. A large reduction in the uncertainty of the initial conditions may yield only a small reduction in the uncertainty of the final state of the system.

The Lyapounov exponents characterize the exponential separation in time of two initially close trajectories. Researchers often try to test for the existence of chaos in a system by calculating these exponents. However, experience shows that the detection of chaos is seldom straightforward, and these calculations may lead to erroneous conclusions.

FIELD-THEORETIC PRINCIPLES

At the beginning of this chapter, we emphasized the importance of field theory as a framework or *Weltanschauung* for thinking about the complexities of reality.

Following are some rather informally stated principles that policymakers and decisionmakers should bear in mind because a policy action can be viewed as a perturbation of a field that may induce surprising, unpredictable, and counterintuitive behaviors:

1. System-fields show different kinds of equilibrium-related conditions including equilibrium, disequilibrium, multiequilibrium, and nonequilibrium. These conditions imply different kinds and degrees of stability and controllability.
2. Under multiple-equilibrium conditions, the system exists in two or more states, which may differ in longevity, and at least one of which may be unknown. When a critical threshold is surpassed, the system may flip surprisingly into another equilibrium condition; that is, it may be captured by another basin of attraction. Stock market crashes and fisheries depletions are examples.
3. Under nonequilibrium conditions the system is hypersensitive to internal fluctuations, that under "normal" conditions would have been damped, and to external perturbations and noise. Scientific revolutions (Kuhn, 1970) are examples.
4. Systems evolving or driven far from equilibrium increasingly approach critical or bifurcation points, beyond which lie structural change (e.g., nonequilibrium phase transitions). The worldwide collapse of Communism, the *real* message of which must not be lost to the policymakers of Capitalism, provides a salient example.
5. As a system-field evolves, the elements may become increasingly correlated, eventually encompassing the entire field. This may improve the viability of the system under stable conditions but render the system most vulnerable to perturbing forces and fluctuations under conditions approaching instability. The world economy and most ecosystems are examples of evolution to a metastable state (cf. Figure 1.6).
6. An evolving system-field is an order parameter or collective macrolevel structure, that has been generated by interactions at the more microlevel and that, with mutual or cyclic causality, constrains behavior at that microlevel. The politico-economic-religious power structure of the West exemplifies this principle.
7. Attractors in phase space may make the future state of the system unknown or unknowable, and the future evolution and behavior of the system may be partly or wholly unpredictable. Anthropogenic perturbations of climate and weather are illustrative.
8. The state of all systems, even that of simple ones, must necessarily be uncertain because of combinations of natural variability, randomness, and sensitivity to forces and the sensory, perceptual, and cognitive limitations of human observers. Even a minuscule amount of uncertainty may make the difference between entirely different evolutionary pathways. Attempts to determine just how many individuals in a population of an endangered species constitute a genetically viable population provide an example.

9. The state of the system changes in ways that are not transparent to the observer. Thus, the same policy or decision may on one occasion have no effect because the internal interrelationships and mechanisms have shifted to swallow up the policy action, whereas on another occasion the action may induce great instability. Many system dynamics models show this phenomenon.
10. The same system elements can be configured into qualitatively different kinds of systems which are separated by instabilities. Hawk/dove political configurations provide an example.

A number of principles can also be stated more formally (expanded from De Greene, 1990b). *When the system-field evolves or is driven far from equilibrium, then:*

Principle One. The system becomes hypersensitive to small fluctuations that under "normal" conditions would have been damped, and the frequency of related fluctuations may increase.

Principle Two. The system becomes hypersensitive to external perturbations and fluctuations (noise). Because of critical slowing down, the system takes longer and longer to recover from these external forces.

Principle Three. Microlevel changes may be indicative of incipient structural change even though the system appears to be "normal" at the macroscopic level.

Principle Four. Recovery from undesirable or pathological states through time and structure reversal is impossible.

Principle Five. New self-organizations will transfer additional entropy (disorder) to the external environment.

FAITH, IDEOLOGY, SCIENCE, AND RELIGION

In this and succeeding sections of this chapter, the field-theoretic framework and the constructs of systems theory carry over to systems thinking and to further considerations of the problems that face policymakers and decisionmakers in a world of transformation.

In various chapters of this book, the authors have been highly critical of mainstream approaches to science, particularly the behavioral and social and environmental sciences that relate to the applications of technology. We have been quite concerned with the limitations of the Newtonian paradigm of science and society. We have emphasized that the mainstream approaches, coupled with the Newtonian paradigm, represent *an* approach to science and society. They do not and cannot represent *the* only, truthful, and everlasting viewpoint, however dominant a *Weltansicht* they may presently be.

In many ways the practice of institutionalized science has led to the degeneration of science. The symptoms are many and are associated with the exhaustion of fundamental ideas and the movement toward excessive confirmation, as discussed in

Chapter 1. Common symptoms include lack of imagination and timidity in the funding agencies, mindless continuation of research in known areas involving slight changes in settings and parameters, reservation of the rewards of career advancement and money for only the ideologically faithful, rejection of new ideas with unbridled hostility, misuse of science and technology for purposes of institutional power, difficulty getting nontraditional work published, and publication of trivia and even falsified data and results stemming from pressures to produce. In short: much of science is ingrown, highly conservative, and context free, but capable of intense emotional arousal in defense of the *status quo*.

This decay, accelerated over the past two decades, has of course been noted by many astute observers. Burbridge (1992), in a succinct but powerful essay, states that Big Bang theory rests on many untested, and in many ways untestable, assumptions and has become bandwagon thought that reflects faith as much as objective truth. But nonconformists are punished by denial of publication and of telescope viewing time. Peer review has become a form of censorship. Burbridge wagers that no young researcher would be willing to jeopardize his/her scientific career in academia by writing an essay such as his own. Forrester (1987) notes that the recognition of nonlinearity can destroy the simplicity and universality by which research papers are judged, and that junior faculty members enter the world of nonlinearity at the risk of their promotions. Bauer (1992) discusses the myth of the scientific method, itself, and the misleading impacts of the false belief on popular thinking. He rejects terms like "scientific fact." He contrasts "frontier science," which must necessarily change, with "textbook science," which is relatively static, and emphasizes that the public does not understand the difference. Interestingly, considering our discussions of systems decline in Chapter 1, Bauer likens present institutionalized science to the Roman Catholic Church at the beginning of the Reformation.

Because mainstream economics plays such a powerful role today in policymaking worldwide, it warrants additional comment. Kuttner (1991) views economics as rigid and politically correct. To be credentialed to practice economics, one must accept a *theology:* that individualized selfishness aggregates to the collective good, and that buying, selling, and consuming, without outside interference, lead to optimal outcomes. To become a tenured economist at a prestigious university, or to achieve an equivalent position at an influential think tank, requires that one be "economically correct." The economics profession ensures that people with incorrect views do not get the "union cards" necessary even to begin practice.

The "free market" is thus also a myth, and the concept of general equilibrium is wrong in the real world. Ingrao and Israel (1990) review the history of general equilibrium theory. In light of the discussion of the Newtonian paradigm in Chapters 1 and 2, the attempt to pattern economics after mechanics is especially pertinent. Later work in economic theory followed axiomatic mathematics which led to a theory that apparently can not be validated by real market behavior. Basically, the theory holds that there is a price for each good for which supply equals demand simultaneously in all markets. This resembles a static equilibrium in physics. Other equilibria are assumed to exist between consumers and producers. Critics have noted that the theory does not explain how the purported equilibria came into being. The unrealistic nature of pricing, especially in social and environmental terms, is also well recognized.

Confusion in the policy applications of economic theory is quite evident at the time of this writing. Gross National Product (GNP) and Gross Domestic Product (GDP) are only approximate measures of the performance of an economy. These measurements contain a good deal of subjective judgment and fudging. Also governments do not like to look bad. A business-cycle recession (cf. Chapter 1) is often defined as zero or negative growth of GNP or GDP for two successive quarters. But it is unclear whether and when the (US) economy recovered from the "recession," acknowledged by economists to have begun in 1988 or in 1990. GNP/GDP appear to show "recovery," but organizations continue to lay off people, unemployment and underemployment continue to increase, and many kinds of jobs continue to disappear. Further, savings are a major source of new capital, and the US has been criticized for its low savings rate. Thus, the economy, although considered by many to have "recovered" from the "recession," is still seen to be sluggish. The problem is perceived as one of "jumpstarting," and this has been attempted by the Federal Reserve Board by lowering interest rates in the hope of encouraging a debt-ridden and despondent populace to spend and consume more. When this policy failed the first time, the Federal Reserve Board lowered interest rates still further, hurting savers and especially older people and others on fixed incomes.

Returning to the main theme of this section, science and technology are artifacts of humanity and as such reflect human limitations as well as human capabilities. Science, technology, ideology, and religion are artifacts that, in much usage, are seen as opposed to one another. In some cases they are seen as equally "valid" but parts of different and irreconcilable spheres. Occasionally, there is an absurd and even pathetic arrogance, as when a given area (say, the mind) is defined as outside the realm of science. In reality, science, technology, ideology, and religion have much in common: *they are all dependent on faith*, faith in human sensations, perceptions, thinking, and doing. Thus, the terms are in many ways interchangeable. Religion does not necessitate the existence of one or more deities. Therefore, we can acknowledge that, in addition to the usually recognized theistic religions, science is a religion, technology is a religion, Capitalism is a religion, Communism is a religion, and so forth. Each religion has its own set of beliefs and reconfirming rituals. But there are no absolutes.

For further reading consider Kafatos and Nadeau (1990), who argue for a complementarity of consciousness and the conduction of science. A new way of understanding the universe could reduce much of the conflict between science and religion. The journal, *Rationality and Society*, contains articles relevant, pro and con, to themes of this book.

Finally, the field-theoretic constructs discussed in the first part of this chapter and the evolution of structures discussed in Chapter 1 apply to the dynamics of the great religions, which are order parameters. As times get worse, people become more conservative. Conservative thinking acts as a mean field, damping nonaverage behavior and needed variety. Policymaking now and in the foreseeable future cannot be rationally abstracted from the religious dynamics. Policymaking faces formidable challenges in such areas as inter- and intrareligious conflict, population control, environmental protection, sustainable development, and abortion. Many of these areas are obviously quite touchy. But the alternative to their consideration is most likely collapse of the nature discussed earlier. I shall conclude with these remarks. Part of the past competitive evolutionary success of *Homo* was due to the aggressiveness of the genus and to the capability of parents, as with most genera, to bear excess offspring, so that at least some would survive to reproduce. But past evolutionary success may mean future evolutionary failure.

THE ACCEPTANCE OF NEW PARADIGMS

We have noted that, when times get worse, people become more conservative. A number of authors have noted that scientists are basically conservative and have not welcomed scientific revolutions. And conservatism can be seen in a great many areas of endeavor today.

Let us briefly review the ideas of Kuhn (1970). The deeply conservative scientists, once indoctrinated into the prevailing paradigm, strongly resist new views. But scientists can never fully understand the real world nor even the works of one another. One can therefore speak of differing versions of "reality," wherein the same words may be used with very different meanings by different scientists. A paradigm is basically structured out of scientists' *subjective* observations and interpretations. The adherence of scientists to the given paradigm is the equivalent of religious faith (a situation we discussed above).

Exceptions to the prevailing paradigm, called anomalies, may be noted. Eventually, the number of anomalies mount up to a scandal, and a scientific revolution occurs. Most scientists switch to the new paradigm, which now becomes normal science. Scientists now devote their activities, called mopping up, to solving "puzzles," which extend and reinforce the newly accepted paradigm but do not challenge it.

Lightman and Gingerich (1992) discuss the beginning of anomalies. They state that anomalies are recognized as such only after they have received compelling explanations within a new paradigm, a process of retrorecognition. Previous to recognition the particular facts are ignored or accepted as givens. These authors also point to the extreme conservatism of science. In invoking Leon Festinger's theory of cognitive dissonance, they note that the familiar is more comfortable than is the unfamiliar and that inconsistencies in belief are uncomfortable. Scientists are motivated to reduce dissonance and achieve consistency. By these processes, the inadequacy of current theory can be avoided. Consider again our discussion of the cognitive/emotional models of defense policy analysts in Chapter 1.

A scientific revolution is creatively destructive. Paradigm shift can be viewed as a nonequilibrium phase shift. It is the rather sudden restructuring of the collective cognitive or macropsychological field. The originators of the new paradigm are fluctuations or mutations of the "great person" kind. The mutants attract persons of like minds, and the ensemble represents a nucleation. The nucleation battles the existing paradigm which represents the larger system or environment. The old paradigm may damp the emerging new paradigm for a period of years (as appears to be the case in much of behavioral and social science), but eventually, if the anomalies and scandals are strong enough, the old paradigm collapses in a scientific revolution and is replaced by a new paradigm.

The personal experiences of a number of creative individuals, the human interest stories that should punctuate even the most objective textbook material, are illuminating. Ilya Prigogine worked on the fringes of chemistry for some 20 years before receiving the Nobel Prize in 1977. In a 1983 interview with *OMNI*, Prigogine noted that all his ideas had been poorly accepted, and that the reaction of well known scientists was almost uniformly negative. An intuitive person, who observes that science is much less objective than is commonly assumed, he uses mathematics only as a communications tool.

Forrester (1990; Chapter 7 of this book) describes the hostility generated by his *Urban Dynamics* and by the *Limits to Growth* models. These works hit tender nerves in the mainstream economics and social science communities, and the ideas are still threatening. Emery (author of Chapter 6) and his colleagues found no publisher willing to accept their early findings about sociotechnical systems. Mandelbrot (1983), whose own fractal geometry is the subject of much controversy, reviews the careers of a number of mathematicians and scientists who received recognition only late in life. He writes (pp. 391-392):

> "....a life story that is interesting to tell is rarely the reward (or is it a punishment?) of those who keep to the mainstream of science... Entry into the mainstream is postponed, and in many cases it is even posthumous...Mavericks' work frequently exhibits a peculiar freshness."

Lotfi Zadeh (*IEEE Spectrum*, March 1991, p. 28) mentions the "downright hostility" in the US toward his fuzzy-set theory or fuzzy logic. Zadeh's theory was opposed by eminent mathematicians. On the other hand, fuzzy logic has been widely accepted in Japan, where it has been implemented in the design of a number of control systems ranging from that of the Sendai subway to those of several kinds of consumer goods, including camcorders, cameras, washing machines, vacuum cleaners, and microwave ovens. Fuzzy logic is analogous to the human capability to reason in approximate, as opposed to just absolute, terms. There are important differences among ethnic order parameters or collective consciousness. The Japanese are more tolerant of ambiguity than are Americans as a whole. This appears to be a very important factor in the decline of American competitiveness and of American hegemony discussed in Chapter 1. I emphasize that quick fixes (e.g., Total Quality Management), borrowed from the Japanese, can exacerbate this situation.

Finally, in this section we have discussed the conservatism of scientists and also alluded to the increased general conservatism that occurs during the Kondratiev-recession phase. With regard to the former, many more individual experiences from past centuries could have been noted. For example, in 1847 Helmholtz's paper on the law of conservation of energy was rejected for publication by the editor of the *Annalen der Physik* and had to be published in pamphlet form (Richtmyer and Kennard, 1942). With regard to the latter point, conservatism is not limited to science. Michener (1992) describes the great difficulties now facing young writers and poets in America, as well as the hostility that often greeted his own writing career.

Some Comments on the Status Quo

As discussed above and in Chapter 1, a new paradigm is seldom if ever accepted initially, and people usually reject paradigm change because of deep cognitive and emotional investments in the *status quo*. And ruling elites have almost always had power and economic investments in the status quo. Thus, even if a new paradigm makes sense to most people and they have a "gut feeling" that it is right, paradigm change and derived policymaking will almost certainly be rejected by the ruling elites until change becomes inevitable. This is part of the policy reality discussed in Chapters 1, 5, and 6. Referring back to the field-theoretic framework, we note that inevitability is linked to nonequilbrium situations, instability, and hypersensitivity to fluctuations and perturbations.

The maintenance of the national power structure is synopsized by Galbraith (1991). He believes that the military-industrial complex is engaged in a massive escape from public purpose that is abetted by a process of concealment. Thus, mainstream economic theory (discussed above) and instruction conceal economic reality, and political attitudes and instruction conceal political reality. Galbraith believes that the managers of large organizations maximize compensation and power, not for the organization or for dispersed and essentially powerless stakeholders, but for themselves. Acquisitions, mergers, and leveraged buy-outs are the outcomes of struggles for power and position by those who seek to control and manage. The military-industrial complex has an autonomy of its own and serves, not social purposes, but its own ends. Democracy and the free market are just covering façades, behind which power-driven activities are pursued with little questioning. See also De Greene (1990c). The exorbitant level of US executive salaries, vis-à-vis executive and company performance, and some 60-70 times employee wages, is a symptom of this deteriorated situation that is currently attracting much attention.

The kinds and frequency of articles and discussions in the popular media can serve as important indicators of the moods of the times. They can be viewed as microlevel fluctuations that signal that all is not well with the prevalent order parameter. This chapter cites several. Duffy (1991) indicates that, rather than providing genuine solutions to the many societal problems, both liberals and conservatives in order to maintain power offer false choices that divide voters. (Duffy mentions the existence of a New Paradigm Society.) Goodwin (1991), a long-time participant and observer in the Washington scene, believes that there are no signs of courage in the present US political structure. He sees hope only in a political movement outside the Republican/Democratic structure. The movement should have agenda and demagogic skills to convince people that it means business. It could lead to a Jeffersonian revolution from within.

GUIDED STRUCTURAL CHANGE IN THE WORLD SYSTEM FIELD

Systems science can provide a basis for true freedom and true choice, coupled with a greater responsibility for the greater whole of Nature and society, and the realization of true democracy. These goals can be accomplished through an emphasis on spontaneity, fluctuation, novelty, variety, appropriate amplification, and self-organization; while recognizing the limits to growth, stability, and survivability, and the dangers of massive, discontinuous, and catastrophic restructuring.

We tend to overestimate greatly our capabilities to control Nature and society through science; although such Newtonian thinking seemed logical at a time when absolute monarchs controlled people (Prigogine and Stengers, 1984). A dynamic field of forces cannot be controlled, but it can be perturbed with nonlinear and structural effects that may be undesirable. Simple, seemingly rational decisions may induce nonrational results in the form of order parameters that defy understanding. The distinction is management within the field, not of the field (De Greene, 1990b).

For complex systems capable of self-organization, the best management strategy is usually: *leave them largely alone*. Destructive consequences can result from overmanaging an organization, a society, an ecosystem, or a mind.

Creativity and innovation require spontaneity and freedom from external control. Learning must stem from self-discovery. Educational policy in many countries has created an atmosphere that actively discourages these processes. Means are rationalized, rather than outcomes helped. In the US fixation on lesson plans, learning objectives, coming up with *the* correct answers, and standardized teacher assessments, coupled with tight control of the classroom environment by layers of administrators, has practically guaranteed the decline of education. One result may be a paucity of critically thinking adults, given the severity of evolutionary changes. See also the approaches to creativity and learning in Chapters 2, 5, 6, 7, and 8.

Most policymakers recognize that education, along with other areas, is in trouble. Unfortunately, the problem(s) is usually seen as a money problem, and a problem that can be remedied by an even greater dose of the given ideology. The emphasis on money, on the consumption of material things as a primary means of achieving satisfaction, on material standard of living, and on a universal level of standard of living is a serious obstacle to guided structural change. The reasons are psychological, social, and environmental. Numerous studies in organizational psychology have indicated the paramount importance of the needs for individual autonomy, competence, and self-actualization. These studies have revealed the inherent or intrinsic value of work itself. A review of the history of great artists, musicians, philosophers, and scientists reveals much the same thing. These great people created out of the joy of creating. They did not work primarily for money although most realized the necessity for some monetary support to avoid penury. Consider: Einstein developed many of the ideas for his theories of relativity while working at a "bread-and-butter" job in the Swiss patent office in Bern. Van Gogh sold only one painting (although ironically his paintings sell for millions of dollars today). Schubert, over his entire lifetime, made the equivalent of less than $US 3000 (not adjusted) for his compositions; much of the time he was only a step from starvation.

The neglect of values and needs other than the monetary has a deleterious effect on the workplace, the community, and the family. In the workplace Tayloristic/Fordian designs are counterproductive and have contributed to the decline in American competitiveness, as was discussed in Chapter 1. In the community and family, abetted by ceaseless advertising, people have been brainwashed to substitute material goals for spiritual goals. Massive consumption has become an end in itself. Here "spiritual" means love in the family, pride in workmanship, cooperation in the community, and feeling of unity with Nature.

The ideology of continued materialistic growth has a certain Messianic fervor. Anything that can be developed for financial gain should be developed. Numerous programs of highway construction, dam building, forest exploitation, and agribusiness both in the developed and underdeveloped worlds provide examples. Although superficially representing "progress," these activities can be immediately destructive of viable systems and trigger further devolution. Small farming and subsistence agriculture are disappearing as ways of life in many parts of the world. Conversion of natural areas to land for livestock raising and cash crops destroys ecosystems, threatens tribal peoples with extinction, and drives small farmers to the cities, where their numbers swell the fetid slums. Crime, prostitution, drug and alcohol abuse, and general hopelessness increase markedly as people all over the world lose contact with their roots. Moreover, it seems increasingly unlikely that the very high resource use per capita in the West, particularly in the US, compared to that in the underdeveloped world can be justified and maintained much longer without some rather serious consequences downstream.

Clearly, here are areas par excellence that call for a systems-based approach to policymaking. Referring again to the unreliability of measures like GNP and GDP, we note that the UN System of National Accounts (SNA) shares the same deficiency. Repetto (1992) discusses the fact that a country can practice widespread environmental destruction without affecting its measured income. "Impoverishment is taken for progress" (p. 94). Economic growth as measured by GDP is overstated when the value of natural resources is ignored, and adjustment for depreciation of such assets presents a far less rosy picture.

Guided structural change will be difficult in the absence of an understanding of the momentum that present ideological systems possess and of power dynamics. Power, like the intrinsic motivation of a job or task, is a nearly universal human need. Power motivation can be expressed in harmless, constructive, and destructive ways. Surmounting a physical or psychological obstacle, like climbing a mountain or developing an insight, can be a beneficial expression of competence and power. Power expressed as leadership in helping people to solve problems and to realize their potentialities is probably necessary in any society. But power over people and Nature for power's sake, and for exaggerated personal gain associated with others' losses, must be viewed as pathological. Also, concepts like international balance of power, accepted as givens, appear now to require serious rethinking.

Are We Near One of the Great Bifurcation Points of History?

Discussion in various parts of this book, particularly in Chapters 1 and 8, provide an affirmative answer to this question. Great bifurcations or transformations have occurred in the past, as around the year 1500, and they most certainly will occur again.

Either we can prepare the proper receptive field and then recognize, choose, and amplify those fluctuations and nucleations that represent the basic values, needs, and goals of most people and the survivability of the natural environment, or we can let the world system run open loop in the long run while reinforcing the dominant ideological power structure in the short run. Fortunately, there are a number of recognized and beneficial fluctuations that can be amplified via policy actions. Several arise in the area of conservation and are familiar to most readers. These include such practices as cutting down discardable waste at the source, recycling matter, designing equipment to be more energy efficient, reducing reliance on fossil fuels, developing new sources of energy (e.g., solar), and changing life styles and daily habits. Other fluctuations can take the form of grassroots efforts that challenge the stability of the ideological order parameter. Of top priority is the elimination of war in any form in the manner presented by Chestnut in Chapter 3. Carl von Clausewitz' *On War* would then be relegated to the realm of quaint, but obsolete 19th-Century thinking.

The alternative pathway beyond the bifurcation point presents great dangers and possible catastrophic collapses in the face of unmanageable complexity (cf. Chapter 1). Added to the examples already offered, consider further the collective cognitive/emotional climates of the various nations. Although different in some ways, these climates have much in common. Despite the blessings touted by our rulers, these are sad times. The kinds of changes taking place defy easy comprehension, and most people find the uncertainty about the future to be highly unsettling. When people are confronted by sustained uncertainty, ambiguity, and anxiety, they often return to the familiar while rejecting the different. Xenophobia, religious fundamentalism and intolerance, and political conservatism are increasing in much of the world. And too many nations are still willing to go to war.

These and other areas are highly critical for reasons that are not always obvious. For example, intolerance must be met with designs that provide hope for everybody; designs must not polarize the races, the faiths, the sexes, and the ages. Tolerance can not be induced by legal proscriptions. Forced lose-lose choices like loggers versus spotted owls, jobs versus redwoods, and jobs for men versus jobs for women show the poverty of the original policy thinking.

Frustration has from time immemorial been relieved by aggression. Prior to the Persian Gulf Crisis and War of 1990-1991, the boredom, loneliness, superficiality, and general frustration and anxiety of the everyday life of millions of people in the West, perhaps especially in the US, had induced a highly correlated field of tensions that needed just one more spark to set it aflame. The US-pressured UN war against Iraq greatly oversimplified reality, provided apparently clear cause-effect relationships, reinforced the myth of the technological solution, and restored (albeit temporarily) a mood of optimism of the "if we can send man to the moon" variety. Studies showed that the more people relied on television for their major source of "news" (i.e., the more ignorant they were), the more they supported the war. The more educated the people, the more critical they were. The war was one of the rare ones fought by the advanced countries to occur during a recession. Richardson in Chapter 10 provides an interpretation of the systemic factors associated with this cynically motivated event. Readers might like to compare the ethics and morality of responses to Iraqi aggression to those of Western responses to Serbian aggression a year later. Of course, military aggression is in the eyes of the beholder.

FINAL REMARKS

Most approaches to policymaking and decisionmaking are context-free and assume an ideological terra firma on which advice to policymakers and decisionmakers is based. In this book the following are considered to be dynamically interacting parts of an evolving system-field (De Greene, 1991b) which generates limits to the adaptability of human systems (De Greene, 1981) and exhaustion of given paradigms (De Greene, in press): (1) the dominant ideology(ies), (2) the approach to policymaking and decisionmaking, (3) the objects of policymaking and decisionmaking, (4) the given policymakers and decisionmakers, and (5) the nature of the evidence. Further, whole systems transformation (Chapter 8) means that it is not just parts of the system that are now being transformed.

Some succinct recommendations are now offered to people concerned with the policy process as just defined:

1. Avoid projecting your own cognitive/emotional state onto other people, times, and places.
2. Always try to anticipate externalities and side effects of any policy action.

3. Do not formulate policy independently from related policies.
4. Do not view any monetary, human, technological, or natural resource as unlimited.
5. Do respect human continuity with Nature in evolution.
6. Avoid making hypocritical, cynical, and manipulative policies.
7. Look for early-warning signs (fluctuations) that may presage imminent structural change.
8. Do not count on reversing evolutionary change.
9. Be aware of the natural processes by which systems age.
10. Seek situations wherein goals and policies can achieve multiple benefits, yet look for parsimony and design for local self-sufficiency.

REFERENCES

Amit, Daniel J., 1978. *Field Theory, the Renormalization Group, and Critical Phenomena.* New York: McGraw-Hill.

Arnowitt, Richard, and Pran Nath (Eds.), 1976. *Gauge Theories and Modern Field Theory.* Cambridge, MA: MIT Press.

Atkins, Peter W., 1986. *Physical Chemistry.* Third edition. New York: Freeman.

Bauer, Henry H., 1992. *Scientific Literacy and the Myth of the Scientific Method.* Urbana, IL: University of Illinois Press.

Boslough, John, 1989. "Searching for the Secrets of Gravity." *National Geographic,* 175(5), 563-583.

Burbridge, Geoffrey, 1992. "Why Only One Big Bang?" *Scientific American,* 266(2), 120.

Casti, John L., 1979. *Connectivity, Complexity, and Catastrophe in Large-Scale Systems.* New York: Wiley.

Contemporary Physics Education Project, 1990. *Standard Model of Fundamental Particles and Interactions.* Wall Chart. Berkeley, CA: Lawrence Berkeley Laboratory.

De Greene, Kenyon B., 1978. "Force Fields and Emergent Phenomena in Sociotechnical Macrosystems: Theories and Models." *Behavioral* Science, 23(1), 1-14.

De Greene, Kenyon B., 1981. "Limits to Societal Systems Adaptability." *Behavioral Science,* 26(2), 103-113.

De Greene, Kenyon B., 1982a. *The Adaptive Organization: Anticipation and Management of Crisis.* New York: Wiley.

De Greene, Kenyon B., 1982b. "Anticipating Critical Reconfigurations in World Societal Fields." *IEEE Transactions on Systems, Man, and Cybernetics,* SMC-12(5), 694-704.

De Greene, Kenyon B., 1987. "US and USSR Adaptation to a Turbulent-Environmental Field of Forces: Avoidance of Catastrophic Flip to the

EMERGENCY State." *Technological Forecasting and Social Change*, 31(2), 189-200.

De Greene, Kenyon B., 1989. "Micro-Macro Interrelations and the Order Parameter Concept in the Field Theory of Societal Systems." *Systems Research*, 6(4), 277-288.

De Greene, Kenyon B., 1990a. "The Turbulent-Field Environment of Sociotechnical Systems: Beyond Metaphor." *Behavioral Science*, 35(1), 49-59.

De Greene, Kenyon B., 1990b. "Nonlinear Management in Technologically-Induced Fields." *Systems Research*, 7(3), 159-168.

De Greene, Kenyon B., 1990c. "The Suicide of Western Capitalistic Democracy: Can It Happen? Is It Happening?" *World Futures*, 30(1-2), 17-40.

De Greene, Kenyon B., 1991a. "Rigidity and Fragility of Large Sociotechnical Systems: Advanced Information Technology, the Dominant Coalition, and Paradigm Shift at the End of the 20th Century." *Behavioral Science*, 36(1), 64-79.

De Greene, Kenyon B., 1991b. "Evolving System-Fields, versus Context-Free, Futures Forecasting: Some Thoughts on Controllability." *Systems Research*, 9(2), 59-76.

De Greene, Kenyon B., 1992 (in press). "The Growth of Exhaustion," *European Journal of Operational Research.*

Duffy, Michael, 1991. "Looking for the Radical Middle." *TIME*, (May 20), 60-61.

Einstein, Albert, 1950. "On the Generalized Theory of Gravitation." *Scientific American*, 182(4), 1-5.

Emery, Fred E., 1977. *Futures We Are In.* Leiden, The Netherlands: Martinus Nijhoff.

Emery, Fred E., and Eric L. Trist, 1973. *Towards A Social Ecology: Contextual Appreciation of the Future in the Present.* New York: Plenum.

Forrester, Jay W., 1987. "Nonlinearity in High-Order Models of Social Systems." *European Journal of Operational Research*, 30, 104-109.

Forrester, Jay W., 1990. *The Beginning of System Dynamics.* Cambridge, MA: Massachusetts Institute of Technology, Sloan School of Management.

Galbraith, John K., 1991. "The Sting of Truth." *Scientific American*, 264(5), 136.

Gilmore, Robert, 1981. *Catastrophe Theory for Scientists and Engineers.* New York: Wiley.

Gleick, James, 1987. *Chaos: Making a New Science.* New York: Viking.

Goldstein, Joshua S., 1988. *Long Cycles: Prosperity and War in the Modern Age.* New Haven, CT: Yale University Press.

Goodwin, Richard N., 1991. "The People Can Do Better Than Two Parties in a Pod." *Los Angeles Times*, Metro/Valley Section, (May 16), B15.

Grebogi, Celso, Edward Ott, and James A. Yorke, 1987. "Chaos, Strange Attractors, and Fractal Basin Boundaries in Nonlinear Dynamics." *Science*, 238(4827), 632-638.

Grossmann, Siegfried, and Gottfried Mayer-Kress, 1989. "Chaos in the International Arms Race." *Nature,* 337(February 23), 701-704.

Ingrao, Bruno, and Giorgio Israel, 1990. *The Invisible Hand: Economic Equilibrium in the History of Science.* Cambridge, MA: MIT Press.

Haken, Hermann, 1983. *Synergetics: An Introduction.* Third edition. New York: Springer-Verlag.

Haken, Hermann, and Michael Stadler (Eds.), 1990. *Synergetics of Cognition.* New York: Springer-Verlag.

Kafatos, M., and R. Nadeau, 1990. *The Conscious Universe: Part and Whole in Modern Physical Theory.* New York: Springer-Verlag.

Kuhn, Thomas S., 1970. *The Structure of Scientific Revolutions.* Second edition. Chicago: University of Chicago Press.

Kuttner, Robert, 1991. "Economics Has No Time for Heretics." *Los Angeles Times,* Opinion Section, (April 17), M5.

Lewin, Kurt, 1938. *The Conceptual Representation and the Measurement of Psychological Forces.* In Contributions to Psychological Theory I, pp. 1-247. Durham, NC: Duke University Press.

Lewin, Kurt, 1951. *Field Theory in Social Science: Selected Theoretical Papers.* Chicago: University of Chicago Press.

Lightman, Alan, and Owen Gingerich, 1992. "When Do Anomalies Begin?" *Science,* 255(5045), 690-695.

Mandelbrot, Benoit M., 1983. *The Fractal Geometry of Nature.* New York: Freeman.

May, Robert M., 1974. *Stability and Complexity in Model Ecosystems.* Second edition. Princeton, NJ: Princeton University Press.

May, Robert M. (Ed.), 1981. *Theoretical Ecology: Principles and Applications.* Second edition. Boston: Blackwell.

Michener, James A., 1992. *The World Is My Home: A Memoir.* New York: Random House.

Mosekilde, Eric, Javier Aracil, and Peter M. Allen, 1988. "Instability and Chaos in Nonlinear Dynamic Systems." *System Dynamics Review,* 4(1-2), 14-55.

Nicolis, Grégoire, and Ilya Prigogine, 1977. *Self-Organization in Nonequilibrium Systems: From Dissipative Structures to Order Through Fluctuations.* New York: Wiley.

Nicolis, Grégoire, and Ilya Prigogine, 1989. *Exploring Complexity: An Introduction.* New York: Freeman.

Prigogine, Ilya, 1980. *From Being to Becoming: Time and Complexity in the Physical Sciences.* San Francisco: Freeman.

Prigogine, Ilya, and Isabelle Stengers, 1984. *Order Out of Chaos: Man's New Dialogue with Nature.* New York: Bantam.

Prigogine, Ilya, and Michèle Sanglier (Eds.), 1987. *Laws of Nature and Human Conduct.* Brussels: Task Force on Research Information and Study on Science.

Repetto, Robert, 1992. "Accounting for Environmental Assets." *Scientific American*, 266(6), 94-100.

Richtmyer, F. K., and E. H. Kennard, 1942. *Introduction to Modern Physics*. Third edition. New York: Mc Graw-Hill.

Rummel, R. J., 1987. "A Catastrophe Model of the Conflict Helix, with Tests." *Behavioral Science*, 32(4), 241-266.

Saperstein, Alvin M., and Gottfried Mayer-Kress, 1988. "A Nonlinear Dynamical Model of the Impact of SDI on the Arms Race." *Journal of Conflict Resolution*, 32(4)(December), 636-670.

Schaeffer, Robert K. (Ed.), 1989. *War in the World-System*. New York: Greenwood.

Schieve, William C., and Peter M. Allen (Eds), 1982. *Self-Organization and Dissipative Structures: Applications in the Physical and Social Sciences*. Austin, TX: University of Texas Press.

Schwab, Adolf J., 1988. *Field Theory Concepts*. New York: Springer-Verlag.

Thom, René, 1975. *Structural Stability and Morphogenesis: An Outline of a General Theory of Models*. Reading, MA: Benjamin.

Weidlich, Wolfgang, and Gunter Haag, 1983. *Concepts and Models of* a *Quantitative Sociology: The Dynamics of Interacting Populations*. New York: Springer-Verlag.

Weiss, Paul, 1939. *Principles of Development: A Text in Experimental Embryology*. New York: Holt.

Wilson, Kenneth G., 1979. "Problems in Physics with Many Scales of Length." *Scientific American*, 241(2), 158-179.

Zeeman, E. Christopher, 1977. *Catastrophe Theory: Selected Papers 1972-1977*. Reading, MA: Addison-Wesley.

CHAPTER FIVE

SCHOOL FOR RULERS

by Yehezkel Dror

CONTEMPLATING THE TABOO: RULERS MUST LEARN

The theories, speculations, ideas, conjectures, propositions, and proposals set forth in the various chapters of this book add up to a clear message: radical upgrading in policymaking capacities is essential in order to reduce the probability of tragedies and to increase the probability of continuing human achievements in terms of evolving values.

This situation is the result of three interactive and mutually reinforcing factors: First and most important of all, thanks to science and technology human, choice has a very large and constantly growing potential for significantly impacting on large-scale futures, including the future shape of humanity and its very survival. Second, the difficulty of issues is sharply increasing, with hyperturbulence (McCann and Selsky, 1984), jumps and mutations up to metamorphosis, land phase transitions characterizing many critical domains, combined with scarcity of any promising options even for problems which are relatively stable and persistent. Third, human expectations from collective action are very high (as already well recognized by Etzioni, 1968), imposing heavy burdens on societal problem handling and governmental decisionmaking.

To meet the challenges posed by the juxtaposition of these and additional factors, radical upgrading of human choice abilities and propensities is required, ranging from production of innovative types of knowledge to changing the value basis of much of public choice (up to the demands of Küng, 1991). From the large set of thus needed transformations, I elect to devote this chapter to a subject which brings out the full measure of needed breakthroughs and of the necessity to shatter taboos and engage in counter-conventional design of novel institutions, namely, a *School for Rulers*. This proposal exposes and considers needed elevations of the policy-cogitation knowledge and capacities of top-level decisionmakers and decision formers, in the sense of "thinking abilities" in a broad meaning of that term.

The *central thesis* of this chapter is that Rulers (both in the strict sense of that term and in an expanded one), including of course democratic ones, must learn a lot in order to perform well and that institutional architecture is required to bring this about. Four premises serve as foundations for this assertion: One, Rulers matter a lot; two, Rulers must know a lot, in addition to meeting other criteria; three, as a matter of fact nearly all Rulers lack essential knowledge, in particular policy-cogitation competence; and, four, carefully designed learning institutions are needed to supply Rulers with required knowledge and can do so.

These premises and their conclusion are anathema to true believers in the withering away of the state and other hierarchical power structures, if not through the dictatorship of the proletariat, then thanks to the hidden hand of the market. Worse, the proposition that Rulers must acquire much knowledge contradicts a fundamental ten-ant of simplistic democracy which postulates that being elected is a sufficient condition for being anointed as Ruler and performing well in that position. It also confronts structural inertia and vested interests; and rejects sanguine views of the capacities and assured permanence of liberal democracy (as put forth, for instance, in Fukuyama, 1992).

All the more so, thinking about what is taboo and iconoclasm toward accepted idols are essential if we want to break out of the mental cages confining capacities to govern, and if we want to open the way for radical upgrading of our ability to advance human values towards the 21st Century.

To explore my subject within the confines of a chapter, I will proceed by first explaining and supporting my four premises and main thesis. Then, some of the forces resisting and proscribing the proposition that Rulers must learn will be confronted. After clarifying main hurdles, demolishing some of them, and discussing ways to go around those that cannot be breached by words, the main task of this chapter will be taken up in the form of an architectonic design of a School for Rulers. In it the substantive policy-cogitation knowledge needed by Rulers will be manifested and ways to teach it will be indicated. Finally, some bridges between present realities and minimal requirements will be erected, in the form of operational recommendations for upgrading the knowledge of Rulers.

In line with the contemplative and reflective nature of this essay, its preliminary formulation, and its departure from mainstream thinking, references—however extensive they may appear to be—are kept to what I regard as a bare minimum, essential for providing some illustration and indicating some of the literature which should be required readings and study material for top decisionmakers. Extensive bibliographies are supplied in other writings of the author which fully elaborate the issues which can be handled only indicatively within the confines of this chapter (Dror, 1986; Dror, 1988a; Dror, in work a; Dror, in work b). In particular, an encompassing view of required redesigns of governance, of which the School for Rulers is but an element, though a central one, which serves as context and basis for the present chapter, will be presented in a Club of Rome Report on *Governance for the 21st Century* being prepared by the author (Dror, in work c).

PREMISES AND THESIS

This section discusses the four premises mentioned above.

Rulers Matter a Lot

The first premise is that Rulers matter a lot. This is true whether we construct the term "Ruler" in a strict sense, including heads of states and comparable office holders, such as the President of the European Commission and the Secretary-General of the United Nations. Or whether the scope of the concept of "Rulers" is somewhat expanded to include other top-level decisionmakers, such as senior ministers and other senior politicians on national levels, heads of sub-national entities such as the *Lånder* in Germany and the states of the USA, and also chief executive officers of very large corporations and of non-governmental non-market organizations such as major parties.

My discourse in this chapter applies to an even broader stratum of senior policymakers and policy formers and candidates for such positions subsumed under the term "Ruler," with suitable adjustments easy for the reader to envisage. However, to sharpen my argumentation, I nominally direct my analysis at Rulers in the strict sense of that term, all the more so as they are a prime target for knowledge-upgrading. But my having in mind a broader, though still relatively small, stratum of high policy elite makes more practical sense of my proposals. After all, it is hard to imagine acting heads of state going to a school, however useful this might be. But fast-rising politicians and aspirants to top-level positions may well do so if suitable opportunities and incentives are provided. Still, I prefer to focus myself on Rulers, not only as a didactic device bringing out the full claims of my position, but because of the real necessity urgently to upgrade the knowledge of top-level policymakers, however difficult doing so may be given present realities.

Despite humanistic, liberal, and egalitarian hopes, Rulers, also in the broadest sense of that term as used here, have significant impacts on reality. Without going into meta-historic debates on historic determinism versus underdetermination of the future by the past, it seems clear that however constrained Rulers are, they do enjoy a significant space of free movement, which can be further expanded by them if they know how to do so.

Empirically, it is enough to survey the contemporary scenery and recent history in order to make a strong case for the impact potential of Rulers. On a deeper level, modern theories of history (e.g., Gellner, 1988) and of change (e.g., Boudon, 1982; Hallpike, 1988), *inter alia* as presented in some of the chapters of this book, clearly perceive future shaping processes as largely nondeterministic, with choice having a significant though constrained and in part indeterminate impact potential.

Overstating the case somewhat (because changing trajectories is more a matter for decision chains than of single choice events), historic spurts characterizing our epoch (and also other periods in history, such as the 17th Century) produce situations often fitting the image posed by the Confucian disciple Hsüntze: "Yang Chu weeping at the crossroads said, 'Isn't it here that you take a half step wrong and wake up a thousand miles astray?'."

Emerging global conditions may well multiply such situations, posing both far-going opportunities and dismal dangers. Given circumstances which approximate crossroads in time, top decisionmakers have significant possibilities to influence historic trajectories for better or worse, sometimes by serving as "butterfly factors," to use a chaos theory metaphor (Gleick, 1987, pp. 9-31), but often also by significantly influencing future systems states more directly and deliberately.

Exaggeration must be avoided. Rulers operate under many external (as well as internal and internalized) constraints, including many very rigid ones. Their decision impact potential is further narrowed and spoiled by inadequate personal abilities, limited knowledge and resources, and unavoidable dependency of Rulers on gate guards and mind keepers. Still, Rulers are one of the human groups distinguished by superior power to impact on the future (though other groups, such as scientists and prophets, exert much more influence on human fate but in very different ways).

Rulers Must Know a Lot

Moving on to the proposition that Rulers must know a lot, the first point to note is that normatively we would like Rulers to meet many other criteria, such as moral and political ones. The fact that I deal in this chapter with knowledge, or to be more exact with some forms of knowledge dealing with policy-cogitation, does not imply that this is the only and most important gap in the qualities of Rulers. But it is an important one which I think is more amenable to improvement. However, the issues of required virtues (in the sense of MacIntyre, 1984) and character (in the sense of Kupperman, 1991) of Rulers, as well as of their values, must be faced—leading to even more far-going proposals which I will discuss at some other opportunities (Dror, in work a).

It is possible to adopt a normative position stating that, as a value in itself, Rulers should know a lot. Alternatively, one can demand more knowledge from Rulers because of an axiomatic assumption that knowledge will upgrade morality. But these are not the justifications in which I ground the second proposition. Thus, I do not claim that Schools for Rulers will necessarily stop evil Rulers and advance outstanding ones, though the point could be made tentatively that a Hitler and Stalin would not be able to pass through a School for Rulers (see the discussion of their makeup in Bullock, 1991), while a De Gaulle would have been encouraged and helped to avoid some mistakes (compare Lacouture, 1990 and Lacouture, 1991).

While I regard myself as a nova-Platonist in the sense of accepting the position that Rulers must learn a lot, and while I follow in Plato's footsteps by supporting the idea of schools for Rulers, my reasons are not normative nor do I think that more knowledge of whatever type assures higher morality. I accept as an empiric fact that many Rulers who were very ignorant in terms of the knowledge I have in mind have had tremendous impact, also for the better. And I see no logical, psychological, and functional reason, nor do I know any empiric evidence, leading to the conclusion that knowledge improves morality, even if the contents of morality could be made non-relative which also raises serious problems.

My proposition that Rulers must know much and, in particular, should acquire policy-cogitation competence, stems from two other lines of reasoning. The first is that changes in policy *problématique*, to use Club of Rome terminology, as presented in the different chapters of this book, make explicit and carefully studied and considered policy-cogitation knowledge, in the sense expounded in this chapter, increasingly essential for Rulers; because under conditions of hyperturbulence, other bases of knowledge, such as experience-based insights and much of normal disciplinary learning, are increasingly irrelevant at best and often misleading. The second justification, as also brought out by some of the parts of this book and to be demonstrated in the main body of my chapter, is that novel knowledge which is truly salient to the tasks of high-level policymaking in a period of change spurts and phase transitions is becoming available, though slowly and as yet inadequately so.

Nearly All Rulers Lack Essential Knowledge

Let me add another main reason for requiring more policy-cogitation knowledge from Rulers; namely, that the usual decision quality of Rulers as demonstrated throughout history and on the contemporary scene alike, with deviations for better and worse, becomes increasingly inadequate. To pinpoint the argument: thanks to the increasing power of policy instruments, as supplied mainly by science and technology, the costs of the "normal" amount of errors in decisionmaking by Rulers will increase by orders of magnitude, up to the specter of human-made irreversible catastrophes. Therefore, determined efforts to break out of the historic range of the quality of decisions by Rulers are imperative.

This assessment is supported by all that is known on decision patterns of Rulers, whether we look on decision outcomes or open up the black box and examine the quality of decisionmaking processes (as well illustrated by Hybel, 1990; Vertzberger, 1990; and Heller, 1992, to pick just three recent illustrations from a very large and growing literature). While some Rulers are at some phases of their career near-geniuses, they too make serious errors. And most Rulers are far removed from being geniuses, to put it mildly, their decisions being saturated with many quality spoilers such as motivated irrationalities (Pears, 1984).

Therefore, even if we limit our aspirations to the minimal though substantive ambition of reducing grievous and consequential errors, radical changes in the quality of choice by Rulers are essential. Accordingly, even though good ideas for upgrading decisions by Rulers are very scarce and difficult to implement, every possible avenue must be tried. Ergo, assuming that relevant knowledge exists and is in fact underutilized by Rulers, as hopefully shown in this chapter and in this book as a whole (and see Dror, in work b), upgrading their knowledge is a promising idea which should be pressed. There are good reasons to hope that it may help while at worst it cannot do much harm.

In a recent book Rosenau (1990, ch. 12) claims that nowadays policy elites enjoy highly enhanced analytical skills and improved tools for decisionmaking. If this were true my third premise that in fact Rulers are ignorant of relevant knowledge would be falsified. But the opinion that Rulers are increasingly equipped with policy-cogitation knowledge is indubitably in error.

My own studies (Dror, in work d) of twenty-five offices of Rulers fully support the conclusion in nearly all of the literature dealing with the matter, that there is no change for the better in the policy-cogitation knowledge base of top decisionmakers, with the exception of some narrow policy domains. Thus, sometimes professional staffs are established, but just as often they are closed down (e.g., Blackstone and Plowden, 1988; Campbell and Wyszomirsky, 1991; and Williams, 1990). Whatever is known on "pathways to power" (e.g., Dogan, 1989) demolishes hopes that selection processes of Rulers assure that they know more. If one adds that policy-cogitation knowledge as needed by Rulers is not readily accessible, cannot in the main be derived from learning by trial and error, is often beyond the reach of "common sense" and frequently is incongruent with it, and can be acquired at present only at few locations, then the inescapable conclusion is that Rulers cannot avoid lacking essential knowledge. Thus the third premise is validated both empirically and analytically.

Rulers Need Schools

This conclusion will be further supported when the scope and nature of policy-cogitation knowledge needed by Rulers is clarified through presentation of a design of a School for Rulers. This design will also clarify and strengthen the fourth proposition, namely, that carefully designed learning institutions can supply Rulers with required knowledge and are essential for doing so.

Assuming that a strong case has been made for the first three premises and that the fourth one is provisionally accepted, the central thesis of this chapter that Rulers must learn a lot should be judged as having been advanced from the status of a daring conjecture to that of a strong hypothesis. Nevertheless, not wishing to fool myself or the reader, let me clearly admit that serious doubts remain, especially as to the availability of policy-cogitation knowledge and the possibilities of teaching it in ways actually influencing for the better choice behavior by top decisionmakers. These doubts can only be resolved experimentally, by setting up institutions for conveying policy-cogitation knowledge to Rulers, having a significant sample of Rulers go through them, and then estimating impacts if any on decision behavior, with iterative improvement of learning processes and further assessment of results.

This is not feasible. At best it is a lengthy process which presupposes that Schools for Rulers are first set up, with what is up to being tested being assumed to be proven. To try to break out of this closed loop it is necessary to make at least a strong prima facie case for setting up learning institutions for Rulers, even if on lesser a scale than a School for Rulers. This I hope to do by presenting an idealized design of a School for Rulers. But, before taking up this daunting task, the ground must be cleared of major obstacles to seriously considering such ideas and implementing them at least in part.

IDEOLOGICAL AND STRUCTURAL HURDLES

Proposals to set up learning institutions for Rulers are not on the agenda, neither of practical politics nor of political science. Therefore their merits and demerits are hardly discussed and the obstacles standing in their way are implicit rather than explicit. Still, examination of political philosophy discourse and political reform proposals, together with some field experience of mine with presentation of proposals on learning institutions for senior policy elites, permit identification of four main types of hurdles, two ideological and two structural.

Ideological Hurdles

The main ideological barrier is rejection as inherently incorrect of the very notion that Rulers need knowledge and should be urged to engage in structured learning. A second ideological counter-position is that, even if Rulers need additional knowledge and organized learning, this conclusion should not be publicly admitted and acted upon because it carries too high a cost in undermining axiomatic assumptions essential for democracy. The two structural barriers are institutional inertia and opposition by vested interests.

Every human position is ultimately grounded in some fundamental posture, whether looked at as cultural biases or as metaphysical stances. Thus, the endeavor to upgrade the quality of choice by Rulers, instead of say trying to get rid of Rulers altogether, accepts hierarchical ways of life (on the lines discussed in Thompson, Ellis, and Wildavsky, 1990) as essential or at least unavoidable in the foreseeable future. And the effort to improve choice by supplying Rulers with knowledge implies rejection of mysticism, astrology, *Führer*-ideologies, and other counter-knowledge beliefs.

On these *Weltanschauungen* all I can do here is to clarify where I stand and work out some of the implications of my fundamental position, including a priori rejection of any argument that Rulers have access to alternative superior insights and therefore do not need better this-world knowledge.

The alternative argument that Rulers acquire adequate knowledge on their way to power and from experience requires some further consideration. The ideological position here is that Rulers do not need advanced knowledge as proposed by me, because common sense meets most of their needs and necessary expert knowledge is supplied by civil servants and other technicians.

This is not the place to take up again the Madison Papers debate and to search for a new middle ground between Jefferson and Hamilton which fits the changing nature of problems facing Rulers. But, as a historic entry into this issue, let me point out a paradox. The idea that Rulers need complex knowledge was widely accepted till the advent of democracy, as evidenced by the long history of Mirrors for Princes as a main genre of writings, from early Accadian and Chinese texts, through *The Kautiliya Arthasastra* to *The Prince*. With the beginnings of modern scientific knowledge, including the evolution of a broadly conceived "noble science of politics" (Collini, Winch, and Burrow, 1983), writings with advice for senior politicians went out of fashion (with distinguished exceptions, such as Neustadt, 1990; and, differently, Matus, 1987; Schultze, 1992). Still, in the country where modern democracy evolved, the United Kingdom, it was strongly believed that education for leadership was essential, this being a main rationale of the public schools (McCulloch, 1991) and also of parts of University teaching (Symonds, 1991).

This acceptance of the necessity of training for rulership has evaporated or at least gone into hiding. In the absence of intellectual histories focusing on this development as a whole, I can only speculate on its reasons. What may have happened, especially with the advent of mass democracy on the one hand and the ascendance of anti-elitist ideologies on the other, is that the population was assumed to elect suitable politicians and that special knowledge requisites were regarded as anti-egalitarian and as contradicting the universal right to be elected. Modern democratic theory continues to stick to the same conclusion, thanks to a mix between trust in the electoral sense of the people, emphasis on functions of Rulers ostensibly not depending on knowledge such as value judgment, and disparagement of "specialist" knowledge as compared with "generalist" common sense and wisdom matured by experience.

Modern egalitarianism, trust in the masses, suspicion of knowledge establishments, views of knowledge as conditioned by dominant ideologies and cultural biases, and similar positions add to opposition against regarding knowledge essential for Rulers.

My counterpoints have already been presented when I examined the proposition that Rulers need knowledge. In short, in view of the demanding nature of policy issues and the high expected costs of errors on the one hand, and the availability of knowledge which may help in better policymaking on the other hand, upgrading the knowledge qualifications of Rulers is essential. Because of the importance of Rulers, which is indeed on the increase in modern mass democracies, it is not realistic to assume that other components of governance will compensate for ignorance on their part. Even worse, if civil servants and other non-elected staffs monopolize policy-cogitation knowledge, it will often be either ignored in top-level decisionmaking; or the actual influence of elected politicians on the contents of critical choice will be reduced (on lines of the "Yes Minister" and "Yes Prime Minister" television parody), impairing democracy much more than demanding knowledge from elected Rulers would ever do.

Having breached, I hope, the first barricade, we reach now the argument that requiring knowledge from Rulers contradicts basic tenants of democracy and should therefore be rejected as too costly in terms of higher values, even if correct in terms of decisionmaking quality. This line of thinking requires close examination, all the more so as I share democratic value and do not propose a new kind of meritocratic utopia or dystopia (Young, 1958).

Short shrift can be made nowadays of the argument that recognizing the importance of structured knowledge and learning by Rulers may undermine the image of democracy, all the more so after the collapse of Communism; however, this argument may have played a latent historic role. But more serious lines of thinking must be confronted. If indeed a real contradiction exists between central tenets of democracy and upgrading knowledge of Rulers, then a harsh dilemma is faced: On the one hand we may undermine democracy by requiring knowledge from Rulers, while on the other democracy may be weakened by impairing its capacity to handle problems as a result of leaving Rulers under-equipped with knowledge. But luckily this is a chimera and not a real dilemma.

Recognition of the need to augment the knowledge of Rulers (and readers are reminded of the broader meanings of this term to which this chapter is directed) does not imply any changes in the central norm of political democracy that elections select the Rulers. There is no incongruity between this fundamental norm and efforts to get Rulers to acquire more knowledge. As illustrated at the end of this chapter, quite a number of ways to provide easy access by Rulers to knowledge and to motivate them to utilize such opportunities, which in no way intervene with the democratic nature of political processes, can be provided. And stronger measures which may be advisable, such as exposing candidates for elections to top positions to public cross examination which discloses their ignorance and knowledge, however unconventional, in fact augment rather than erode fundamental canons of democracy.

Still, some will claim that the very demand that Rulers acquire and use knowledge will strengthen the elitist nature of politics and policymaking. I agree that my proposals are elitist in nature and stand in stark opposition to extreme egalitarian conceptions of democracy, such as selecting Rulers by lottery (e.g., Emery, 1987). But in my view policymaking in any case is, and in the foreseeable future, will and must remain elitist, hand in hand with grass-roots participation in decisions on local and also national matters and some measures referenda (of very doubtful results, as study of the extensive use of referenda in Switzerland shows).

The real question is not whether policymaking will be elitist or not, but what are the qualifications and compositions of the elites and how are they selected, rotated, and legitimized. The emphasis on knowledge requirements hopefully changes some characteristics of the policy elites for the better, but does not interfere with their democratic selection, rotation, and legitimation and does not cause policymaking to be more closed and guided by a smaller number of persons than it is in any case.

Many additional problems emerge. Upgrading the knowledge of Rulers is a trans-incremental and even radical reform which should be related to the retrofitting of governance as a whole. Thus, more knowledgeable Rulers require and are able to handle more knowledgeable civil servants. More important for our present concerns, increasingly knowledgeable politicians and ignorant publics manipulated by mass media do not go well together and may undermine democracy in its substantive meanings. Therefore, upgrading the knowledge of Rulers should be accompanied by policy enlightenment of the public at large. Indeed, in addition to being an important improvement by itself, augmenting the knowledge of Rulers can also serve as a locomotive of change and shock intervention (on lines with Hirschman, 1986), bringing about other urgently needed upgrading of the capacity to govern. But I am leaving such implications of a systems view of governance, with Rulers being a critical component but not more than one element among many, to another opportunity (Dror, in work c).

Some of the arguments against requiring more knowledge from Rulers may well mask vested interests of political and administrative establishments. Even more insidious is, I think, institutional inertia, which prevents serious consideration of innovative ideas of the type proposed in this chapter (and in this book as a whole) and which leads to totally ignoring the problems of knowledge by Rulers (compare Dahl, 1989). Requiring specified learning by Rulers deviates radically from the path built into the structure of main political and governmental institutions (on these concepts, see North, 1990) and rebuts many assumptions accepted as "obvious." This is also true with respect to large parts of the academic community (compare Lindblom, 1990) and further complicated by the political functions of scientific knowledge (e.g., see the important discussion by Ezrahi, 1990). Therefore, not only are ideas demanding knowledge and learning by senior politicians opposed, but they are prevented from surfacing and being seriously considered.

Structural Hurdles

Ideological contentions can be demolished, though not removed, with better arguments. But institutional resistance and inertia will not give way to words. The barriers to upgrading knowledge of Rulers should be seen in all their strength, so that realistic ways to overcome them can be worked out.

Implementation issues are not central to the purposes of this chapter, which is directed at presenting an idea for reflection and not at proposing a program of action. Still, to counteract hasty rejection of my basic thesis as completely outside the realm of the possible and therefore not worthy of serious consideration, let me add a few words on possible realization strategies.

The idea of a School for Rulers can serve as a guideline for crafting less radical but still useful recommendations which may be more feasible, as I will illustrate in the final part of this chapter. Having good ideas ready so as to avail oneself of rapidly passing windows of opportunity, which may be provided by accidents of crisis and enlightened power holders, is another advisable strategy.

Albeit, building up an epistemic community (in the sense of Holzner and Marx, 1979, p. 108, as further developed in Haas, 1990, but not with the connotations of Michel Foucault) convinced of the merits of the idea and committed to advancing it in one form or another, as may be feasible in shifting situations, is a first essential step for breaking through or going around the barricades hindering upgrading knowledge of Rulers. In order to move ahead in such directions, the contents of proposed policy-cogitation knowledge for Rulers and methods for learning them must be clarified; so as to make at least a prima facie case and so as to provide a basis for shared thinking by the emerging community of scholars, professionals, and practitioners committed to the challenges presented in this book.

DESIGN OF A SCHOOL FOR RULERS

Instead of proceeding element by element, discussing different dimensions and categories of knowledge necessary and useful for Rulers, I advance by presenting an idealized School for Rulers, starting with some principles and then presenting its main constituents and features, adding additional comments on its overall rationale at convenient locations while dealing with its various components. In this way I hope to present the holistic idea of a School for Rulers and thus further to clarify the basic thesis of this chapter, though for sure the ideas as a whole and the different elements need much more elucidation, elaboration, explanation, and justification.

For convenience and to indicate some of the recommended sequences and flows, I divide the School program among four years of study. But this is not intended to reflect a strict order because overlaps and iterations characterize the course of learning as a whole.

Before moving to the design, some delimitation of my task is necessary. I focus on policy-cogitation knowledge, which can be portrayed in diverse ways as knowledge relevant to thinking on policy thinking, as meta-brain principles for improving policy-relevant cognitive processes on the main brain level, and more. Of course, such policy-cogitation knowledge must be supplemented by more specific knowledge relevant to policy issues as a whole and in particular domains. The two levels of knowledge are interdependent and should be taught in tandem. Still, the core of the proposed School for Rulers is constituted by policy-cogitation knowledge, with emphasis on "knowledge-in-action" (Schön, 1983). An alternative way to look on the main contents of the proposed School for Rulers is to view it as focusing on "statecraft professionalism," with emphasis on approaches and modes of appreciation and thinking, mixed with subjects related to specific statecraft issues. But, rather than pursuing here such language games, however necessary for other purposes, the School for Rulers design will serve to clarify what I have in mind on a pragmatic (in the true sense of that term) level adequate for the world of discourse of this chapter.

Policy versus Politics

However, one boundary must be clarified. As stated, I concentrate on policy-cogitation, as distinct from other aspects and dimensions of the job of a Ruler which require dissimilar modes of reasoning and diverse types, levels, and forms of knowledge. Differentiation between various facets of rulership is difficult. We must keep in mind that at least some parts of policy-cogitation are very hard to differentiate from political calculus (as evidenced by the absence in most languages of a semiotic distinction between "politics" and "policy"). Also, there is a lot of overlap between the two, as evidenced by the attention to political feasibility in policy analysis (e.g., Majone, 1989). This is demonstrated strikingly by the amalgamation of treatment of policy and politics in the main writings of Machiavelli, who unquestionably is the founder of modern statecraft professionalism, including important parts of policy-cogitation.

Albeit, the School for Rulers as designed here does not deal with knowledge for gaining and maintaining political and personal power, however cardinal to Rulers such knowledge may be. Rather, it concentrates on knowledge relevant to policy-making separate from personal and political power considerations, without ignoring the importance of the latter.

On Professionalism

The School concentrates on policy-cogitation professionalism, in a broad sense of that term but without taking up other conceptions of "wisdom" (e.g., Sternberg, 1990). Going beyond Schön (1983), I define professionalism as including both positive and negative capacities. Positively, professionalism involves the ability to build bridges between abstract thinking and pure theories on one side and what are perceived as concrete issues and substantial problem domains (whatever their phenomenological nature may be) on the other. Professional knowledge in this main sense requires an ability to operationalize knowledge, including very abstract knowledge which in part is far removed from direct experience. It includes approaches, methods, and orientations as well as generalized and particular information.

In its negative faculty, professionalism comprises doubts about the validity of surface impressions and what appears as "obvious" or "common sense" causal connections, with a lot of skepticism (Hookway, 1992). In the words of M. I. Finley (Finley, 1986, p. 73) when dealing with historic generalizations and explanations, which directly apply to policy-cogitation as a whole, "professionalism is, among other things, the habit of mind, which comes from experience with historical study and reflection, that sends up a warning light every time one makes such a connection."

Professionalism in both its parameters operates on all levels of knowledge, explicit, tacit, and intuitive, though some self-awareness is always required in order to subject professional judgment to critical evaluation by oneself and others.

In its nature, professional knowledge is grounded in philosophy of judgment and action, while in contents it should be grounded as far as possible in philosophy of science. The difference is that professionals (including Rulers in this context) must make decisions, including the default option, within time constraints. They should try to base judgment on the best available knowledge, but are neither bound nor able to follow the criteria of science with the "when I don't know at the required level of validity I reserve judgment" norm.

Even with somewhat relaxed standards of praxis, the crucial and difficult question which we now reach is whether we have sufficient policy-cogitation knowledge available to justify the claim that there is knowledge that Rulers should learn and acquire. My assertion that such knowledge is available, even though inadequately so, and that more can be developed if a concerted effort is made, will as already promised be put to the critical test of a mental experiment in the form of designing a School for Rulers.

To embark on the promised critical mental experiment, let us assume that some enlightened creature from outer space is deeply impressed by our self-accreditation as experts in "The Systems-Based Approach to Policymaking" and by the various chapters of this book, thinking or deluding himself that here he has discovered a way to help what calls itself *Homo sapiens* to get out from its mess. He gives us plenty of gold to set up a School for Rulers and guarantees a supply of intelligent and willing students, who are assured a fast climb to ruling positions if only they finish the school with distinction. The question we now face in earnest is, Do we really know what to teach and how?

My way to try to answer this question is to work out a design of a School for Rulers. Then we can see whether the best we can do and teach does demonstrate that enough knowledge is available or can be produced in the foreseeable future to justify the endeavor, or whether we remain with a school made of sand which the first puff of critical evaluation will blow away.

I will leave it to the readers to supply the winds of criticism and then check what remains of the School and its derivatives. My task is to construct the School for Rulers, though in skeleton version, and thus put the main thesis of this chapter, and perhaps of this book as a whole, to what I regard as a preliminary decisive test.

Overall Design Approach

A main principle underlying the School as a whole is what can be called an Massachusetts Institute of Technology (MIT) orientation and not a trade school doctrine. The School aims at providing understandings, perspectives, approaches, frames, and generic knowledge related to policy-cogitation in a broad sense of that term. Combined with building up learning habits and learning tools, this approach tries to provide knowledge, skills, craftsmanship, and routines which will not become quickly obsolete in an epoch of rapid ultrachange (that is, change in the patterns of change itself) and hyperturbulence.

A second and concomitant feature comprises the already mentioned professional aims of the School, with emphasis on applying abstract thinking and general theories to specific contexts and concrete needs. This implies that knowledge should not only be explicit but should be *internalized* on all levels of awareness, intuition, and action, leading to semi-automatic patterns, habits, and practice, up to "Finger-tip sense" (my inadequate translation for the German term *Fingerspitzengefühl*).

A third principle of the School is to try to help students to develop personality features related to policy-cogitation and necessary for assuring high-quality policy-making. These include, for instance, capacity to exit oneself, to take a cold view while surrounded by the white-hot heat of labyrinths of power, and to welcome remonstrations by knowledgeable advisors. More difficult to achieve is immunization against characteristic diseases of Rulers (going back to "Rulerscraze" as first discussed by Tacitus), such as a growing distance from reality, subjugation to mind keepers, and self-images which get more and more distorted the longer and the more successful one serves as a Ruler. Still, some resistance to these congenital dangers of rulership can perhaps be inoculated by sensitization to the pervasive and pernicious effects of being a Ruler and habituation to countervailing habits on mind and action.

But ambitions for the Schools, however far going, have their limits. While moral reasoning and value analysis is an important part of the curriculum, I do not realistically expect the School as proposed here to produce more moral Rulers. Care must and can be taken to prevent the instrumental rationality orientations characterizing large parts of the subject matter from transferring what are basically utilitarian, if not manipulative, attitudes from the legitimate domain of consequentialism in policymaking, given responsible goals and value contexts, into the political and power games of public office and into goal setting itself. But as to raising the moral standards of Rulers—this is quite a different endeavor which needs to be taken up, but is beyond the scope and ambition of Schools for Rulers (a point at which, as noted, I sharply depart from Plato and his Academy).

A fourth principle of the School is multiplicity of approaches and, in particular, a combination and even synthesis of quantitative and humanities approaches and modes of policy-cogitation. To comprehend intricate realities and to handle them, it is necessary to combine and integrate numerical with literati perspectives and to apply them in all their variety combinationally to major policy issues. Thinking-in-history is a main dimension of doing so, but not the only one. Artistic, semiotic, literary, and other imagery are also needed to perceive more of complexity, as well as to deal with other requirements of improved policy-cogitation such as construction of realistic visions as policy-compasses.

In one way or another students must be encouraged and helped to apply a variety of modes of imagining, thinking, and calculating to policy issues. I am not quite sure how to meet this need by structured teaching and courses and therefore leave this open for further thinking. But, not to make it too easy on myself, let me at least indicate here some possibilities that may encourage thinking in terms of nonlinearity, uncertainty, indeterminacy, hyperturbulence, ultradynamics, and chaos, as discussed in other chapters of this book. Some additional ones, such as mental exercises, will be presented later.

What is aimed at is to get students to think, analyze, create, intuit, decide, and act on the basis of full assimilation and incorporation of nonlinearity, uncertainty, and the other above-mentioned features of reality in our epoch. This requires, for instance, overcoming difficulties in conceptualizing the idea of "chance" and reconciliating deterministic causal thinking, which seems to be hard-wired into our brains, with indeterminacy as a form of causality. To move in this demanding direction, students must study and understand strict treatments of complexity, such as in the works of Ilya Prigogine (Prigogine, 1980; and Nicolis and Prigogine, 1989) and others (e.g.,

Bacchus, 1990; Campanella, 1988), as well as chaos theory (Gleick, 1987 and Stewart, 1989 suffice for a School for Rulers, as distinct from training of policy professionals). However, this is far from enough; philosophical (e.g., Suppes, 1984) and literary treatments have a lot to contribute to the comprehension of fate (e.g., Nussbaum, 1986), chance (e.g., Auster, 1990; Gelven, 1991), and chaos (e.g., Hayles, 1990); all this should be supplemented by cultural perspectives (e.g., Douglas and Wildavsky, 1982; Wildavsky, 1988) and interpretative approaches (e.g., Brown, 1987).

But, as stated, this is only quarter-baked thinking on a very demanding issue which is fundamental to the idea of the School for Rulers and pivotal for its success or failure.

As a fifth and last but overriding principle let me reiterate that the School focuses on policy-cogitation and not the political facets of Rulership, despite the difficulties of this distinction as mentioned above.

These five principles constitute the edifice within which the various building blocks of the School are located. They provide the setting for presenting, exploring, and pondering some of the main dimensions, elements, and features of the idea of a School for Rulers.

Terms of Study

The School for Rulers, as outlined here, has a study program of about four years (as distinct from a school for policy professionals, which requires two or three years more), with individual adjustments according to the baggage which students bring with them, which may require additional time for learning and unlearning.

Studies are intense and full time, all year round. To reduce fatigue, learning methods are diverse. But, all in all, students are undergoing a strenuous and penetrating multi-dimensional learning experience utilizing their mental resources to the utmost while developing and strengthening them.

Entrance Prerequisites

The School should take in students who have some experience with life, in addition to having finished college and, preferably, graduate studies, and who are around the ages of twenty-five to twenty-seven and more. Mid-career learning programs are also needed and serve as a possible alternative for parts of the School and a supplement to it; but my design here deals with the idea of the School for Rulers, on which other programs, diluted or augmented, are to be based.

The pre-entry experience of students should be diverse, with no preferences with regard to college and graduate study subjects. But priority is given to candidates who also have street-level experience, as a safeguard against their later distance from grass roots which is inherent in their designated positions. Students who lack such experiences should supplement their background before entering the School or by taking a special leave for gaining street-level and grass-root experiences after their second year of studies.

An effort should be made to get students with multiple socioeconomic backgrounds reflecting diverse ethnic and cultural groups and including females and males in more or less equal numbers. But no relaxation of standards on admission is acceptable.

The School is very demanding. Recognizing the pluralistic facets of cognitive abilities, including, *inter alia*, differences between more analytic and more creative capacities, and possible tensions between more convergent and more divergent thinking, still only "the bests" (in plural) should be admitted, however multiple the criteria may be.

Care has to be taken *not* to rely on personality testing which looks for mainstream "normality." We want the very exceptional student too (compare Simonton, 1984; and classic Lange-Eichbaum and Kurth, 1927), even if this entails a cost in terms of other scales. One personality feature to be insisted on is an "open mind" (in the sense of Rokeach, 1960), or at least a strong potential in that direction. Neurotic and narcissist personalities are definitely not to be excluded as long as they are not clearly pathological, though an effort should be made to screen out sadists and all kinds of mental monsters. (See also the discussion of the cognitive/emotional models of policymakers and decisionmakers in Chapter 1).

Languages, Tools, and Literacy

Minimum languages and tools to be acquired before entering the School program proper include at least one international language in addition to English, good numeracy, and easy usage of diverse computer software (but not programming).

To this must be added science and technology literacy, as a categoric imperative for senior decisionmakers. What I have in mind is something like a habit of reading regularly publications of the type of *Scientific American*, understanding well most of it, and being able to deliberate knowingly on the policy implications of at least half of it. More important, required is understanding of the modes of thinking underlying science and technology. But this understanding can often be best acquired apropos consideration of ongoing science and technology endeavor, if suitably presented. Indeed, taking into account the crucial though delicate and largely ignored roles of Rulers in protecting and guiding, and perhaps restraining, science, as put forth sharply by Nietzsche (see Eden, 1983, for a very pertinent treatment), much more may be required, but a first essential step is to eliminate science and technology illiteracy.

The requirement of science and technology literacy illustrates serious deficits in knowledge of present Rulers. Science and technology is one of the three most important variables changing the nature of the world and impacting on most policy issues (the other ones being value transformations, as discussed, e.g., in Chapter 8, and demographic shifts). Therefore, science and technology ignorance is a serious handicap for policymakers. But, as a matter of fact, science and technology literacy is scarce in most governance elites, and the need for it is far from being recognized. As science and technology literacy is not hard to acquire with suitably conceived learning methods, this point is I think a strong though simple demonstration of the validity of my basis thesis that Rulers must learn and that specific institutions are needed to assure required learning.

Base Disciplines

Base disciplines, to be learned during the first year of studies if not acquired earlier, include economics, social sciences, and public law. The emphasis in learning these subjects, in accordance with the principles presented above, should be on ways of thinking, central concepts, and main theories, together with main "nuts and bolts" (Elster, 1989), *not* on detailed monographic information. Therefore, with effective learning methods and well prepared materials, one year of intensive work should be enough to provide access to main base disciplines so as to permit in the following years their application to policy issues, combined with further study of them.

The situation is different with respect to history. Thinking-in-history (going far beyond Neustadt and May, 1986; compare Nolte, 1991) is essential for improved policy-cogitation, including achieving profundity in handling complexity. This requires a sense for the ebbs and flows of history which one can develop only by extensive reading of history and by life-long thinking about history and in terms of history.

The first year should provide a gate to historic thinking, by study of different approaches to history and the debates on the uses and misuses of history. More importantly, students who lack appreciation of the importance of history, and who are not in the habit of reading history as a major leisure-time activity should be stimulated and inspired to become ardently interested in history.

Knowing, Understanding, and Outlooking Reality

The second year should largely be devoted to knowing and understanding reality within a long-term time perspective. Given that the present is just a thin slice in time and that policymaking is always future oriented, outlooking (if I may coin the verb "to outlook") is a main theme of comprehending reality.

What I mainly propose is not knowledge of manifest facts, though this too is required (and amazingly lacking among senior policymakers, as revealed in some surveys of mine). Rather, the studies are directed at understanding *deeper* layers of dynamic reality, together with realizing the barriers to getting at reality, and understanding the constructionist nature of reality images.

A good way to take up this subject, which illustrates novel approaches needed for the School as a whole, is via the history and problematic nature of security intelligence. There is plenty of case material available on the many failures and few successes of security intelligence (a good introduction is Laqueur, 1985). These serve as an anvil to which to apply the hammers of epistemology, cognitive sciences, organization theory, etc., so as to understand the barriers hindering more correct appreciation of realities.

Having developed a sense for the problems of estimating reality, students move into substantive theories on the dynamics of their societies and on the global situation, trying to upgrade their cognitive maps and reference theories on the level of facts, understanding of dynamics, and epistemological doubts.

As noted, all examination and understanding of reality involves a temporal dimension, returning us to history on the one hand and leading us to outlook (I suggest shunning the term "prediction" which seems to imply that we can foresee the future) on the other. Utilization of main outlook methods and study of their rationales, potentials, and limitations, should therefore accompany assessment of situations and be integrated into it.

Here, we return to thinking-in-history, the question being what are the time frames to be taken into account. French *Annales* historiography (Burke, 1990), as largely shaped by Fernand Braudel (e.g., Braudel, 1980), constitutes an essential antidote to the preponderance of short-term political and "human time" thinking and the dominance of surface phenomena in much of policymaking and in the concern of most Rulers. Indubitably, inculcating into students a strong commitment and firm habit of applying long time frames to pressing issues and of trying to move into deeper layers of the stream of change, is a major requirement and possible contribution of the School.

The Future Between Necessity, Contingency, Chance, and Choice

The second year should culminate in one of the peaks of the study program; namely apprehension of the future as being a dynamic product of interacting necessity, contingency, chance, and choice. Nonlinear models, catastrophe and chaos theory, study of historic mutations and contemporary upheavals, long-wave approaches and Kondratiev phenomena, dissipative-structure theory, deterministic philosophies of history—all these perspectives, metaphors, views, and theories should be summed up in a deep understanding and sense of change processes, with evolutionary potentials (on line with Hallpike, 1988) being perceived as a dynamic mix between linear determinism, underdetermination, and non-necessitarianism (on the latter concept, see Unger, 1987). To achieve such comprehension on all levels of knowledge, multiple languages, and imagery are essential as already noted.

Of course, students are free to reach their own conclusions throughout their studies and should be encouraged to do so. But I think that the world view best fitting policymaking is one of the future being between necessity, contingency, chance, and choice. The mix among them and its dynamics are uncertain, but it seems clear, as also brought out in other chapters of this book, that our epoch tends to mutations, shifts, and jumps up to phase transitions.

This raises grave issues about the potentials to choice. For sure the impact of choice is getting stronger, because of the already mentioned growing power of policy instruments. But pervading hyperturbulence raises serious doubts about the balance between expected and unexpected, desired and counter-desired outcomes.

Improved policy-cogitation knowledge is directed at increasing capacities to achieve desired impacts through choice under ultradynamic conditions, but this capability is not assured. Furthermore, upgraded policy-cogitation knowledge, such as concerning the nonlinear nature of ongoing historic processes, also explicates doubts on our abilities to direct the impact of choice on complex mixtures between necessity and chance.

This has profound implications not only for the contents of policy-cogitation knowledge, which must become much more "uncertainty sophisticated" as mentioned below, but also for the personality features of our students. While persons who aspire to rulership are usually not susceptible to Hamlet's syndrome, still full awareness of the fragile nature of choice may impair needed will-power and reduce the determination to try and to mold at least some elements of the future. The recommended mental and philosophic posture is one of "Principle Hope" (adapted from Ernst Bloch), with maximum efforts to use human intelligence and knowledge to impact positively on the future through better considered choice. But also needed is a good dose of late stoicism, the *Meditations* of Marcus Aurelius being essential material for study, contemplation, and mental exercises as suggested below.

Policymaking Realities and Pathologies

As a special part of understanding reality, the second year also includes a course on the realities of decisionmaking, as explored in decision psychology, group dynamics, organization theory, bureaucratic politics studies, etc. This is relatively a straightforward subject. But here, too, doubts should accompany knowledge. Thus, the cultural bases of policymaking realities should be examined, such as special features of decisionmaking in Japan and unique characteristics of choice in true believer groups and societies. The issues of "dominant ideologies" and social processing of reality also deserve attention in this context, as in other ones.

The study of policymaking realities should adopt the perspective of a top decisionmaker who wants to avoid main pathologies characterizing the process and to improve choice behavior. Therefore, special attention is devoted to main diseases of decision processes (compare Hogwood and Peters, 1985), such as groupthink (Hart, 1990), ways to reduce them, and improved management of decision processes as a whole.

Policy Analysis Methods and Techniques

Part of the second and of the third year are devoted to familiarizing students with main policy analysis methodologies and techniques, such as modeling, gaming, cost-benefit analysis, operations research, decision analysis, implementation monitoring and evaluation approaches, etc. (Consider, for example, the system dynamics methodology discussed in Chapter 7). But the main purpose of the School should be kept in mind, differentiating it from advanced training of policy professionals. Senior decisionmakers must understand main policy analysis methods and techniques, know their potentials, and be alert to their limitations and pitfalls. But they are not professionals in policy analysis and therefore do not have to acquire detailed skills in using the methods and techniques.

For example, they should know what a Delphi study is and be alert to its underlying and often doubtful assumptions, such as those of the validity of subjective probabilities. But they do not need to know how to do design and run a Delphi study.

Grand Policy Thinking

What I call grand policy thinking stands in the center of the third year and constitutes the main basis for the clinical studies, as presented below. Grand policy thinking deals with main perspectives and frames for considering strategic choice. This is the main equipment which Rulers require in their role as decisionmakers, and it is the core of the professional dimensions of statecraft. *Therefore, grand policy thinking constitutes the hard kernel of the School.* The subjects of the first and second year lead up to it. And most of the clinical learning which follows aims at grand policy thinking professionalism; that is, application of grand policy thinking to major policy issues—explicitly, tacitly, and intuitively.

It is impossible to present here the contents of grand policy thinking, a subject with which I deal at length in other works of mine (Dror, 1991; Dror, in work b). But some of its main ideas can and must be briefly specified, all the more so as the whole idea of a School for Rulers and the central thesis of the chapter hinge on the existence or at least potential availability of such policy-cogitation knowledge salient to Rulers. Therefore, I move briefly through some main principles of this central grounding of the School for Rulers and this chapter as a whole.

Focus on Grand-Policies. Rulers should be concerned with overall directives, macro policies, grand strategies, policy paradigms, etc. To put the matter in its most ambitious form: grand policy thinking deals with policies aiming at significant impact on the future, up to setting and changing long-term trajectories (compare Kennedy, 1991).

The actual tendency in most of policymaking is to move from ad hoc and limited decisions to aggregative, implicit, and de facto grand policies seldom considered as such, accompanied by declarations of intentions and hopes which have little action implications and even less lasting consequences. Hence, the importance of the mission of Rulers to focus on grand-policies, including selection of policy modalities as discussed below. Grand-policies in turn provide guides, frames, directives, and settings for more limited decisions, while being influenced and reshaped in part by the latter and operationally expressed through them.

Policy Modality Deliberation. In some circumstances the preferable grand-policy is to have none, but instead to proceed by incremental policy changes at most. This is the case when realities are on an ascending curve which can with a high probability be expected to continue. The reverse situation holds when a domain is on a steep decline and approaches disaster, with the risks of very radical policy innovations being preferable to the certainty of failure in their absence. This may be the case with some global issues.

Different endeavors also require quite distinct policymaking modalities. Thus, radical self-transformation efforts, such as in the successor states of the USSR and in East Europe, need quite different policy modalities than do policies in more settled situations, with reformcraft moving into the center of the Rulers' task.

Environmental (in the broad sense of that term) dynamics is another factor to be taken into account when selecting policy modalities. Thus, under conditions of hyperturbulence, as increasingly prevailing, fine policy tuning is frequently a delusion and robust policies are needed instead. Still, action portions need careful determination, in quantitative and qualitative measures, the question being one of critical mass and appropriate scales of intervention. Policy effectiveness is usually not linear, different allotments being required to achieve various results under multiple conditions. In some cases only large-scale efforts can bring about meaningful results, while in others more than a little is waste and sometimes counterproductive. (See the discussions of nonlinearity in other chapters of this book).

A different policy modality choice considers the use of systematic planning, in its diverse forms, as a main way of policymaking, in contrast to reliance on improvisations, up to crisis decisionmaking, or on dispersed and also fragmented and piecemeal decisionmaking. Related, though different, is the choice between more coordinated policies on the one hand and reliance on sharp but narrow selective-radical interventions to achieve desired results by shock effects and "locomotive-of-change" jolts (Hirschman, 1986) on the other.

Another modality choice deals with allocation of policymaking tasks, between different mixes of more decentralization and deconcentration versus more centralization and concentration, with the pure model of a market system and the pure model of a hierarchical command system serving as polar antipodes. Here, policy-modality deliberation overlaps institutional concerns, as discussed as follows, and has also intense relations with ideologies and value systems.

Without going into the many additional facets and forms of policy modalities, one conclusion seems clear: Rulers must think in terms of policy modalities and devote much of their effort to choosing better ones.

Structure and Institution Concern. A main feature of better policy-cogitation by Rulers is intense concern with structures and institutions, as constraints, as objects for policies, as influenced by policies even if not aimed at, and as setting paths for policies (North, 1990). Policy reasoning in terms of societal architecture, constitution setting (also in the broad sense of that term as referring to societal structures, see Giddens, 1984), shaping of politics via institutions (March and Olsen, 1989), social and economic institution building, structural adjustment, macro-organizational design and redesign (e.g., Nystrom and Starbuck, 1981, to be applied to broader structures) should be central to the concerns of Rulers and explicitly handled by them.

This requires studying institutional limitations, dynamics, and processes (March, 1988; Olsen, 1982) and exploring possibilities for changing, up to *de novo* designing new institutions—leading to reformcraft.

Reformcraft. The culmination of structure and institutions reconsideration as a central concern for grand policy thinking is reached when broad systems reforms are the main mission of Rulers. This is nowadays drastically the case in different ways in the successor states of the USSR, the Republic of South Africa, Eastern European countries, and China; to a significant though lesser extent in many developing countries; and, in other but very significant ways, in the European Community area.

These instances demonstrate the central position which radical, guided systems transformations occupy in contemporary policymaking. Additional needed societal transformations (as emphasized in some of the chapters of this book) further add to the importance of such endeavor. Hence, the crucial importance of reformcraft knowledge (e.g., Dror, 1989) as a main component of grand policy thinking for Rulers. This is all the more important as individual top-level decisionmakers do often fulfil determinative roles in such processes.

Consideration in Terms of Rise and Decline of Nations and Success and Failure of Great Enterprises, Including Policies Aiming at Global Futures. Grand policy analysis should use as its broadest frame of thinking historic and theoretic conjectures concerning the rise and decline of nations, the long-range impacts of revolutions, successful and failing development endeavors, and the fate of other "Great Enterprises" (in the sense of Wakeman, Jr., 1985), with special attention to needed grand policies oriented at overall global regimes and futures (e.g., G. Porter and Brown, 1991; Wagar, 1992). Close study of outstanding policy successes, such as Singapore (e.g., Sandhu and Wheatley, 1989), together with failures, such as much of Latin America, integrates this subject with thinking-in-history and many of the other subjects within a concrete context.

This principle fully reveals the ambitious and demanding nature of policy-cogitation knowledge in general and of grand policy thinking in particular. Our understanding of the variables shaping the fate of nations, etc. is rudimentary. At the very least, to gain some access to this subject, students have to engage in extensive readings in history and in social change theory and to work a lot to clarify to themselves what is known or at least knowable and what is unknown and perhaps unknowable. And students have to learn how to design and implement high-aspiration and even heroic policies on the basis of a mixture between some knowledge, many conjectures and speculations, and a large amount of ignorance. (The topic of the rise and decline of ideas, economies, technologies, nations, and civilizations receives major emphasis in Chapter 1. For example, the concept of the Kondratiev phenomenon is considered to be a major organizing principle).

Alternative Overall Futures, Grand Designs, and Realistic Visions. In order to help and overcome the enslavement of policymaking to current pressures and imminent adversities, and in order to provide long-range compasses and dynamic goal directions, longer-range alternative overall futures, up to grand designs and realistic visions, are required. They are needed so as to link future oriented and intergenerational values to present decisions.

Construction of scenarios backward in time, from alternative futures, grand designs, and realistic visions to the present, is a related essential grand policy thinking method, supplementing and counter-balancing the usual practice of thinking from the here and now into the future. This should also be done in respect to "nightmares" and very negative futures, as a basis of policies aiming to reduce their realization probability.

The evolution of the European Community can serve as an outstanding example of realistic visions in action (as illustrated by Delors, 1992), to be studied closely by students for lessons to be drawn.

Policy Thinking in Terms of Interaction with Ultradynamic Environments. Grand policy thinking considers issues in terms of active interaction with very dynamic, ultradynamic, and hyperturbulent environments which in part are responsive to our policies, as contrasted with the tendency often to underrate the transmutations and reactions of contextual and target systems.

In dealing with interactive and, in part, competitive and sometimes malevolent and intrusive environments, the appropriate strategy mix among pre-adjustment, post-adjustment, efforts to shape or at least to influence the environments, and self-isolation from the environments is a main issue for grand policy thinking, with special attention to dialectic and nonlinear interaction sequences (e.g., Mitroff, Quinton, and Mason, 1983), competitive relations (e.g., M. Porter, 1990), and more. Global policies dealing with possible heat-trap effects and policies trying to preserve national cultures against international mass communication overloads illustrate the range of issues where interaction analysis with dynamic environments is essential within grand policy thinking.

Focus on Critical Choice within Coherent Perspectives. Every nation and other entity faces a limited number of more critical choices, in the sense that they have a greater impact potential on the future (compare Janis, 1989). Contrary to rampant tendencies to allocate most of their inherently very limited attention, thinking, and energy resources to what seems pressing, Rulers should focus on the really critical choices. Therefore, grand policy thinking involves efforts to identify critical choices and put them at the top of decision agendas with open-ended formulations which do not serve as blinders for policy-cogitation.

Critical issues should be handled within broad systems perspectives nested in coherent policy sets (a point strongly made by Haas, 1990). But critical variables should be concentrated on as policy instruments, rather than clinging to illusions of "comprehensive planning."

Handling of Deep Complexity. This principle has already been discussed, but must be reiterated as crucial for grand policy thinking. Handling of deep complexity faces a main lacuna in knowledge: *No overarching conceptions of complexity which meet the needs of policymaking are available.* Indeed, the very notion of deep complexity may in large parts be unspecifiable, at least with presently available epistemological and phenomenological concepts. We may well ultimately run here into congenital limits of the human brain.

But, still, much can and should be done to upgrade handling of deep uncertainty by top decisionmakers. Thus, to add to what has already been said, one of the main requirements for handling deep uncertainty is perceiving, accepting, and processing *contradictions*. Following some trends in modern logic, grand policy thinking must embrace and handle contradictions, whether taking the form of dialectic processes and ironies of history, or of contradictory policy requirements, or of cognition of contradictions in estimations, and more.

Related is the need to accept and utilize counterintuitive dynamics, such as "micromotives and macrobehavior" (Schelling, 1978), and to overcome the already underlined mental trapping of "common sense," such as cultural conditioning (Geertz, 1983, ch. 4). To put it more bluntly than before: Common sense is no good for handling complexity, which often is counterintuitive. Concomitantly, simplification maxims of the Occam's Razor type and Mach's Principle of Economy, interpreted as a logical rule (Einstein, 1950, page 40 of 1990 reprint) are not to be automatically adopted in policy-cogitation. Rather, the multi-variance of forces and factors and their often circuitous interaction, as better depicted in Tolstoy's *War and Peace* than in most of the professional literature, characterize deep complexity and should be accepted, understood, and steered.

Less demanding is the recommendation to adopt multiple perspectives (as proposed in Linstone, et al., 1984, but going far beyond it), with main issues to be considered from diverse points of view, both in terms of theories and models and in terms of types of actors. Tougher is the more advanced requirement to apply multiple disciplinary perspectives and a variety of disciplinary perspectives. Most inclusive of the ideas mentioned here for better handling of deep complexity is the already discussed necessity to apply a combination of different languages, modes of reasoning, images, symbols, principles, frames, approaches, orientations, methods, methodologies, and tools. To put this principle metaphorically into the form of a modified law of requisite variety: very complex approaches are essential for handling deep complexity—including the capacity to recognize fundamental simplicity often hiding behind facades of pseudo-complexity (compare Slobodkin, 1992, to be contrasted with Nicolis and Prigogine, 1989. Attempts to deal with deep complexity via a field-theoretic framework are made in Chapter 4 of this book).

Policy-Gambling. In the face of pervasive uncertainty, much of it built into societal and physical processes, decisions are in essence gambles. When uncertainties involve the very forms of alternative futures and the very dynamics of change, as they usually do, decisions are fuzzy gambles with unknowable and in part indeterminate payoff functions, involving *inter alia* unavoidable proliferation of unexpected and often undesired consequences. When critical decisions are faced, Rulers have no choice but to "gamble with history" (Dror, 1990).

Such *a policy-gambling perspective, fundamental to the already presented view of the future as between necessity, contingency, chance, and choice, constitutes in my view the single most important policy-cogitation insight and knowledge which students should acquire and absorb at the School.* The fact that it is bitter, hard to accept, and tragic in more than one sense makes the conception of policy-gambling all the more cardinal and essential, with much attention to be devoted to it.

But of course, our aim is not to make Rulers feel bad, though we will not dodge such results when useful for improving performance. Understanding of the true nature of choice as fuzzy gambling should lead to ways to upgrade policy-gambling. This involves professional assimilation of better protocols for handling the uncertainties dominating much of choice (Dror, 1988b and literature quoted there). Not less important, though in many respects harder to achieve, the error-propensities characterizing human choice in the face of uncertainty (e.g., Kahneman, Slovic, and Tversky, 1982) must be diagnosed and counteracted.

Addressing Crisis Decisions. With crises being built into the very fabric of hyperturbulence, crisis decisions constitute a very important mode of policymaking providing both unique windows of opportunity and Pandora Boxes. Therefore, crisis decisionmaking must be addressed by grand policy thinking, all the more so as Rulers fulfil determinative roles in crisis choice.

Attention to crisis decisionmaking is recognized as all the more important once simple conceptions of crisis management such as aiming at the restoration of the *status quo ante* are rejected, both as impossible and as neglecting singular opportunities (Godet, 1985); and, even more so, when crisis instigation is recognized as sometimes a preferable policymaking modality.

This view makes crisis decisionmaking an essential part of grand policy thinking for Rulers. Study of relevant professionalism (e.g., Comfort, 1988; Lagadec, 1990; Lagadec 1991; and Rosenthal, et al., 1989) is essential as a part of the School, with a lot of crises exercises serving as a main teaching and learning method.

One of the main lessons to be conveyed is that in order to improve unavoidable crisis improvisation, crisis decisionmaking must, paradoxically, be grounded in deep policy thinking. Unless main policy issues are considered in depth *before* a crisis occurs, policy thinking infrastructures essential for guiding crisis decisionmaking are lacking. Therefore, crisis decisionmaking improvement is integrated into upgrading of policy-cogitation as a whole, though it has some special features, such as design and running in of compressed-time decision modes.

Meta-Policymaking Concerns. Meta-policymaking, in the sense of policymaking on policymaking, including structures, processes, personnel, and guiding principles (Dror, 1983), is a main activity of Rulers, however unaware they often are of it and however ill prepared they are for it. Unavoidably, the behavior of Rulers shapes decisionmaking styles, usually without thought being given to such effects of their behavior.

Thus, their willingness to listen to contrary opinions, their support of or indifference towards think tanks, their demand to get options or preference for a single-dimensional recommendation, the extent to which they require concise but fully worked out staff chapters and read them or their liking for short briefings where rhetoric abilities shape choice—all these determine policymaking qualities above and beyond the machinery which Rulers head. If one adds critical decisions which many countries face on rebuilding their machinery of government as a whole, including its decision-preparatory and decisionmaking dimensions, then the importance of meta-policymaking knowledge for Rulers becomes clear.

This does not imply that the School can and should go into details. But main issues of decision process improvement within a broader setting of "how governments think or do not think" are a significant part of grand policy thinking for Rulers.

Feasibility Concerned but not Overbound. A word on feasibility mapping and implementation testing serves to wind up this short excursion into the universe of grand policy thinking. Feasibility is taken into account and implementation problems are considered, but as challenges rather than only or mainly as constraints. Instead of the adage that "politics is the art of the possible," statecraft should be advanced as "the capacity to make the necessary and desirable possible."

In Brief Summary. I hope these few observations provide at least some sketch of the contents of grand policy thinking, as a central part of the curriculum of the School for Rulers. Outside the mental experiment in which we are engaged, the look at some facets of grand policy thinking is intended to serve as at least prima facie proof that knowledge in policy-cogitation, which justifies the thesis the Rulers must learn and can serve as a basis for such learning, is available.

Value Analysis and Moral Reasoning

Supplementing grand policy thinking is value analysis and moral reasoning.

Let me repeat that I do not expect to make students more moral in the sense of reshaping their superegos. But learning main principles of moral reasoning (e.g., Boyce and Jensen, 1978), studying the nature of value judgment (e.g., Hecter, Cooper, and Nadel, in press), and exploring the main dilemmas of value choice as dealt with in political philosophy and parts of jurisprudence may help to clarify one's values and improve capacities to face the bitter choices overloading Rulers.

Related is study of main political and social ideologies, including some theology as well as main streams of utopian thinking, as well as political philosophy with attention to emerging new ideas (e.g., Kymlicka, 1990) in addition to the classics. The culmination of these studies is the already mentioned design of realistic visions as policy-compasses (as well as support-mobilizing instruments), combining examination of the evolutionary potential of situations, so as to remain within the boundaries of the perhaps possible, with translation of alternative and preferred values and ideologies into operational action maps.

The Ruler's "I"

A different central constituent of the program is thinking and contemplation on the vocation of rulership, to be taken up both in the third and in the last year. Studying the literature on Rulers and leaders, historic as well as analytic, psychobiographic, and theoretic, is the easy part of this undertaking. Literary treatments of rulership, as well as other artistic depictions, can add important dimensions which also do not pose special obstacles.

A selection from classical and the few modern Mirrors for Princes (e.g., Hodgkinson, 1983; and Wriggins, 1969), together with select diaries and memoirs by Rulers, serves to introduce the main thrust of this component of the School; namely, analysis of and contemplation on the subjective meanings, existential projects, and inner spaces of being a Ruler, in the context of external functions, tasks, and missions and within a broad philosophy of life, duty, and "calling." In other words, development of the "I" (compare Glover, 1988) of a Ruler is the demanding and also enigmatic task of this part of the curriculum.

For example, consideration of the relationships between power and knowledge, including development of a self-image that is in part a rulership-professional, in part a politician, in part an entrepreneur and change agent, and as a whole human being, is an early step in the endeavor. A good starting point for doing so is the classical essay of Max Weber on "Politics as a Vocation," in comparison with his essay on "Science as a Vocation" (as discussed, for instance, in Eden, 1983, especially chs. 5 and 6). Thinking through the subjective meanings of having power, the temptations it poses, and the opportunities for better or worse it opens leads into the innermost and most important layers of the Ruler's "I".

To bring out fully the unconventional nature of this aspect of the School, which should be in a relation of creative tension with the more explicit knowledge-oriented studies, let me mention here one of the essential methods for helping students to develop their Ruler's "I", even if by doing so may expose me to quite some misunderstandings and accusations. I think the School should employ mental exercises modeled in conception, though not in contents and objectives, after the Spiritual Exercises of Ignatius, the founder of the Jesuit Order (Puhl, 1952).

To give an inkling of what I have in mind (to be fully developed in my book *The Path of the Ruler: A Mirror for Top Decision Makers*, Dror, in work a), one can have students spend two or three days in a hospital seeing giving birth and dying, brain operations, the desperation of AIDS patients, and the woeful hopes of children undergoing kidney transplants. Then, a day is devoted to individual contemplation on implications of the experience for one's image as being a Ruler, to be summed up in group deliberations and an essay.

Already mentioned exercises involve contemplation on the *Meditations* of Marcus Aurelius. But more demanding mental exercises are needed, such as imagining various temptations facing a Ruler, contemplating alternative responses, considering weaknesses of will one may experience, and forming determination on the way to choose. Challenging possibilities in this direction are illustrated by taking up Thucydides' *History of the Peloponnesian War*, studying it in terms of statecraft and choice and in terms of moral dilemmas, and then running a series of contemplative and gaming exercises. For instance, the personality of Alcibiades as discussed in some recent literature (especially Forde, 1989) provides an excellent basis for considering in-depth both professional and moral statecraft problems.

But mental exercises (also relevant are some methods for mind training for the martial arts, see Layton, 1990), however essential, should be undertaken only if suitable mentors who can guide students in such an activity are available. This is all the more essential in order to avoid any possible slippage into various forms of quasi-mystic "transcendental" contemplation, which are strictly outside my conception of a School for Rulers.

Other personality development activities are available to help with crystallization of the student's Ruler's "I". Thus, to add to what has been said on needed abilities to withdraw from situations, to exit oneself, and to take a cold look at hot circumstances, also essential for learning is the ability to change one's mind even in regard to strongly held convictions. A mixture of studying the theory of such topics, working them out in group sensitivity sessions, and, if as mentioned above suitable mentors are available, dealing with them in solitudinous contemplation combined with mental exercises can serve to meet such needs, with further thought to be given to the possibility of offering didactic psychotherapy to students (but not depth analysis, which can impair drives needed by Rulers).

The effort to help students to work out their "I" as Rulers brings out a central feature of the school which is obvious and has been mentioned above, but deserves to be further explicated and accentuated. The success of the School depends critically on the quality of the teaching and mentoring staff. For sure, preparing mental exercises, guiding them, and integrating them into the quite different main body of teaching requires outstanding personalities. However, relatively more conventional curriculum subjects also require unusual teachers and mentors because of the clinical learning modes which are central to the professional aspirations of the School.

Clinical Learning

The proposals above reveal the dependency of the idea of a School for Rulers on *novel* learning approaches. Leaving aside the complex of problems of personality development, advancement of grand policy thinking and facilitation of professionalism in statecraft require clinical learning (leaving aside apprenticeship possibilities, which go beyond the confines of this chapter).

Cases, exercises, and projects should be introduced from the start of the first year through the second year via systematic study. Changing the learning modalities, parts of the third and nearly all of the fourth year are devoted to clinical learning, by moving through a variety of situations and policy problem sets realistically reflecting the challenges of a Ruler. These are taken up then with the help of all that has been learned and additional knowledge to be searched for, with opportunities to add theoretic and factual knowledge as needed in the context of the clinical learning.

Incoming mail baskets, crises simulation games, budget preparation exercises, national critical choice selection, global projects—these should be vicariously experienced and handled professionally in individual and group learning situations. It is through such surrogate clinical (in the sense of clinical years in medical education) practice that professionalism, in the already mentioned senses of a capacity to build bridges between abstract knowledge and concrete issues while avoiding superficial conclusions, can be advanced and the School can achieve its main aim.

There is an additional objective to be mentioned here which can be achieved via the clinical learning period. No main policy domain should be strange to the graduates of the School for Rulers. Familiarity with different policy fields posing challenges to Rulers (Sawhill, 1988 illustrates such a concern) can be accumulated by selecting cases, exercises, and projects so as to deal with major policy domains and the relations between them, supplemented with extensive readings.

Various additional ways of learning can be of help, such as study tours, internships, and individual guided readings (essential to inculcate reading habits which should last a life time). But the main need is for clinical learning as the chief road to professional growth within the limitations of the School.

To enable effective clinical learning, teaching materials have to be carefully prepared, requiring both a lot of hard work and much creativity. But more important, again, is the quality of the teaching staff. Guiding professional growth through work on simulated rulership issues, requires teachers to combine a lot of theoretic and factual knowledge with much practical experience in high-level policy issues. In addition, teachers must have outstanding didactic skills.

Rulership Equipment and Skills

Better insights into the realities of rulership and the problems of handling concrete issues serve as bases for a fourth-year course on equipment and skills. Apparently mundane subjects, such as staffs, decision support systems, interface with think tanks, etc., are revealed as essential for success, as are communication skills. Some of the needed skills are developed during the first years of studies, some are practiced during clinical learning, and all are summed up towards the end of the course of studies.

BACK TO REALITY: PRACTICAL RECOMMENDATIONS

No harm is done by leaving additional features of the School for Rulers vague because, however much I regret it, establishment of such a school is not near—though I am convinced that the time for such schools (as well as of other ways to screen and upgrade Rulers) is sure to come, hopefully before a very high price is paid thanks to prevailing under-preparation and under-qualification of top decisionmakers. Let us, therefore, leave our mental experiment behind, hoping that the design of a School for Rulers has served its main purpose to concretize the idea that Rulers must learn and to justify it.

Assuming that the thesis that Rulers should learn is by now accepted, at least provisionally so, the next question is what to do now in the absence of a benefactor from outer space. The answer I propose is that the ideas expressed through the school design can serve as a basis for a number of operational recommendations, which are diluted in comparison to the School for Rulers, but can be very useful, are feasible, and may if successful pave the road towards more ambitious endeavor of the type of a School for Rulers.

Four recommendations, presented in bare outline, will serve to illustrate such possibilities.

Write Texts for Rulers and Prepare Other Self-Learning Tools for Them

It is amazing, as already pointed out, that the literary genre of Mirrors for Princes has been nearly completely abandoned, with very few texts being prepared for high-level public decisionmakers. This omission is all the more striking when compared with the proliferation of books written for business decisionmakers.

Leaving aside other very important implications of this state of affairs, such as the possibly involved downgrading of public service as compared to business careers with all the very serious dangers of such as myth and preference, the recommendation is clear: Encourage writing of texts for Rulers (in the broadest sense of that term) and write such books (as attempted by this volume and by Dror, in work a; and Dror, in work b).

But books are only one, though an essential, tool for learning by Rulers. Suitable computer programs, video tapes, and carefully designed games provide additional underutilized possibilities to provide Rulers, and Ruler-aspirants, with learning instrumentation which can combine the useful with the pleasant.

Improve On-the-Job Learning

Much more complex is the subject of on-the-job learning by Rulers. Much such learning does take place, even though inadequate and often misleading as already pointed out. To improve on-the-job learning the whole advisory system of Rulers has to be revamped, with one of the main tasks of advisory staffs of all types being mentoring and teaching. This is a cardinal subject of tremendous potential though not an easy one to handle. As I deal extensively with staffs for Rulers and related ideas, with focus on encouraging and facilitating learning by Rulers, in another publication (Dror, in work d), I do limit myself here to mentioning this very important possibility.

But one point must be added. The scarcity of suitable briefing and display facilities in nearly all offices of Rulers which I examined, and serious weaknesses even in what are regarded as very advanced display facilities such as in defense operations centers, for instance in presenting and analyzing uncertainties and complexity, indicate the scope of innovations needed in order to enable Rulers to process advanced policy studies and learn from them.

Concomitantly, novel types of decision support systems fitting the needs of top decisionmaking (what I call DSSR = decision support systems for Rulers) are essential for upgrading the interface between advanced professional policy studies and top-level decisionmakers, so that the latter can really benefit from them. Suitable DSSR, completely absent at present despite the verbiage of advertisements and the ill-informed efforts of interested companies, can also encourage and serve self-learning by Rulers.

Provide Intensive Workshops and Courses

Quite a number of senior decisionmakers want to engage in learning assuming it is relevant to their needs as they see them but have very little time available. Having been approached by such persons, I was hard put to recommend to them where to go. It is easy to send them to some business executive course, some senior public service seminar, or some issue-oriented exchange of opinions. But these do not fit the needs of learning policy-cogitation (in addition to sensitive issues, such as possible leakage of ignorance revealed by the participant when serious learning takes place).

Therefore, organizing of intense and short learning opportunities for senior public policymakers is a must, with workshops of five to ten days as a starter. Intense one- to three-month seminars are a good second step, up to National Policy Colleges (see Dror, 1986, pp. 293-4), where mixes of different policymakers and policy formers spend say three months studying main policy issues with heavy doses of input of policy-cogitation knowledge.

Grant Sabbaticals

As my last illustration here let me propose that granting of Sabbatical leave, that is, a year or two with full payment, should not be a privilege of academics only. Senior policymakers, including elected politicians, need and deserve such opportunity, as long as the Sabbaticals are devoted to bona fide learning.

I have only hinted at some possibilities to facilitate learning by Rulers in useful and feasible ways, having neither discussed some very relevant experiences (such as at some party schools, defense colleges, government executive courses, the *Curso Alta Direccion* run by ILDIS in Caracas for senior Latin American government decisionmakers, etc.), nor presented additional ideas (such as special short modules on select subjects, for instance, science and technology literacy). But the four examples suffice I think to make the point that we are not in a situation of either realizing at once an impossible idea, such as the School for Rulers, or doing nothing. Many learning opportunities which may succeed can be created for Rulers in the broad sense of that term. Failure of such efforts cannot be precluded. But not trying to move in such directions is, I think, not only a serious neglect but a moral crime.

There remains one major imperative which is presupposed in this chapter; namely, that what I call policy-cogitation knowledge will continue to be developed, so that a substantive and solid basis for learning by Rulers is built up. The responsibility for doing so is imposed on us, policy professionals, and not on politicians and other policy elites.

EPILOGUE

Let me, finally, look again at my proposal by taking up main studies of emerging policy problems, global, regional, and national. A clear conclusion from such studies is that *present policies are inadequate for handling main predicaments and opportunities.* For instance, revised evaluation of the classical Club of Rome Report *Limits to Growth* (Meadows, Meadows, and Randers, 1992) raises the specter of collapse. The meaning of the nuclear revolution is far from being realistically faced (Jervis, 1989, with needed revisions). Major issues of powershifts are getting more threatening (Toffler, 1990). The overall diagnosis is one of inadequate handling of increasingly demanding issues (King and Schneider, 1991). And, a comprehensive view of *Change: Threat or Opportunity* (Kirdar, 1992), as well as a broad survey of problem perceptions (Union of International Associations, 1991) clearly leads to the conclusion that in order to utilize the opportunities and contain the threats faced towards the 21st Century *radical policy innovations are essential.*

Even if some of these estimations are discounted as unduly pessimistic and as disregarding much progress, such as in democratization and economic modernization, still any reasonable situational estimate of the dynamics of the situation will, I think, lead to the conclusion that unprecedented problems and opportunities are faced, in respect to the implications of science and technology, ecological issues, global equity, diffusion of mass-lethal weapons, demographic imbalances, mass media influences, possible new diseases, international regimes, and more. In short, top decisionmakers face what can be viewed metaphorically as labyrinths (see Doob, 1990, especially ch. 3) full with traps and abysses and also with a few marks indicating ways out into very promising though unknown futures.

Given such a situation, for sure radical upgrading in ability to find one's bearings and to choose the way to proceed are essential. Such upgrading in collective choice capacities require, as one of their main elements, improvements of the qualities of top decisionmakers. One of the relatively easier ways to do so is to enable, encourage, push, and maybe oblige high-level policymakers to engage in relevant studies, in addition to being elected. Ergo, the proposal for a School for Rulers comes to answer an imperative of the situation.

Claims that this idea is unrealistic are correct when we look into the rear mirror, but they neglect to look forward into the labyrinths into which we are moving. Given the policy labyrinths, neglect of the possibilities to upgrade navigational abilities with the help of Schools for Rulers, realized in phases as proposed above, is an act of gross negligence and perhaps of self-destruct, and not of "realism" in any true sense of that term.

REFERENCES

Auster, Paul, 1990. *The Music of Chance.* New York: Viking.

Bacchus, Fahiem, 1990. *Representing and Reasoning with Probabilistic Knowledge: A Logical Approach to Probabilities.* Boston: MIT Press.

Blackstone, Tessa, and William Plowden, 1988. *Inside the Think Tank: Advising the Cabinet, 1971-1983.* London: Heinemann.

Boudon, Raymond, 1984. *Theories of Social Change.* Cambridge, UK: Polity.

Boyce, William D., and Larry Cyril Jensen, 1978. *Moral Reasoning.* Lincoln, NB: University of Nebraska Press.

Braudel, Fernand, 1980. *On History.* Chicago: University of Chicago Press.

Brown, R. H., 1987. *Society as Text: Essays on Rhetoric, Reason, and Reality.* Chicago: University of Chicago Press.

Bullock, Alan, 1991. *Hitler and Stalin: Parallel Lives.* London: Harper Collins.

Burke, Peter, 1990. *The French Historical Revolution: The "Annales" School 1929-89.* Oxford: Polity.

Campanella, Miriam (Ed.), 1988. *Between Rationality and Cognition: Policy-Making Under Conditions of Uncertainty, Complexity, and Turbulence.* Torino, Italy: Albert Meynier.

Campbell, Colin, and Margaret Jane Wyszomirsky (Eds.), 1991. *Executive Leadership in Anglo-American Systems.* Pittsburgh, PA: University of Pittsburgh Press.

Collini, Stefan, Donald Winch, and John Burrow, 1983. *That Noble Science of Politics: A Study in Nineteenth-Century Intellectual History.* Cambridge, UK: Cambridge University Press.

Comfort, Louise K. (Ed.), 1988. *Managing Disaster: Strategies and Policy Perspectives.* Durham, NC: Duke University Press.

Dahl, Robert A., 1989. *Democracy and its Critics.* New Haven, CT: Yale University Press.

Delors, Jacques, 1992. *Our Europe: The Community and National Development.* New York: Verso.

Dogan, Mattei (Ed.), 1989. *Pathways to Power: Selecting Rulers in Pluralistic Democracies.* Boulder, CO: Westview Press.

Douglas, Mary, and Aaron Wildavsky, 1982. *Risk and Culture: An Essay on the Selection of Technical and Environmental Dangers.* Berkeley,CA: University of California Press.

Doob, Penelope Reed, 1990. *The Idea of the Labyrinth: From Classical Antiquity Through the Middle Ages.* Ithaca, NY: Cornell University Press.

Dror, Yehezkel, 1983. *Public Policymaking Reexamined.* Supplemented edition New Brunswick, NJ: Transaction Books.
Dror, Yehezkel, 1986. *Policymaking Under Adversity.* New Brunswick, NJ: Transaction Books. 1988 paperback.
Dror, Yehezkel, 1988a. "Visionary Political Leadership: On Improving a Risky Requisite." *International Political Science Review*, 9(1), 7-22.
Dror, Yehezkel, 1988b. "Uncertainty: Coping with It and with Political Feasibility." In: Hugh J. Miser and Edward S. Quade (Eds.), *Handbook of Systems Analysis: Craft Issues and Procedural Choices*, pp. 247-281. New York: North-Holland.
Dror, Yehezkel, 1989. "Memo for System-Reforming Rulers." *Futures*, 21(4), 334-343.
Dror, Yehezkel, 1990. "Faithful Decisions as Fuzzy Gambles with History." *The Jerusalem Journal of International Relations*, 12(3), 1-12.
Dror, Yehezkel, 1991. *Grand Policy Analysis.* Presidential Address, Policy Study Association Annual Meeting, Washington, DC, August.
Dror, Yehezkel, in work a. *The Path of the Ruler: A Mirror for Top Decision Makers.*
Dror, Yehezkel, in work b. *On Ruling: A Text for Top Decision Makers.*
Dror, Yehezkel, in work c. *Governance for the 21st Century.* Club of Rome Draft Report.
Dror, Yehezkel, in work d. *Staffs for Rulers.*
Einstein, Albert, 1950. "On the Generalized Theory of Gravitations." *Scientific American*, April. Reprinted in Special Scientific American Issue, *Science in the 20th Century*, 1991, pp. 40-45.
Elster, Jon, 1989. *Nuts and Bolts for the Social Sciences.* Cambridge, UK: Cambridge University Press.
Emery, Fred E., 1987. *On Learning to Democratize.* Draft. Canberra, Australia: Australian National University, Centre for Continuing Education.
Etzioni, Amitai, 1968. *The Active Society: A Theory of Societal and Political Processes.* New York: Free Press.
Ezrahi, Yaron, 1990. *The Descent of Icarus: Science and the Transformation of Contemporary Democracy.* Cambridge, MA: Harvard University Press.
Finley, Moses I., 1986. *The Use and Abuse of History.* London: Hogarth Press.
Forde, Steven, 1989. *The Ambition to Rule: Alcibiades and the Politics of Imperialism in Thucydides.* Ithaca, NY: Cornell University Press.
Fukuyama, Francis, 1992. *The End of History and The Last Man.* New York: Free Press.
Gellner, Ernest, 1988. *Plough, Sword and Book: The Structure of Human History.* Chicago: University of Chicago Press.
Geertz, Clifford, 1983. *Local Knowledge: Further Essays in Interpretive Anthropology.* New York: Basic Books.
Gelven, Michael, 1991. *Why Me? A Philosophic Inquiry into FATE.* De Kalb, IL: Northern Illinois University Press.
Giddens, Anthony, 1984. *The Constitution of Society: Outline of the Theory of Structuration.* Cambridge, UK: Cambridge University Press.
Gleick, James, 1987. *Chaos: Making a New Science.* New York: Viking.

Glover, Jonathan, 1988. *I: The Philosophy and Psychology of Personal Identity*. London: Allen Lane.

Godet, Michel, 1985. *Crises Are Opportunities*. Montreal: Gamma Institute Press.

Haas, Ernest B., 1990. *When Knowledge is Power: Three Models of Change in International Organizations*. Berkeley, CA: University of California Press.

Hallpike, C. R., 1988. *The Principles of Social Evolution*. Oxford, UK: Clarendon.

Hart, Paul T., 1990. *Groupthink in Government: A Study of Small Groups and Policy Failure*. Amsterdam: Swets & Zeitlinger.

Hayles, N. Katherine, 1990. *Chaos Bound: Orderly Disorder in Contemporary Literature and Science*. Ithaca, NY: Cornell University Press.

Hecter, Michael, Lynn A. Cooper, and Lynn Nadel (Eds.), in press. *Towards a Scientific Understanding of Values*. Stanford, CA: Stanford University Press.

Heller, Frank (Ed.), 1992. *Decision-making and Leadership*. Cambridge, UK: Cambridge University Press.

Hirschman, Albert O., 1986. *Rival Views of Market Societies*. New York: Viking.

Hodgkinson, Christopher, 1983. *The Philosophy of Leadership*. Oxford, UK: Blackwell.

Hogwood, Brian W., and Guy B. Peters, 1985. *The Pathology of Public Policy*. Oxford,UK: Oxford University Press.

Holzner, Burkhart, and John H. Marx, 1979. *Knowledge Application*. Boston: Allyn & Bacon.

Hookway, Christopher, 1992. *Skepticism*. London: Routledge.

Hybel, Alex Roberto, 1990. *How Leaders Reason: US Intervention in the Caribbean Basin and Latin America*. Oxford, UK: Blackwell.

Janis, Irving L., 1989. *Crucial Decisions*. New York: Free Press.

Jervis, Robert, 1989. *The Meaning of the Nuclear Revolution: Statecraft and the Prospect of Armageddon*. Ithaca, NY: Cornell University Press.

Kahneman, Daniel, Paul Slovic, and Amos Tversky (Eds.), 1982. *Judgment Under Uncertainty: Heuristics and Biases*. Cambridge, UK: Cambridge University Press.

Kennedy, Paul (Ed.), 1991. *Grand Strategies in War and Peace*. New Haven, CT: Yale University Press.

King, Alexander, and Bertrand Schneider, 1991. *The First Global Revolution: A Report by the Council of The Club of Rome*. New York: Simon & Schuster.

Kirdar, Üner, 1992. *Change: Threat or Opportunity*. Five Volumes. New York: UNDP.

Küng, Hans, 1991. *Global Responsibility: In Search of a New World Ethics*. New York: Crossroad.

Kupperman, Joel J., 1991. *Character*. Oxford, UK: Oxford University Press.

Kymlicka, Will, 1990. *Contemporary Political Philosophy: An Introduction*. Oxford, UK: Clarendon.

Lacouture, Jean, 1990 and 1991. *De Gaulle, Vol. I. The Rebel, Vol. II. The Ruler*. London: Collins Harvill.

Lagadec, Patrick, 1990. *States of Emergency: Technological Failures and Societal Destabilization*. London: Butterworth-Heinemann.

Lagadec, Patrick, 1991. *La Gestion Des Crises: Outils de Réflexion à l'Usage des Décideurs.* Paris: McGraw-Hill.
Lange-Eichbaum, W., and W. Kurth, 1927. *Genie Irrsinn und Ruhm: Genie-Mythus und Pathographie des Genies.* Munich: Ernst Reinhard, 1979 reprint.
Laqueur, Walter, 1985. *A World of Secrets: The Uses and Limits of Intelligence.* New York: Basic Books.
Layton, Clive, 1990. *Mind Training for the Martial Arts.* Norfolk, UK: Kime Publishing.
Lindblom, Charles, 1990. *Inquiry and Change: The Troubled Attempt to Understand and Shape Society.* New Haven, CT: Yale University Press.
Linstone, Herald A., et al., 1984. *Multiple Perspectives for Decision Making: Bridging the Gap Between Analysis and Action.* New York: Elsevier.
MacIntyre, Alasdair, 1984. *After Virtue.* Second edition. Notre Dame, IL: University of Notre Dame Press.
Majone, Giandomenico, 1989. *Evidence, Argument, and Persuasion in the Policy Process.* New Haven, CT: Yale University Press.
March, James, 1988. *Decisions and Organizations.* Oxford, UK: Blackwell.
March, James, and Johan Olsen, 1989. *Rediscovering Institutions: The Organizational Basis of Politics.* New York: Free Press.
Matus, Carlos, 1987. *Adiós Señor Presidente.* Caracas: Pomaire/Ensayos.
McCann, Joseph E., and John Selsky, 1984. "Hyperturbulence and the Emergence of Type 5 Environments." *Academy of Management Review,* 9(June), 460-470.
McCulloch, Gary, 1991. *Philosophers and Kings: Education for Leadership in Modern England.* Cambridge, UK: Cambridge University Press.
Meadows, Donella H., Dennis L. Meadows, and Jørgen Randers, 1992. *Beyond The Limits: Global Collapse or a Sustainable Future.* London: Earthscan Publications.
Mitroff, Ian I., Harold Quinton, and Richard O. Mason, 1983. "Beyond Contradictions and Consistency: A Design for a Dialectic Policy System." *Theory and Decisions,* 15(June), 107-120.
Neustadt, Richard E., 1990. *Presidential Power and the Modern Presidents: The Politics of Leadership from Roosevelt to Reagan.* Revised edition. New York: Free Press.
Neustadt, Richard E., and Ernest R. May, 1986. *Thinking in Time: The Uses of History for Decision Makers.* New York: Free Press.
Nicolis, Grégoire, and Ilya Prigogine, 1989. *Exploring Complexity: An Introduction.* New York: Freeman.
Nolte, Ernst, 1991. *Geschichtsdenken im 20. Jahrhundert: Von Max Weber bis Hans Jonas.* Frankfurt am Main: Propylän Verlag.
North, Douglass C., 1990. *Institutions, Institutional Change and Economic Performance.* Cambridge, UK: Cambridge University Press.
Nussbaum, Martha C., 1986. *The Fragility of Goodness: Luck and Ethics in Greek Tragedy and Philosophy.* Cambridge, UK: Cambridge University Press.
Nystrom, Paul C., and William H. Starbuck (Eds.), 1981. *Handbook of Organizational Design. Vol. 1: Adapting Organizations to their Environments; Vol. 2: Remodelling Organizations and Their Environments.* Oxford, UK: Oxford University Press.

Olson, Mancur, 1982. *The Rise and Decline of Nations: Economic Growth, Stagflation, and Social Rigidities.* New Haven, CT: Yale University Press.
Pears, David, 1984. *Motivated Irrationality.* Oxford, UK: Clarendon.
Porter, Gareth, and Janet Welsh Brown, 1991. *Global Environmental Politics.* Boulder, CO: Westview Press.
Porter, Michael E., 1990. *The Competitive Advantage of Nations.* London: Macmillan.
Prigogine, Ilya, 1980. *From Being to Becoming: Time and Complexity in the Physical Sciences.* New York: Freeman.
Puhl, Louis J., 1951. *The Spiritual Exercises of St. Ignatius.* Chicago: Loyola University Press.
Rokeach, Milton, 1960. *The Open and Closed Mind.* New York: Basic Books.
Rosenau, James N., 1990. *Turbulence in World Politics: A Theory of Change and Continuity.* Princeton, NJ: Princeton University Press.
Rosenthal, Uriel, Charles T. Michael, and Paul T. Hart (Eds.), 1989. *Coping With Crisis: The Management of Disasters, Riots and Terrorism.* Springfield IL: Charles C. Thomas.
Sandhu, Kernial Singh, and Paul Wheatley (Eds.), 1989. *Management of Success: The Moulding of Modern Singapore.* Singapore: Institute of Southeast Asian Studies.
Sawhill, Isabel V., 1988. *Challenges to Leadership.* Washington, DC: The Urban Institute.
Schelling, Thomas C., 1978. *Micromotives and Macrobehavior.* New York: Norton.
Schön, Donald A., 1983. *The Reflective Practitioner: How Professionals Think in Action.* New York: Basic Books.
Schultze, Charles L., 1992. *Memos to the President: A Guide Through Macroeconomics for the Busy Policymaker.* Washington, DC: The Brookings Institution.
Simonton, Dean Keith, 1984. *Genius, Creativity, and Leadership.* Cambridge, MA: Harvard University Press.
Slobodkin, Lawrence B., 1992. *Simplicity & Complexity in Games of the Intellect.* Cambridge, MA: Harvard University Press.
Sternberg, Robert J., 1990. *Wisdom: Its Nature, Origins, and Development.* Cambridge, UK: Cambridge University Press.
Stewart, Ian, 1989. *Does God Play Dice? The Mathematics of Chaos.* Cambridge, MA: Blackwell.
Suppes, Patrick, 1984. *Probabilistic Metaphysics.* Oxford, UK: Blackwell.
Symonds, Richard, 1991. *Oxford and Empire: The Last Lost Cause?* Oxford, UK: Clarendon.
Thompson, Michael, Richard Ellis, and Aaron Wildavsky, 1990. *Cultural Theory.* Boulder CO: Westview Press.
Toffler, Alvin, 1990. *Powershift: Knowledge, Wealth, and Violence at the Edge of the 21st Century.* New York: Bantam.
Unger, Roberto Managabeira, 1987. *Part I of Politics, a Work in Constructive Social Theory: False Necessity—Anti-Necessitarian Social Theory in the Service of Radical Democracy.* Cambridge, UK: Cambridge University Press.

Union of International Associations, 1991. *Encyclopedia of World Problems and Human Potential.* Two volumes. Third edition. New York: Bowker/Saur.

Vertzberger, Yaacov Y., 1990. *The World in Their Minds: Information Processes, Cognition, and Perception in Foreign Policy Decisionmaking.* Stanford, CA: Stanford University Press.

Wagar, W. Warren, 1992. *The Next Three Futures: Paradigms of Things to Come.* London: Adamantine Press.

Wakeman, Frederic, Jr., 1985. *The Great Enterprise: The Manchu Reconstruction of Imperial Order in Seventeenth-Century China.* Two volumes. Berkeley, CA: University of California Press.

Wildavsky, Aaron, 1988. *Searching for Safety.* New Brunswick, NJ: Transaction Books.

Williams, Walter, 1990. *Mismanaging America: The Rise of the Anti-Analytic Presidency.* Lawrence, KS: University Press of Kansas.

Wriggins, Howard W., 1969. *The Rulers' Imperative: Strategies for Political Survival in Asia and Africa.* New York: Columbia University Press.

Young, Michael, 1958. *The Rise of Meritocracy (1870-2033).* London: Thames & Hudson.

CHAPTER 6

POLICY: APPEARANCE AND REALITY
by Fred E Emery

INTRODUCTION: OVERVIEW OF 'PROBLEMS' AND 'SOLUTIONS'

"Gracious me! I've been talking prose for the last forty years and have never known it". That exclamation was by M Jourdain in Moliere's *Le Bourgeois Gentilhomme*. I and my fellow applied social scientists, with no notable exceptions, would have to confess the same ignorance, with regard to policy, not prose, for the last forty years. We have been that way since Lerner and Lasswell announced the emergence of 'The Policy Sciences' in 1951, in a book with that title. In policy situations we have gone on doing what we know best to do as social scientists. We have consistently evaded the question of what it is about the 'policy sciences' that make them different from the other social sciences. We have been successful in that evasion, I suggest, because we have been sensitive to, and responded appropriately to, changes in context as we move from private to public organizations and from executive to administrative levels. In attributing the reasons for different practices to these contextual differences we have been guilty, I further suggest, of the fallacy of misplaced concreteness. In what follows I will argue that the significant differences are conceptual, not merely contextual. I will go on to suggest how we might now try to realize Lasswell's vision "that the policy science orientation... will be directed toward the knowledge needed to improve the practice of democracy ... to affirm the dignity of man, not the superiority of one set of men" (Lerner and Lasswell, 1951, p. 10). That next step takes us beyond what we mean by policy to challenge what is meant by science when people talk of policy, or economics, as a science. This issue was also discussed in Chapter 2.

CONCEPTS AND DEFINITIONS OF POLICY

Policy *formulation* must reflect the complexity and uncertainty of the environment within which choices are being made. We are not here referring to the choices that are made with respect to which goal or objective one should pursue: that is the domain of strategic planning and mission formulation. Our concern here is with forming policies. We are referring to a bias that an organization wishes to introduce with respect to a whole range of choice of means that might be made by those acting as agents of the organization; or by persons and organizations that are to some degree dependent on that organization. What is common to both formation of strategies and policies is that both involve choices made by purposeful individuals or organizations.

Adoption of a policy should presuppose that viable plans have been made to enable the 'bias' to operate in relevant areas of decisionmaking. Unless ways can be identified so that the agents have to make choices that conform to organizational policy, e.g., follow-ups and inspections, then the policy is no more than wishful thinking, or window-dressing. With respect to independent parties, adoption presupposes that their life-spaces have been modified in ways that lead them more often than not to make the desired choices. In this latter case there often is no clear-cut measure of effective adoption. To continually insist that there is a public policy in the absence of such evidence of adoption is to invite the so-called boomerang effect (as with many inhibitory policies, e.g., anti-smoking).

The above two statements presuppose a view of policymaking that is not commonly held. A 'good' policy statement would normally be held to be one which is simple and unambiguous with a clearly stated purpose that obviously serves some higher social or organizational good. The implementation of such a good policy statement could then be left to any person of common-sense to work out. 'Policy adoption' is simply not a problem if the policy is straight-forward and stated 'loudly and clearly'. In this view policy formulation is very much like formulating an advertising theme, and this may be why governmental and corporate policymaking is so frequently associated with large Public Relations and media budgets. We shall try to explore the differences between these two views of policy. Unfortunately, we run into difficulties when we ask the first and obvious question of what makes a statement a policy statement.

In fact, the striking feature about the whole field of policy studies is the lack of definition of what it is that distinguishes policy studies ('policy sciences'!) from political science, economics and applied social sciences. A common practice is to provide a simple postulate of what a policy is and then, without analysis or justification, proceed to attribute whatever meanings suit the matters under discussion. Equally common is the practice of simply ignoring the issue on the assumption that, if the author says he is writing about policy issues, then they are policy issues (see Meehan's critique, 1985). Wildavsky, at least, in his classic study of *The Art and Craft of Policy Analysis* (1979) tries to confront the issue, admits that he and faculty colleagues have failed to resolve the matter:

> "How can you teach (or write a book about) a subject if you can't say what it is?
>
> "At the Graduate School of Public Policy in Berkeley, I discouraged discussions on the meaning of policy analysis. Hundreds of conversations on this slippery subject had proven futile, even exasperating, possibly dangerous... Yet... students did learn to do (policy) analysis... What was it, then, that could be learned but not explained, that all of us could sometimes do but that none of us could ever define (at least to anybody else's satisfaction)?" (p. 2)

Their teaching experiences led them to conclude that this is a subject that is "better taught backward" (p. 3).

For the moment let us just suggest that the success of 'teaching backwards' is probably due to the same factors that made the case-study method of Harvard Business School so attractive when we knew so little about what constituted good management. The case-study method, insofar as it works, cannot do so from inductive generalizations but can only do so in that some general principles are intuited that can then be put to the test. We hope, in what follows, to spell out what it is that might be so intuited. Hence our primary interest, for the moment, is in the definition of policy/its conceptualization. Table 6.1 provides some dictionary definitions.

If the leading practitioners and scholars cannot define policy can some help be gained from the way the word is used in society at large? This is not an unrealistic ploy as in everyday life, as distinct from academia, it is often permissible to call a spade a spade.

Table 6.1: Some Dictionary Definitions of Policy

Webster's International, 1898.

". The settled method by which the government and affairs of a nation are, or may be, administered.

. The method by which any institution is administered.

. Management or administration based on temporal or material interests, rather than on principles of equity or honour; hence worldly wisdom; dexterity of management; cunning; stratagem.

. Prudence or wisdom in the management of public and private affairs; wisdom; sagacity; wit."

Webster's International, 1964.

". A definite course or method of action selected (as by a government, institution, group or individual) from among alternatives and in the light of given conditions to guide and usually determine present and future decisions.

. A specific decision or set of decisions designed to carry out such a chosen course of action.

. Such a specific decision or set of decisions together with the related actions designed to implement them.

. A projected program consisting of desired objectives and the means to achieve them (formulation of policy)."

Oxford English Dictionary, Compact Edition, 1971.

". In reference to conduct or action generally: Prudent, expedient, or advantageous procedure; prudent or politic course of action.

. A course of action adopted and pursued by a government, party, ruler, statesman, etc: any course of action adopted as advantageous or expedient (the chief living sense of 'policy')."

Macquarie Dictionary, 1981.

". A definite course of action adopted as expedient or from other considerations.

. A course or line of action adopted and pursued by a government, ruler, political party or the like."

In the last two sets of dictionary definitions, the most modern, the repeated reference is to "a course of action", even a "definite course of action". These references blur the distinction between policy and means-end planning, strategy and problem-solving. Each of these last three terms can be taken to imply a course of action that is adopted or pursued. However the *OED* reference to prudent procedures seems to group together courses of action in terms of the qualities they have as means, as such, rather than their qualities as means-to-an-end. The *OED* definitions abjure terms like efficacy and efficiency which would better describe courses of action as means-to-an-end.

The apparent confusion in public usage of the term 'policy' does not, however, appear in the definitions provided by Webster.

It will be noted that the definitions of Webster's, 1898, are focussed on the quality of means (methods) not ends, and that the criteria for judging the selection of means (methods) are notions of prudence, wisdom, sagacity, wit. (The explicit exclusion of equity and honour as criteria possibly reflects the decades of corruption of public policy formulation, in late nineteenth century USA, by the 'robber-barons' and their ilk.) Webster's, 1964 edition, reflects the massive growth, in the twentieth century, of "the administration of people". It distinguishes between policies as potential guides to "present and future decisions", plans for implementation, and formulated policy (program) as the combination of both features.

It is necessary to merge the 1898 and 1964 definitions proffered by Webster's:

. A policy is a principle which is intended to guide the choice of means (methods) in the pursuit of objectives.

. A formulated policy is one which specifies how that principle will be implemented in given conditions.

Policy Compared to Strategy

It will be noted that a policy is not what is usually understood to be a strategic plan. Four distinctions need to be made.

A *strategic plan* typically depicts a strategic objective, to be achieved within a specified period, and a hierarchy of sub-goals whose achievement will ensure the achievement of the objective, e.g., to land a man on the moon by year X, to seize x% market share in the next financial year, to graduate in law in five years. A *policy*, however, is typically one step removed from direct pursuit of an objective. A policy typically seeks to produce a change in the social environment such that certain kinds of ends are more or less easier to achieve, e.g., that there is less discrimination against female employees, that there is more equity in educational opportunities, or that fewer people are motivated to violence. The bench-mark for judging the worth of a policy is already given by the existing levels of such phenomena, and has only to be pointed to by the policy maker, not invented, as in goal-setting; past social experience indicates the rate of change in the occurrence of such phenomena that might be expected from policy interventions and thus serves to measure progress in the implementation of policy.

A strategy is expected to create both the *necessary AND sufficient* conditions for the achievement of the objective. At best a policy can only establish some of the *necessary* conditions for the pursuit of some strategic objectives - a producer-product relation not an analogue of cause-effect relation (Ackoff and Emery, 1972).

With the achievement of its strategic objective an organization has usually managed to move itself to a more propitious place in its environment. By contrast, a policy typically seeks to change the environment or organizational culture within which objectives are pursued. By such changes it seeks to establish conditions that are necessary, although not in themselves sufficient, for the achievement of other objectives.

In pursuit of strategic objectives we see an interdependent set of activities that are convergent on the objective. The objective acts as the system principle and the relation between the parts tends to be multiplicative. Failure in one step may undermine the whole whilst success may make the other steps easier. In the implementation of a policy we see a divergent set of activities. Those activities add up to determine the extent of implementation but are largely subject to local determination rather than system determination. Failure in one area does not have any necessary implications for success or failure in other areas.

The distinction between strategy and policies appears to be the same distinction as that we make in military affairs between strategy and doctrine. The military have had no problem in distinguishing strategy from doctrine; for example, doctrines for the employment of armour from the strategies which those indoctrinated forces implemented. In civil affairs the distinction was less apparent as government administrators concerned themselves with how policies evolved, and corporate leaders concerned themselves with corporate strategies. The professionals advising these two worlds have tended to live in separate professional communities. Selznick tried to bridge the gap with his book *Leadership in Administration* (1957). He was ahead of his time.

DISTINCTIVE PROBLEMS OF POLICY FORMULATION

The differences between the formulation of strategy and the formulation of policy are not trivial differences: different concepts are required to explain or predict the different dynamics. And evaluation must be done in different ways.

Governing bodies are typically involved in both strategic planning and in policy formulation and often, in practice, the activities will overlap. Thus, strategic planning may indicate a need for new policies in some areas, and the simultaneous failure of several policies may lead to the formulation of new strategies. Nevertheless, the distinctive problems of policy formulation need to be borne in mind with respect to conceptualization, construction, critical criteria and institutional supports (Meehan, 1985).

Conceptualization

It is wishful thinking to imagine that realistic policies can be arrived at by either deduction or induction. No social science, including economics, offers a theory that adequately represents social reality. Deductions from the theories that we have can only give our arguments a spurious rigour. Similarly, there is no inductive methodology that can be cranked up to yield reliable policies automatically, whether it be survey sampling, experimentation, or computer simulation. We can only proceed in the way that scientific discovery usually proceeds, that is, by retroduction. Given the issue that happens to concern us, traffic accidents, child poverty, or whatever, we can only study that from as many angles as possible until it occurs to us that that phenomenon might be a consequence of some general principle. Thus, for instance, we might be concerned with the dissatisfaction expressed by a particular workforce. However we look at it, the dissatisfaction seems very real and persistent. Then we call to mind the general proposition 'that alienation causes dissatisfaction' and tentatively entertain the hypothesis that these workers may be alienated and that that is what is causing the dissatisfaction that we observe. With this hunch we have begun to define a problem and a direction in which a solution might be found. If we find that there is no way in which these workers can be described as alienated, then the hypothesis has to be discarded and we are back at square one without a problem that a policy might solve.

In our search for hypotheses we have to remain 'domain oriented'. That is, we have to take the phenomenon as it presents itself and search for whatever principle appears to fit the phenomenon regardless of how that phenomenon might be defined by disciplines or professions.

We are typically looking for policies that can be effective for social groupings that are multi-nodal and heterogeneous, and we cannot assume the level of authority that can pertain to uni-nodal, homogeneous groups (Ackoff and Emery, 1972, pp. 227-229).

Construction of Adequate Policy Statements

The most important consideration in the construction of policy statements is precision. Whether rewards or sanctions are used to induce compliance with a policy, the most important consideration is that there should be no fuzziness about what is being rewarded or sanctioned, and no more behaviours should be covered than is essential for the purposes of the policy. Thus, if there is a policy to protect a particular fish population, it should specify that only fish over x-inches should be taken, not that only 'mature' fish be taken or that 'under-sized' fish should be returned to the water. The underlying aim is that introduction of a new policy should make the least possible change in existing conditions of life and the least possible demands on, or threats to, those it is hoped will observe the policy. If an explicit policy is needed, it is because sufficient people have not already seen the sense of behaving in the desired fashion. However, it is rarely the case that people act as they do without some reason. Hence, it should always be assumed that a new policy will arouse opposition and some measures must be planned to minimize this opposition. An 'unfreezing' of existing habits is a prelude to change of habits.

Critical Criteria

At very least a policy should meet the criteria of specifying what actions should follow from observing the policy, of identifying the target population(s), and of producing a measurable change for individuals (not just some superficial change in the behaviour of aggregates or institutions).

Institutional Supports

Inducing a change in behaviour is not enough in itself. Steps need to be taken to support the continuance of the new behaviour. Policymakers have been quick to take advantage of the marketing potential of television. They need to take a further leaf from the book of the modern marketers. Those people know that a successful launch has to be prepared for with careful organization of continued support from branding, pricing, and distribution to outlets. Similarly, with policies there needs to be careful consideration of what institutional supports are going to be needed to sustain the changes in behaviour once they have been induced. The social field needs to be 'frozen' again, so that the new behaviours resulting from the policy come to be seen as natural as the previous behaviours (e.g., wearing of car seat belts, non-smoking in aircraft).

In Brief Summary

The points made about these four issues were meant as serious comment on the practice of policymaking. However, the main point was to illustrate that the concerns of the policymakers draw them into a universe of discourse that is demonstrably different from that of the strategic planners. We did not speak above of identifying strategic objectives, formulating mission-type orders, allocating resources to sub-goal attainment, or of operational, as distinct, from tactical planning.

A SYSTEMS-THEORETIC INTERPRETATION

Enough has been said here, I think, to justify a more formal statement about the differences between policymaking and strategic planning. I shall try to state the differences in systems-theoretical terms because, as I shall try to show later, this links policymaking to some well established and powerful lines of thought. I shall have to take this carefully as, in the past three decades, we have generated great confusion about what is a 'systems-theoretical' statement; and we have generated a corresponding cynicism that such statements are just the traditional arguments dressed up in the new 'buzz' words. See also Linstone's discussion in Chapter 9.

First, if we take L11 to represent the system (or set of systems) with which we are concerned, and let L22 represent the environment, then L12 and L21 can be taken to represent the effects of each on the other. To represent a total system-environment situation at any given time, there must be some lawful statements (L) for each one of sets (11, 12, 21, 22). The L22 can range from placid, randomized environments to turbulent-field environments and possibly, in extremes, to type V, vortical environments (Baburoglu, 1988; Emery and Trist, 1972). See De Greene's discussions in Chapters 1 and 4.

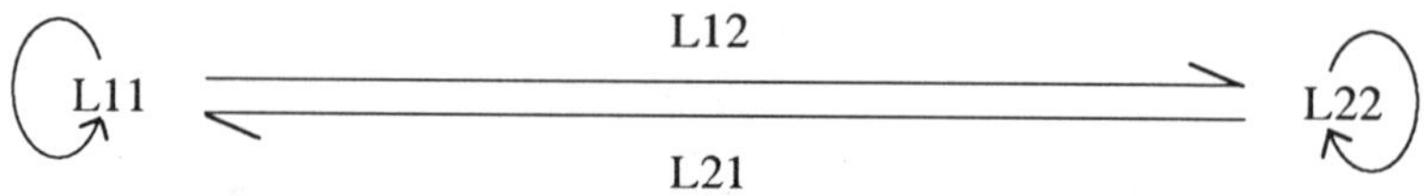

Figure 6.1: Parameters of the System-Environment Situation. The circular arrows around L11 and L22 represent the lawful determining relations between parts of the L11 and the L22 respectively.

Figure 6.1 is an approximate representation of what I take to be the fundamental theorem of systems theory. Bertalanffy's (1950) concept of the 'open system' totally ignored the L22; and Prigogine's (1980) concept allowed for only a minimal specification of the L22 as random, and hence with random bursts of relatively high energy in-flows. The position I took in the early sixties is that changes in the L21 are not all random but can sometimes be determined from changes taking place in the L22. Hence, the concept, implied in Figure 6.1, that *a system can be fully characterized only if we can characterize its environment.* It took a little longer to realize that the symmetry of the figure also implied that *an environment can be characterized only if we can characterize the kinds of systems for which it is an environment.* Either way, the testable assumption is that L12 and L21 are both greater than zero. With this formulation, the environment for living systems is much less than the universe of the physical scientists and much more than the world of the closed systems social scientists (e.g., Parsons, 1937) or Lewin's (1936) "perceived life space". Two of the hoary problems of systems thinking just drop away; that is, what, apart from complexity, distinguishes a system from an aggregate of elements, and how do boundary properties determine the system-environment relation.

In 1956 Hall and Fagan distilled out the classic definition of 'system' as "a set of objects, conceptual or material, with their interrelations". This definition, like the sources from which it was distilled, blithely ignored the point that Angyal had made in 1941. Angyal argued that defining a system requires identification of the system principle which alone explains why these particular items enter into these particular relations; and which alone explains why we can confidently speak of systems as incomplete or having elements or relations that are redundant; and alone explains why, if we find more that one system principle, we always find two or more entangled systems. The concept of 'a system principle' is unacceptable to scientists if this is construed as 'autopoiesis', a sort of Aristotlean essence inherent in the system. Thus, Maturana and Varela (1980) define the organism as a self-regulating system and give the name autopoiesis to the self-regulating process. That is acceptable, but they go beyond that to assert that this is some sort of entelechy, within the organism, that drives the system. For them the L22 is merely a source of perturbations. The metaphor they give (p. 51) is of a pilot flying blind on his instruments without ever knowing what it is that is outside and reflected, presumably accurately, on his instrument panel. Using their assumption the scientific study of organism-environment relations is ruled out. A scientist would insist that only the study of organism-environment relations could lead one to entertain such an hypothesis. It is not a matter to be determined on a priori grounds. In line with the formulation of system-environment relations presented above, we are asserting that, the system principle, as with system goals, is to be found in the special forms of interdependence that exist between the system and its environment. These interdependencies are open to scientific study; they are not speculations about untestable essences. It is this concept that is embodied in the dictum that any organization that seeks to influence its future must first ask itself, "What business are we actually in?" Get this wrong, then the organization's offers of dependency will not be reciprocated; there will be no interdependence, and no future.

The argument represented by Figure 1 also does away with the hoary concept of 'system boundary' as a third type of thing, like a fence or a wall, that mediates between system and environment. Instead, the boundary is represented as a process involving the *interaction* of L12 and L21. It is a negative reflection on systems thinkers that they should have reified the concept of boundary for so long. Militaries long ago recognized that physical obstacles did not constitute boundaries between social systems (armies). The dynamic property of a boundary is that any uninvited crossing has the potential of invoking purposeful reaction. Obstacles covered by a reaction force can become a boundary, but without such covering defensive fire they are simply physical obstacles.

The theorem represented in Figure 6.1 can be better represented as follows.

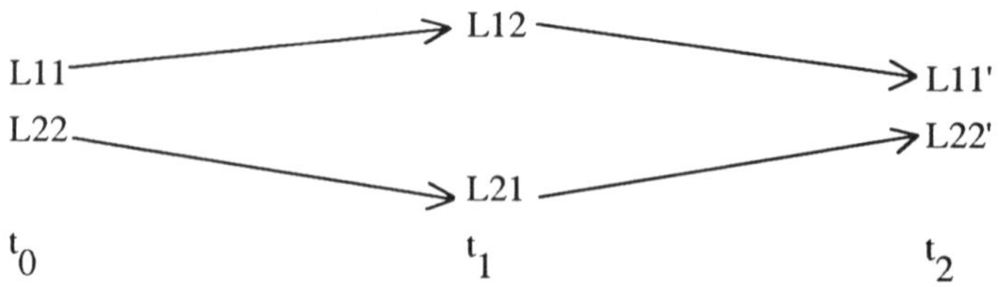

Figure 6.2: Temporal Relation of the Parameters of the System - Environment Situation.

This representation has the advantage of highlighting:

1. that purposefulness is the result of the imbrication of two or more causal strands (we do not resile from the search for causal relations within each strand, and we do not resort to teleology);
2. that time is intrinsic to the transactional relation defined by the four terms. The emergent, and unpredictable, characteristics of the event at t_2 define it as later in time than the events at t_0 (time is extrinsic to the two-term inter-actions that can fully characterize any mechanism. Interactions define only a change in the arrangement of parts; the change could take place in either direction. Hence, only some external criteria can establish which was the prior arrangement). T_0-t_2 constitutes 'the present' for the temporal gestalt defined by the transformation of L11, L22 to L11', L22';
3. that change and novelty are inherent in the transaction because the interaction of the independently determined processes L12 and L21 at t_1-t_2 is a unique, one-off event.

Systems Definitions of Policymaking and Strategic Planning

Having laid this groundwork we are in a position to give a more rigourous, formal definition of strategic planning and policymaking.

If, in *strategic planning*, an appreciation of L22 and the capabilities of the system L11 has revealed that new and more desirable forms of interdependencies are possible, then the plan for implementation is typically:

$$1.\quad (L11, t_0 \rightarrow L11', t_1) \longrightarrow \underline{(L12, t_0 \rightarrow L12', t_1)}$$
$$2.\quad \underline{(L12, t_0 \rightarrow L12', t_1)} \longrightarrow (L21, t_0 \rightarrow L21', t_2)$$

3. to give at t_2 a new steady state level of

$$L11' \underset{L21'}{\overset{L12'}{\rightleftharpoons}} L22'$$

If, in *policymaking*, it is considered desirable to change the behaviours of the system, or set of systems, from L12 to L12', then the plan for implementation is typically:

1. $(L22,t_0 \longrightarrow L22',t_1) \longrightarrow \underline{(L21,t_0 \longrightarrow L21',t_1)}$
2. $\underline{(L21,t_0 \longrightarrow L21',t_1)} \longrightarrow (L12,t_0 \longrightarrow L12',t_2)$
3. to give at t_2 a new steady state level of:

$$L11' \underset{L21'}{\overset{L12'}{\rightleftharpoons}} L22'.$$

Naturally, both strategic planning and policymaking are working within the same model of system-environment relations but, aiming for different kinds of changes, they start from different places and work in different directions. The end result may not appear very different.

This presents us with two theoretical extremes. In practice strategic planning and policymaking are not always such exclusive concerns. I shall return to this point, but for the moment another matter should be considered.

FOUR WORLD HYPOTHESES

The theoretical position outlined above corresponds with Pepper's (1942) world hypothesis of *contextualism*. No apologies are needed for that 'coincidence'. The leading theorists concerned with policymaking, (e.g., Lasswell, 1951; Meehan, 1982; Wildavsky, 1979) have unanimously stressed that any scientifically guided policymaking must accept the rich diversity of interests, concerns, and understandings of the contexts within which need for policy emerges, and within which policy, once implemented, must find a home.

Granted that we have no place for policymaking that is derived from interpretation of sacred texts or the whims of autocrats, then there are only three alternatives to proceeding from contextualism. These are organicism, mechanism and formism.

We could proceed from the world hypothesis of *organicism*; that is, that the State is a living organism, like one big family. Under the influence of Hegel much of the welfare system of Bismarck's Germany appears to have evolved as a form of paternalism. In a family it is usual for the welfare of one member to be positively correlated with the welfare of the other members. In a modern society it is usual to find that at any given time (but not always), the welfare of some members is negatively correlated with the welfare of other members. This is Marx's thesis of class conflict. Much more prevalent these days, and just as deep-seated, is 'dissociation', the perceived lack of any relation between one's own welfare and the welfare of others (Emery and Emery, 1976, pp. 64-71, 109-114). The other side of the modern coin of social exchange is the widespread sensitivity of age, gender, and ethnic groups to anything that vaguely smells of paternalism. Whatever other faults the organicist world hypothesis has (Pepper, 1942, pp. 295-304), it does not adequately describe modern societies and for a policymaker to presume otherwise is to enter a political mine-field.

The *mechanist* world hypothesis has much more relevance to policy-makers. If they have had any tertiary education in the physical or the social sciences (not including economics), they will be deeply steeped in the categories and methodologies that have, since Descartes and Newton, been associated with this model. In the public eye the quintessential scientific activity is the experimental demonstration in the laboratory of what causes what. When policy experts seek help from these sciences, it is likely to come in the forms suggested by this world hypothesis. If the policy experts consult the scientists about Problem B, the answer will take the form 'A causes B' (more or less); therefore, if you change A, then B will change. Despite the usual passing reference to "multiplicity of other causes", this model seems always to encourage tunnel vision and single issue politics. More importantly, the logic of causes and effects seems more relevant to the pursuit of strategic objectives than the formation of policies. The logic of the policy expert is not usually 'if A, then B'. The logic of the policy expert is more appropriately that of 'not A, therefore not B'. A policy is argued on the grounds that if the policy is not in place, then B will not occur; if it is in place, it is possible that B will occur, but still possible, for other reasons, that it will not. See also the discussions in Chapters 1, 2 and 4.

At the practical level it might seem that policy experts need not give much attention to advice from the mechanistic schools in the social sciences, as they have produced so little knowledge that matches the rigour of the physical sciences. This is not so. There are many social scientists who will argue forcefully and persuasively that their sciences can achieve this stature, if only they are given enough resources and encouragement. It is obvious that the policymakers of the Western democracies are not impervious to such persuasion. Often the support given to social scientists is cynical window-dressing. Exercises like the massive US Congress support for James Coleman's (1966) study of *Equality of Educational Opportunity* suggest that sometimes the policymakers do believe that social science can deliver the goods. Or, at least, give that impression to the electorate.

The kind of promise that is held out to policymakers is well illustrated by two of the textbooks on systems theory for social scientists:

> "If we have a full description of the input signals and a full description of the system, it IS possible to derive a full description of the output system. Thus dynamic system analysis remains a determinate discipline. Perhaps it is even more significant that any two of the three descriptions.. can be used to derive the third." (Heise, 1975, p. 225)
>
> "Interpreted in social terms, the theory of linear systems supplies a logical apparatus for treating the dynamics of social processes. The fundamental premise of this logic is that the manner in which some states of society are transformed into other states depends upon the structure of the social system that is involved in this transformation... the structure of a system determines its function. This last assertion is true not just under any set of assumptions but rather is valid under the set of assumptions and interpretations of systems theory developed in this book." (Cortes, *et al.*, 1974, p. ix)

Policy experts could be forgiven for not noticing that these social scientists are not talking about the world in which policies are supposed to work. In both cases the assumption is made that social organizations and societies are mechanical systems but the assumption is hidden behind verbiage. Social scientists then celebrate their 'discovery' of the explanatory power of their fully determinate models. They have, in fact, discovered nothing, but simply demonstrated that a logically consistent model can be constructed on those assumptions. In the first case the assumption is that the systems are incapable of exercising genuine choice. But policy experts have to work with purposeful systems. The responses of purposeful systems are not fully determinate. We cannot rule out responses that are wholly or in part a change in state of the system; for example, when a community gets fed-up with attempts to interfere with their life-style. We cannot rule out novel responses to the situations created by new policies; for example, the criminal responses to the US Prohibition. In the second case the assumption that "structure of a system determines its function" is not an assumption that policy experts should readily accept. Inputs into a social system, including policy inputs, may produce specific outputs or a transformation of structure. Nowadays, policy inputs about freeways and the forests appear to produce both, in indeterminate mixtures. The image of structure, as in a machine, is fundamentally flawed because, in human societies, "organizational structure is a functional structure describing the allocation of choice" (Ackoff and Emery, 1972, p. 222).

Behind the facade of systems theory the policy expert must be able to detect the difference between causal thinking and systems thinking:

> "In causal research the task is to single out from a multiplicity of data pairs of data between which there is a necessary connection [i.e., 'if A, then B']. In system thinking the task is not to find direct relations between members but to find the superordinate system in which they

are connected or to define the positional value of members relative to the superordinate system" (Angyal, 1941, p. 35).

LIMITATIONS OF FORMISM AND FORMAL RATIONAL MODELS

We could proceed from the Formist (Aristotlean) position, the last of the four world hypotheses, of assuming certain postulates about human nature, or the nature of human society, and deduce what must necessarily, and naturally, be good for people. When this is attempted we discover that neither people nor societies quite fit our preconceptions. The temptation is then to discard the people who do not fit our preconceptions. It is easy to understand that policymaking still proceeds in such a fashion in Eire where the medieval influence of Thomism still prevails. It is harder to understand the dominance of economic theory in the policymaking of modern secular societies when that theory proceeds from the same basic world hypothesis. It is approximately half a century since F S C Northrop (in Northrop, 1947) proved that, despite the mathematical trappings, economic theory was not a scientific enterprise. The scientific enterprise is inextricably involved in external reality at the beginning and end of its projects. It starts with reductive assumptions about reality, but these are reviewed and amended as the outcomes in practice differ from predictions. In contrast, economic theory is a logical enterprise. It is a machine that reviews and amends its assumptions in the light of the interest generated by the deductions derived from those assumptions. The development of economic theory has been influenced by neither the empirical relevance of its assumptions nor the empirical validity of its deductions. When theoretical economists definepreferences, markets, equilibrium, etc, they are concerned only with precise nominal definitions; they are not concerned with the scientist's painstaking efforts to determine valid empirical definitions. Thus, the theoretical economists are concerned only with possible worlds, that do not have actually to exist. Then economists do not have to concern themselves, as theorists, with whether the outcomes predicted by their models correspond to any changes in reality. It is enough for them that they prove that, if such and such a world existed, then change X must necessarily (logically) result in Y. The following diagrams from Northrop, Figures 6.3 and 6.4, tell it all.

Northrop was a competent analyst. His paper was read before the Economics Clubs of Yale and Princeton before being accepted for the *Quarterly Journal of Economics* in November 1941. No economic theorist came forward with a rebuttal. Subsequent evidence suggests that the economic enterprise is even more deeply entrenched in its scholastic, non-scientific, enterprise. In his presidential address to the American Economic Association in 1970, Wassily Leontieff was severely critical of the fact that economic theorists did not need to have any knowledge of economic realities. "Unfortunately, any one capable of learning elementary, or preferably advanced calculus and algebra, and acquiring acquaintance with the specialized terminology of economics can set himself up as a theorist" (1971, p.1).

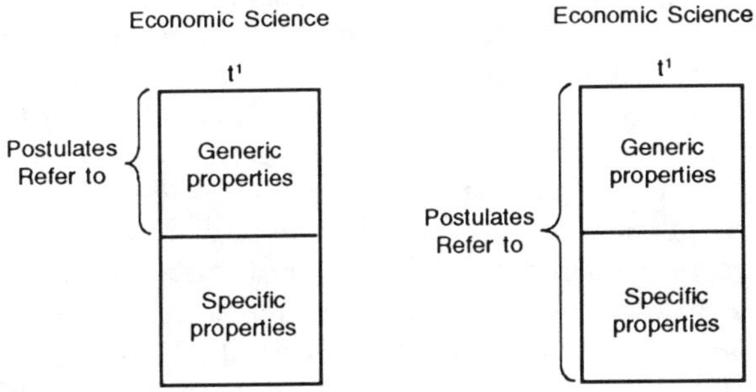

Figure 6.3: The Properties to which the Postulates Refer
Source: Northrop, 1947, p. 252.

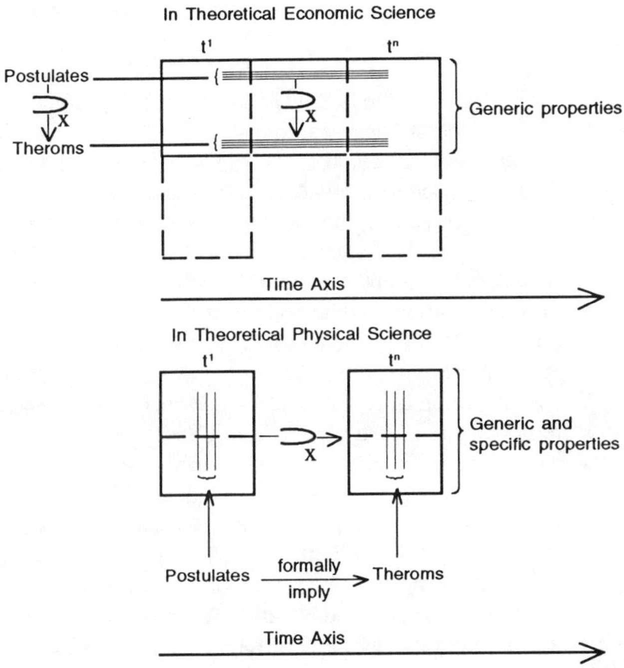

Figure 6.4: The Factors between which the Relation of Formal Implication ($\supset_x$) Joining Postulates and Theorems Holds
Source: Northrop, 1947, p. 252

In 1982 Eugene Meehan published a study based, in part, on an analysis of the contents of the leading US economic journals during 1970-1979. Only five to ten percent of the articles published in those journals over the decade met Northrop's criterion of a scientific contribution; that is, starting from earth and coming back to earth. And many of those few were only dubiously tied to the real world. Even more striking was the apparent lack of any professional pressures toward producing scientific publications rather than self-contained logical exercises. Unlike mathematical exercises, the ones in economic theory, and computer econometric modelling, are replete with words like 'markets', 'preferences', 'goods', and 'values' that have ordinary everyday meanings. However, as used by economists those words have only the special nominal definitions that are arbitrarily chosen by the economic theorist. This lends their work the spurious and misleading appearance of being about the real world of economic transactions. But: "The principal finding of this study is unequivocal. From the policy-maker's perspective, both economic theory generally and econometrics are virtually useless. Their present products cannot be used in reasoned policymaking and their approach to inquiry makes future improvement extremely unlikely" (Meehan, 1982, p. 125). A similar study that I did in the late seventies of ten years of the journal *Operations Research* yielded results that paralleled Meehan's. Operations research had found a home in academia and the published output reflected that change. Over ninety percent of the published articles were exercises in applied mathematics. To repeat the analogy I used at the time, they were like a ship with an engine room but no sharp end and no blunt end with screws and rudder - not exactly a seaworthy condition. The published papers did neither start from an empirical problem nor end with a testable hypothesis. The editor over this period was retiring, and he pointed to the same fault and lamented that all of his efforts to attract properly scientific contributions had come to naught. See also the discussions in Chapters 1 and 2.

The developments in economics, econometrics, and operations research suggest that universities are not inherently structured about the ends that policy experts pursue. Universities, quite universally, are most concerned with the pursuit of academic excellence; and they seem to be in agreement that excellence is best demonstrated in the mastery of the logical interplay of abstract universals, that is, the peak achievement in the Formist (Aristotlean) world hypothesis. (As universities go into the business of educating policy experts, one has to wonder about what protects them from being turned in the same direction as the economists, econometricians, and operations researchers have been turned.) See also the discussion in Chapter 5.

This review of the implications for policy experts of the world hypotheses of organicism, mechanism and formism has necessarily been sketchy. Enough has been said, one hopes, to suggest that serious attention should be paid to the *contextualist* hypothesis.

For the policy expert the biggest obstacle is that the formal logic, which is the final arbiter of arguments in the other world hypotheses, is inadequate in the contextualist world hypothesis. We have pointed to the difficulties that economics, and now operations research, have gotten into with the pursuit of syllogistic perfection. The reason that the formal logic is inadequate is that contextualism takes time, not space, as the basic category; that is, contextualism focuses on time-space, not space-time. It accepts Peirce's dictum, that whilst there can be no motion *in* an instant, there can be motion *at* any instant in time (Peirce, 1982, p. 392). By contrast, in the classical world hypotheses "every state of the world at each particular instant was defined as a huge instantaneous configuration of an enormous number of particles, each of which was moving according to the laws of classical mechanics while preserving its physical identity through time. World history was thus viewed as a continuous succession of such instantaneous configurations" (Capek, 1991, p. 326). On this assumption that everything is determined by everything else, abstract universals could be taken as referring to possible determinants. Democritus is credited with first formulating this assumption ("By necessity are foreordained all things that were and are and are to come."), but it is in the formal logic of Aristotle that this metaphysical view of the world has been enshrined (Harris, 1983, pp. 161-170). The laws of identity and the excluded middle precisely reflect this world of successive, determinate instants. The fact that, by mere assumption, Aristotlean logic ruled out the world view of Heraclitus ("that nothing is but is always becoming") was made plain for all to see by Zeno's famous paradoxes (Peirce, 1982, Vol.1, pp. 390-392; 1984, Vol.2, pp. 254-257).

Contextualism in effect reasserts the Heraclitean world-view and hence necessarily finds formal logic less than an adequate guide. More specifically, it insists on a distinction between formal or abstract universals and material universals. Only those universals can be considered material which can claim some connection with material entities (Feibleman, 1979, p. 198; Peirce, 1984, Vol.2, p. 258). Only material universals can help us to understand the processes of becoming in a field of material entities. In formal logic the major premiss in a syllogism can take the form 'All Ms are P' because M specifies a formal universal, and there are no restrictions on its references and the members of the class thus specified are infinite and can be treated as identical. This abstract emptiness explains why contextualism requires concrete (material) logics. In concrete logics the major premiss cannot take the form of 'All Ms are P' as the universals are material universals. With material universals reference is restricted to the range of members of the designated class (which membership may be extended indefinitely but is not infinite), and the members, whilst similar in some respects, are not and cannot be treated as identical. By the same token concrete logic accepts 'middle terms' and rejects the 'law of the excluded middle'.

Stephen Toulmin (1958) showed us the kind of concrete logic that policymakers need in real-life contexts. Toulmin's thesis is that in real life we are arguing about material universals. Hence, we should not accept any formal logic that substitutes formal universals for material universals, as formal universals do not, and cannot, constitute empirical evidence. They cannot even point us to where we should look for empirical evidence.

Instead, Toulmin argues we must ask of any major premiss: "What evidence does it provide to warrant drawing conclusion C from data D?" If the warrant is not acceptable further backing (B) arguments are needed, and so on (B2, B3...).

Because we are arguing about material classes, whose members are in varying degrees similar but not identical (certainly not identical with respect to the determining contexts within which they find themselves), the conclusion will usually have to be qualified. The qualifications themselves are subject to rebuttals and exceptions.

Diagrammatically these relations are shown in Figure 6.5.

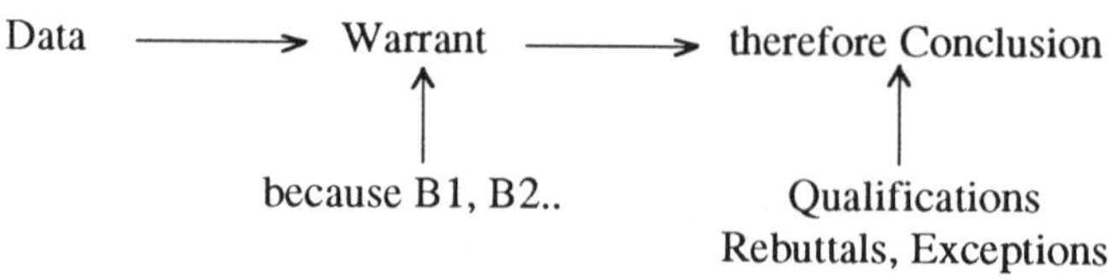

Figure 6.5: The Structure of Arguments

This is not a fully spelt-out concrete logic. For an example of such we have to turn to Feibleman's concrete logic of organizations (Feibleman, 1979, pp. 218-226). As Fischer has pointed out, however, the work of Paul Taylor (1961) on these informal logics clearly identifies a hierarchy of four levels of warranty that a policymaker should require before concluding that "the facts lead to this conclusion" (Fischer, 1990).

POLICY EXPERTISE, AND RATIONALITY REVISITED

However far we succeed in progressing in these directions, we are still left with serious problems arising from the concept of 'policy expert' itself. I have used this term to distinguish between those concerned with the formulation and adoption of policies and to distinguish these two groups from the policymakers. In a democratic society it is assumed that the elected representatives are the policy-makers. In practice the skilled policy experts often seem to be effectively the policymakers. Insofar as this is the case policy experts would seem to be white-anting the democratic processes they are employed to serve.

I suggest that this is so only so far as expertness in policy formulation is seen as identifying the most probably effective course of action. I am referring here to the concept of rationality that economists attribute to 'economic man'. It is not, however, the concept of rationality that best describes the behaviour of human beings and their institutions. Human beings enter choice situations not only with conceptions of objectives they wish to achieve but also with a usually unstated, determination to preserve their own identity and self-respect, and a preference for courses of action that they find enjoyable in their own right. Consider again the discussions of rationality in Chapters 1 and 2.

In the first rigourous concrete logic of 'psycho-logics' Heider concluded with the observation that:

> "Max Weber and others pointed out one kind of rationality in behaviour, namely, the rationality of the means-end relation.
> Choosing the appropriate means to gain an end makes for a 'good', a 'rational' action and we can understand it. However, understandable behaviour often is not of this sort, but is based on the simple configurations of U (cognitive) and L (attitudinal) relations. Since they determine both behaviour and perception, we can understand social behaviour of this kind" (Heider, 1981, pp. 110-111).

This can be explicated further. Rationality for human beings and their institutions is a function of the perceived Probable Outcome, where Probable Outcome is a function of both Probable Effectiveness and Probable Choice, taken jointly. Probable Effectiveness is Heider's U relation and the economists rationality. Probable Choice is Heider's L relation (Ackoff and Emery, 1972, p. 58). What these authors are asserting is that people do not choose between courses of action simply on the basis of which will be most effective in achieving the objective they have in mind. Even with full knowledge of the relative effectiveness of the possible courses of action, they will be biased toward those courses that they like, are familiar with, or seem most in character with their own sense of identity. More than that. The whole set of effective courses of action may be so off-putting or repulsive, by their values and ideals, that a human being might decide the desired end is "not worth the candle".

These considerations should disturb any conceptions of what is, or who could be educated to be, a policy expert. In the inputs for the facts, warrants, and qualifications of Toulmin's (1958) argument, we are inevitably concerned with data, values and ideals. In formulating policy we are concerned with influencing the choices that people make. Hence, we must consider the attitudes, values, and ideals that will enter into those choices. If we ignore these matters, then we can expect policies to fail because people persist in their 'irrational' ways or ignore, subvert, or actively organize to defy the 'irrational' policies of the experts. We can go some way in empirically determining attitudes. I think the expertness of policy experts now should be in 'searching for common ground' with those whose values and ideals are likely to be influenced by the policies that are under consideration (Emery, 1981, Vol.2, Chapter 27). The policy expert would be the one who formulates policies that people perceive, perhaps for the first time, as sensible and desirable.

CHALLENGES AND RECOMMENDATIONS

Policymaking is not, and cannot be, a determinate science. The 'light on the hill' that Lerner and Lasswell lit in 1951 must be dismissed as wishful thinking. Guidelines to policymaking can certainly be devised, but not 'Policy Sciences'. This is no cause for dismay, if we can continue to evolve guidelines that enable us to avoid the disasters in policy formulation and adoption that mark the past. It would, of course, be much more reassuring if we could, more often, produce policies that succeed because the people, whom the policies are supposed to help, actively ensure their success. These are sentiments that, I think, most policy experts would share.

I have been embarrassed, however, constantly, for some decades, with being a social science expert in policy matters. Three sources of embarrassment keep recurring. My knowledge, as a social scientist is conditional, but is treated as definitive; the 'powers that be' whom I am advising make what they wish, within limits, of my advice; people, being purposeful creatures, make what they will of policies that are imposed on them. I am embarrassed because my knowledge is largely about what will not work in human affairs; it is impossible for me to be definitive about courses of action that have not been tried before. I am embarrassed when my work, as a social scientist, is used, in some power-ploy, for ends that I do not approve. I am embarrassed when my research is used to justify reduction in the range of choices open to people.

In 1964, C West Churchman and I sat down to pool our experiences with respect to the second question, the question that Wildavsky (1979) aptly phrased "Speaking truth to power". Churchman brought his experiences as an operations researcher and I had my experiences as a social scientist. We had no difficulty in identifying the three roles that were open to us as scientific advisers. We could wear our university hats and make our observations and give advice from a securely independent position; we could take out 'dual citizenship' by becoming a professional employee of the organization seeking our advice; or, thirdly, we could seek the sanctioning of some body whose interests are seen as encompassing what are divisive within the organization we wish to study and advise. A simple example is the Joint Committee of the employers and unions that guided the research into new managerial policies in Norway, 1962-68. With regard to the independent role we concluded that it "is incomplete because the independent research community has no adequate way of judging whether its focus of interest or its output are of any real concern or value to any other organization" (Churchman and Emery, 1966, p. 79). With regard to the dual role we observed that it "is incomplete because the researcher has no adequate basis for resolving conflicts of interest and values within the organization under study or between it and the research community" (p. 80). The third role seemed obviously desirable if we were to get close to the issues and conflicts and yet not become a captive tool of some factional group. Finding an encompassing third organization was recognized as a serious practical difficulty that might not always be resolved.

In gathering experience with this problem, I continued to grapple with the other two problems of being taken as an expert with determinate knowledge and involving people in formulating and taking responsibility for policies that concern them. As the problems came into focus it seemed to me that a solution would involve:

- conducting some search process whereby the main parties to the proposed change can clearly identify and agree about the ideals the change is supposed to serve and the kinds of paths most in character with their traditions;
- designing a change process which will enable relevant learning to take place at rates appropriate to the demands of time. This process involves the time within which change must occur to avoid intolerable costs of not changing, and the time by which decisions need to be made if adequate resources are to be mobilised;
- devising social mechanisms for participation, whereby the choice of paths will reflect the intrinsic value of these paths for those who will have to traverse them.

The social mechanism that emerged was the *search conference*. This is a temporary planning community, meeting under 'social island' conditions for as many days and nights as seem necessary for their work. The conference and its membership have to be sanctioned by the key operational bodies involved in the issue. Experts may be participants, as resource people, but it is the conference that is sanctioned to give the advise normally sought from experts. Responsibility for adoption and implementation of conference recommendations remains, of course, with the key operational bodies (Emery, 1981). Occasionally situations have been encountered that are so polarized that the concerned parties will not meet together in joint search. These are rare occasions. Most times search conferences have been approved, and met the three aims outlined above, regardless of the issues, and kinds of parties involved (M Emery, 1982).

It seems to me that the search conference, or some such participative mechanism, must replace the traditional role of the policy expert. The policy expert will be one who is equipped to identify when such mechanisms need to be used, and advise on who should be involved and on how the process should be conducted. The role of Plato's guardians simply corrodes our democratic institutions. Emeritus Professor Robert A Dahl, after a lifetime of studying modern democracies, observed that: "If the democratic process is not firmly anchored in the judgments of the demos, then the system will continue to drift toward quasi-guardianship" (1989, p. 338). See also Chapter 5 and the "School for Rulers".

REFERENCES

Ackoff, Russell L, and Fred E Emery, 1972. *On Purposeful Systems*, Chicago: Aldine Press.

Angyal, Andras, 1941. 'A Logic of Systems'. In Fred E Emery (Ed.), *Systems Thinking*, 1, 27-40, 1981, London: Penguin.

Baburoglu, Oguz. N, 1988, 'The Vortical Environment: the Fifth in the Emery-Trist Levels of Organizational Environments', *Human Relations*, 41, 181-210.

Bertalanffy, Ludwig von, 1950, 'The Theory of Open Systems in Physics and Biology', *Science*, 111, 23-29.

Capek, Milic, 1991, *The New Aspects of Time*, Dordrecht, The Netherlands: Kluwer.

Churchman, C West, and Fred E Emery, 1966, 'On Various Approaches to the Study of Organizations' In John R Lawrence (Ed.), *Operational Research and the Social Sciences*, London: Tavistock.

Coleman, James, 1966, *Equality of Educational Opportunities*, Washington, DC: US Government Printing Office.

Cortes, Fernando, Adam Przeworski and John Sprague, 1974, *Systems Analysis for Social Scientists*, New York: Wiley.

Dahl, Robert A, 1989, *Democracy and Its Critics*, New Haven, Yale University Press.

Emery, Fred E (Ed), 1981, *Systems Thinking*, 2, London: Penguin.

Emery, Fred E, and Merrelyn Emery, 1976, *Futures We Are In*, Leiden, The Netherlands: Martinus Nijhoff.

Emery, Fred E, and Eric L Trist, 1972, *Towards a Social Ecology*, New York: Plenum.

Emery, Merrelyn, *Searching*, Canberra, Australia: Centre for Continuing Education, The Australian National University.

Feibleman, James K, 1979, *Assumptions of Grand Logics*, The Hague, The Netherlands: Martinus Nijhoff.

Fischer, Frank, 1990, *Technocracy and the Politics of Expertise*, Newbury Park, California: Sage.

Hall, A D, and R E Fagen, 1956, 'Definition of System', *General Systems*, 1, 18-29.

Harris, Errol E, 1983, *An Interpretation of the Logic of Hegel*, Lanham, Maryland: The University Press.

Heider, Fritz, 1946, 'Attitudes and Cognitive Organization', In Fred E Emery (Ed), 1981, *Systems Thinking*, 2.

Heise, David R, 1975, *Causal Analysis*, New York: Wiley.

Lasswell, Harold D, 1971, *A Preview of Policy Sciences*, New York: American Elsevier.

Leontieff, Wassily, 1971, 'Theoretical Assumptions and Nonobserved Facts', *American Economic Review*, 61, 1-7.

Lerner, Daniel and Harold D Lasswell, 1951, *The Policy Sciences*, Stanford, California: Stanford University Press.

Lewin, Kurt, 1936, *Principles of Topological Psychology*, New York: McGraw-Hill.

Maturana, Humberto R and Franciso J Varela, 1980, *Autopoiesis and Cognition*, Dordrecht, The Netherlands: Reidel.

Meehan, Eugene J, 1982, *Economics and Policymaking; the Tragic Illusion*, Westport, Connecticut: Greenwood Press.

Meehan, Eugene J, 1985, *Reasoned Argument in Social Science; Linking Research to Policy*. Westport, Connecticut: Greenwood Press.

Northrop, F S C, 1947, *Logic of the Sciences and Humanities*, New York: Macmillan.

Parsons, Talcott, 1937, *The Structure of Social Action*, New York: McGraw Hill.

Peirce, Charles S, 1982, *Writings of Charles S Peirce*, 1, Edward C Moore (Ed), Indiana University Press.

Peirce, Charles S, 1984, *Writings of Charles S Peirce*, 2, Edward C Moore (Ed), Indiana University Press.

Pepper, Stephen C, 1942, *World Hypotheses*, Berkeley: University of California Press.

Prigogine, Ilya, 1980, *From Being to Becoming*, San Francisco: Freeman.

Selznick, Philip, 1957, *Leadership in Administration*, New York: Harpers.

Taylor, Paul, 1961, *Normative Discourse*, Englewood Cliffs, New Jersey: Prentice-Hall.

Toulmin, Stephen, 1958, *The Uses of Argument*, Cambridge, England: Cambridge University Press.

Wildavsky, Aaron, 1979, *Speaking Truth to Power; The Art and Craft of Policy Analysis*, Boston: Little, Brown.

Chapter Seven

System Dynamics and the Lessons of 35 Years [1]

by
Jay W. Forrester

INTRODUCTION

The professional field known as system dynamics has been developing for the last 35 years and now has a world-wide and growing membership. System dynamics combines the theory, methods, and philosophy needed to analyze the behavior of systems in not only management, but also in environmental change, politics, economic behavior, medicine, engineering, and other fields. System dynamics provides a common foundation that can be applied wherever we want to understand and influence how things change through time.

The system dynamics process starts from a problem to be solved—a situation that needs to be better understood, or an undesirable behavior that is to be corrected or avoided. The first step is to tap the wealth of information that people possess in their heads. The mental data base is a rich source of information about the parts of a system, about the information available at different points in a system, and about the policies being followed in decision making. The management and social sciences have in the past unduly restricted themselves to measured data and have neglected the far richer and more informative body of information that exists in the knowledge and experience of those in the active, working world.

---

[1] This paper incorporates an edited version of most of the author's "Lessons from System Dynamics Modeling," *System Dynamics Review*, Vol. 3, No. 2, Summer 1987.

System dynamics uses concepts drawn from the field of feedback control to organize available information into computer simulation models. A digital computer as a simulator, acting out the roles of the operating people in the real system, reveals the behavioral implications of the system that has been described in the model. The first articles based on this work appeared in the *Harvard Business Review* (Forrester, 1958; Forrester, 1959). From over three decades in system dynamics modeling have come useful guides for working toward a better understanding of the world around us.

The continued search for better understanding of social and economic systems represents the next great frontier. Frontiers of the past have included creating the written literatures, exploring geographical limits of earth and space, and penetrating mysteries of physical science. Those are no longer frontiers; they have become a part of everyday activity. By contrast, insights into behavior of social systems have not advanced in step with our understanding of the natural world. To quote B. F. Skinner:

> "Twenty-five hundred years ago it might have been said that man understood himself as well as any other part of his world... Today he is the thing he understands least. Physics and biology have come a long way, but there has been no comparable development of anything like a science of human behavior... Aristotle could not have understood a page of modern physics or biology, but Socrates and his friends would have little trouble in following most current discussions of human affairs" (Skinner, 1971, p. 3).

The great challenge for the next several decades will be to advance understanding of social systems in the same way that the past century has advanced understanding of the physical world.

DESIGNING MANAGERIAL AND SOCIAL SYSTEMS

Everyone speaks of systems: computer systems, air traffic control systems, economic systems, and social systems. But few realize how pervasive are systems, how imbedded in systems we are in everything we do, and how influential are systems in creating most of the puzzling difficulties that confront us.

People deal differently with different kinds of systems. Engineering systems are designed using the most advanced methods of dynamic analysis and computer modeling to anticipate behavior of a system when finally constructed. On the other hand, although political, economic, and managerial systems are far more complex than engineering systems, only intuition and debate have ordinarily been used in building social systems. But, powerful system-design methodologies have evolved over the last 50 years.

In designing an engineering system, say a chemical plant, engineers realize that the dynamic behavior is complicated and that the design can not successfully be based only on rules of thumb and experience. There would be extensive studies of the stability and dynamic behavior of the chemical processes and their control. Computer models would be built to simulate behavior before construction of even a pilot plant. Then, if the plant were of a new type, a small pilot plant would be built to test the processes and their control.

But observe how differently social systems are designed. We change laws, organizational forms, policies, and personnel practices on the basis of impressions and committee meetings, usually without any dynamic analysis adequate to prevent unexpected consequences.

"Designing" social systems or corporations may seem mechanistic or authoritarian. But all governmental laws and regulations, all corporate policies that are established, all computer systems that are installed, and all organization charts that are drawn up constitute partial designs of social systems. Such redesigns are then tested experimentally on the organization as a whole without dynamic modeling of the long-term effects and without first running small-scale pilot experiments. For example, bank deregulation and the wave of corporate mergers in the 1980s constituted major redesigns of our economy with inadequate prior consideration for the resulting bankruptcies and unemployment. All systems within which we live have been designed. The shortcomings of those systems result from defective design, just as the shortcomings of a power plant result from inappropriate design.

Consider the contrast between great advances during the past hundred years in understanding technology, and relative lack of progress in understanding economic and managerial systems. Why such a difference? Why has technology advanced so rapidly while social systems continue to exhibit the same kinds of misbehavior decade after decade? I believe the answer lies in failing to recognize that countries and corporations are indeed systems. There is an unwillingness to accept the idea that families, corporations, and governments belong to the same general class of dynamic structures as do chemical refineries and autopilots for aircraft.

People are reluctant to accept the idea that physical systems, natural systems, and human systems are fundamentally of the same kind, and that they differ primarily in their degree of complexity. To admit the existence of a social system is to admit that the relationships between its parts have a strong influence over individual human behavior.

The idea of a social *system* implies sources of behavior beyond that of the individual people within the system. Something about the structure of a system determines what happens beyond just the sum of individual objectives and actions. In other words, the concept of a system implies that people are not entirely free agents but are substantially responsive to their surroundings.

To put the matter even more bluntly, if human systems are indeed systems, it implies that people are at least partly cogs in a social and economic machine, that people play their roles within the totality of the whole system, and that they respond in a significantly predictable way to forces brought to bear on them by other parts of the system. Even though this is contrary to our cherished illusion that people freely make their individual decisions, I suggest that the constraints implied by the existence of systems are true in real life. As an example, we see the dominance of the political system over the individual in the evolution of the U.S. Federal budget deficit. Every presidential candidate since 1970 has campaigned with the promise to reduce the federal deficit. But the deficit has on the average doubled every four years. The social forces rather than the president have been controlling the outcome. How to harness those social forces has not been effectively addressed.

The feedback structure of an organization can dominate decision making far beyond the realization of people in that system. By a feedback structure, I mean a setting where existing conditions lead to decisions that cause changes in the surrounding conditions, that influence later decisions. That is the setting in which all our actions take place.

We do not live in a unidirectional world in which a problem leads to an action that leads to a solution. Most discussions, whether in board meetings or cocktail parties, imply a structure as in Figure 1.

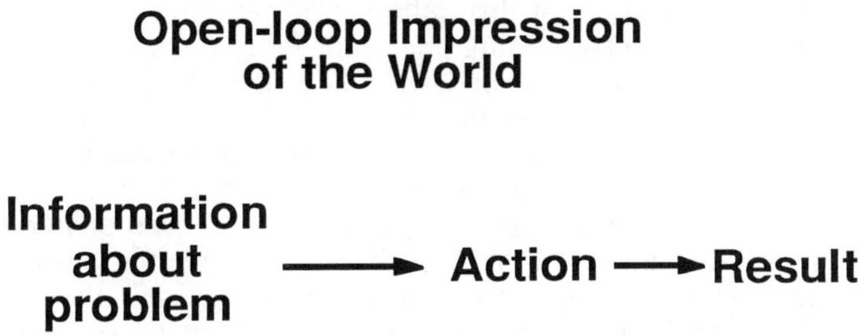

Figure 1

Figure 1 suggests that the world is unidirectional, that the problem is static and we need only act to achieve a desired result.

Instead, we live in an on-going circular environment like Figure 2 in which each action is based on current conditions, such actions affect conditions, and the changed conditions become the basis

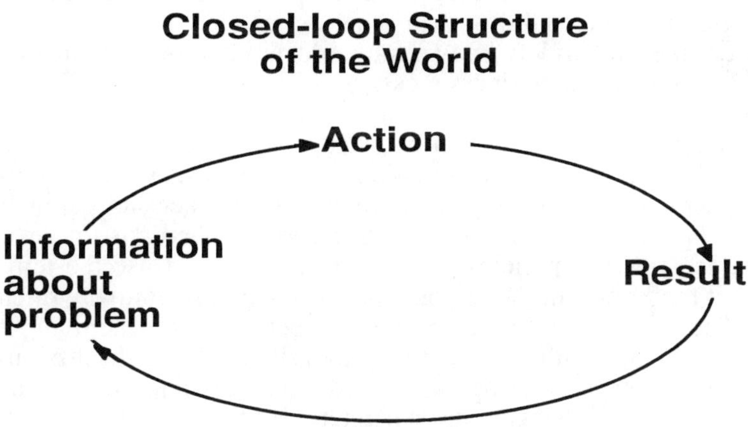

Figure 2.

for future action. There is no beginning or end to the process. People are interconnected. Many such loops are intertwined. Through long cascaded chains of action, each person is continually reacting to the echo of that person's past actions as well as to the past actions of others.

We discovered surprising things in our early work with corporations that we now realize carry over to all social systems (Roberts, 1978, Edward). First, most difficulties are internally caused, even though there is an overwhelming and misleading tendency to blame troubles on outside forces. Second, the actions that people know they are taking, usually in the belief that the actions are a solution to difficulties, are often the cause of the problems being experienced. Third, the very nature of the dynamic feed-back structure of a social system tends to mislead people into taking ineffective and even counterproductive action (Sterman, 1989). Fourth, people are sufficiently clear and correct about the reasons for local decision making—they know what information is available and how that information is used in deciding on action. But, people often do not understand correctly what overall behavior will result from the complex interconnections of known local actions.

In our early system dynamics work we found we could go into a troubled company and uncover the reasons for its problems. The difficulty might be falling market share (Forrester, 1968), or fluctuations in production with employment varying from working overtime one year to having half the work force laid off two years later (Forrester, 1961), or a lower profitability than other companies in the industry. Such difficulties are widely known to employees and the community, and are discussed in the business press.

Such an analysis draws on knowledge about how structure and policy relate to behavior. Information comes primarily from interviewing people in the company about how they make decisions at their individual operating points. Statements describing the basis for decisions are the rules or policies governing action. As I use the term "policy," it represents all the reasons for action, not just formal written policy. These interviews are extensive and penetrating. There might be several sessions with each of many individuals. The discussions range widely from normal operations, to what was done in various kinds of past crises, what is in the self-interest of the individual, where are the influential power centers in the organization, what would be done in various hypothetical situations that have never happened, and what is being done to help in solving the serious problem facing the company.

Talking to a manager usually reveals a clear and comprehensive picture of the rules and conditions driving decisions at that position in the corporation. Then, when talking to another manager about the first manager, the same picture usually emerges. In other words, people see themselves very much as others see them. There is substantial consistency throughout the organization as to the actual operational policies that are guiding decisions. Furthermore, the policies are justified in terms of how those policies are expected to correct the great difficulty that the company is experiencing.

Up to this point, the study of such a company follows the case-study approach to management education. That is, a comprehensive examination of all related parts of the company is made in the context of the problem that is to be solved. But, if left at this point, the weakness of the case-study method would dominate the outcome. A descriptive model of the company would have been assembled, but the human mind is not able to deal with the inherent dynamic complexity of such a situation.

For those who have studied mathematics through differential equations, such a descriptive model is equivalent to a high-order nonlinear differential equation. No scientist or mathematician can solve such a system mentally. Just as with the operation of a chemical plant, only computer simulation methods are capable of revealing the behavior implicit in the structure that can be built from knowledge about the many local decision-making individuals and how they are connected.

After obtaining a description of the important policies, information flows, and interconnections in a company, the next step is to translate that description into a computer simulation model. A simulation model does not involve complicated mathematics but instead is a language translation from the original description to computer instructions. Such a model allows the computer to act out the roles of each decision point in the model and feed the results to other connected decision points to become the basis for the next round of decisions. In other words, a laboratory replica of the company then exists in the computer where one can observe the behavioral consequences of the policies that had been described in the interviews—policies that are intended to solve the company's problem.

To the surprise of those unfamiliar with the devious nature of such dynamic systems, the computer model, based on policies known to people in the company, will usually generate the very difficulties that the company had been experiencing. In short, the policies that were believed to solve the problem are, instead, the cause of the problem. Such a situation creates a serious trap and often a downward spiral. If the policies being followed are believed to alleviate the problem, but, in hidden ways, are causing the problem, then, as the problem gets worse, pressures increase to apply still more strongly the very policies that are causing the problem.

Sometimes, one need not even go into a company to identify the system creating a problem. Many readers will remember the People Express airline. During its early history People Express was spectacularly successful with one of the highest growth rates in the history of American corporations. Don Burr, the founder, was a popular speaker at business schools on the philosophy and policies for corporate success. In 1983, the Harvard Business School published a management case on the history, practices, and success of People Express (Harvard Business School, 1983).

From Don Burr's public speeches and the published case study, Professor John D. Sterman at the MIT Sloan School of Management created a system dynamics simulation model of the People Express corporation. The model represents the propensity to expand the air fleet, the relatively greater difficulty in expanding trained service personnel, and the competitive effects of low fares on other airlines and on the financial position of People Express. The model generates a powerful growth mode followed by sudden failure, just as happened with the actual airline. One discovers from the model that it was implicit in publicly stated policies that the company was doomed to collapse after an initial unusual success.

Such a model provides a powerful way to introduce people to the importance of systems interactions. For example, Sterman used the computerized People Express case as a dramatic introduction to management for the 200 students enrolling each fall for a master's degree in management at MIT. For more than six hours, in the ballroom of the Marriott Hotel, with 100 Macintosh computers, students explored various policies affecting corporate profitability and the rate and stability of growth. They came to appreciate how a mix of apparently reasonable policies can produce rapid unbalanced expansion followed by deteriorating quality and sudden collapse. They also experienced how "obvious" policies for improvement so often do not work.

Misjudgments rather similar to those at People Express lie behind much of the foreign encroachment on American markets in the 1980s. Foreign infiltration was initially blamed by American companies on lower foreign wages and lower product prices. In response, domestic prices were reduced until there were insufficient profit margins to permit fixing the real difficulties, which were usually more in design and in quality of product and service than in price. As so often happens, the domestic failure to compete arose more from mismatched internal policies than from external forces.

A NEW EDUCATION FOR CORPORATE AND GOVERNMENT LEADERS

All of this points to a new kind of education for leaders of corporations and governments. Beyond that, it suggests a new kind of leader for the future. One can now see clearly a kind of education that we might call "enterprise design" or "social system design." And in the future there is a role for the output of such an education, the "enterprise designer." There is no need for a significant difference between education for corporate leadership or for governmental leadership. The following paragraphs are cast in terms of business managers, but they apply equally well to preparation for government.

A fundamental difference exists between an enterprise operator and an enterprise designer. To illustrate, consider the two most important people in the successful operation of an airplane. One is the airplane designer and the other is the airplane pilot. The designer creates an airplane that the ordinary pilot can fly successfully. Is not the usual manager more a pilot than a designer? A manager is appointed to run an organization. Often there is no one who consciously and intentionally fills the role of organizational designer.

Organizations built by committee, by intuition, and by historical happenstance often work no better than would an airplane built by the same methods. Time after time one sees venture capital groups backing a new enterprise in which the combination of corporate policies, characteristics of products, and nature of the market are mismatched in a way that predetermines failure, just as with People Express airline. Like a bad airplane design that no pilot can fly successfully, such badly designed corporations lie beyond the ability of real-life managers.

I first began to glimpse the possibilities of enterprise design in the 1960s when, for its first ten years, I was on the board of directors of the Digital Equipment Corporation. To guide my own position on the board, I developed a system dynamics model of how high-technology growth companies evolve. The model incorporated some 250 variables ranging from physical processes, to managerial goals and leadership characteristics, to interactions among company, market, and competitors. The model exhibited the full range of typical behaviors for such companies from early failure, through limited growth followed by stagnation, to sustained growth with repeated major crises, and on to untroubled growth. These differences emerged in the model as a result of different policies that could be clearly identified in the various companies having the corresponding kinds of behavior. From the model came improved understanding of how corporate policies determine the corporate future (Forrester, 1964).

Management education, in all management schools, has tended to train operators of corporations. But there has been rather little academic attention to the design of corporations. The determination of corporate success and failure seldom arises from functional specialties alone, but grows out of the interactions of functional specialities with one another and with markets and competitors. The policies governing such interactions have not been adequately handled in management education. We need to deal with the way policies determine corporate stability and growth in an intellectual, challenging, quantitative, and effective way. Such management education leads to what I refer to as enterprise design. Such an education would build on three major innovations that have already occurred in this century.

The first innovation was the case-study method of management education as pioneered by the Harvard Business School beginning around 1910. Second was the development of theory and concepts related to dynamic behavior of feedback systems as first developed in engineering at the Bell Telephone Laboratories and MIT in the 1930s and 1940s. Third has been digital computers, especially the recent personal desk-top computers, that permit simulation modeling of systems that are too complex for mathematical analysis.

The first innovation, the case method of management education, has achieved a wide following because it addresses the problems of general management and the interactions among parts of the corporate-market-competitor system. The case method also draws great strength from being based on the full range of descriptive information and managerial knowledge that is available in the actual working world. But the case method has a major weakness. The description of a case captures policies and relationships that together describe a system so complex that it can not be reliably analyzed by discussion and intuition. Such attempts often draw the wrong dynamic conclusions and fail to reveal why corporations in apparently similar situations can behave so differently.

The second innovation, the understanding of the dynamics of feedback systems, has now emerged from engineering to be seen as an organizing concept for human systems as well. Feedback processes govern all growth, fluctuation, and decay. They are the fundamental basis for all change. They allow new insights into the nature of managerial and economic systems that have escaped past descriptive and statistical analysis.

The third innovation, the digital computer, for the first time allows efficient simulation of complex dynamic models. Such simulation is the only known way to determine behavior in complicated nonlinear systems. Practically all managerial and social systems are nonlinear, putting them beyond the reach of traditional mathematical solution. Trial and error experimentation, through computer simulation, provides the only route to improving our understanding of the behavior of such systems.

Bringing these three innovations together offers the potential for a major breakthrough in management education. The combination will permit going far beyond the case-study method of management education by adding a rigorous dynamic dimension to the rich policy and structural knowledge possessed by managers. The difference between present management schools and management education in the future will be as great as the difference between a trade school that trains airplane pilots and a university engineering department that trains aircraft designers. Pilots continue to be needed, and so will operating managers. But just as successful aircraft are possible only through skilled designers, so in the future will competition create the necessity for enterprise designers who can reduce the number of design mistakes in the structure and policies of corporations.

Correct design can make the difference between a corporation that is vulnerable to changes in the outside business environment and one that exhibits a high degree of independence from outside forces. Correct design can improve the stability of employment and production. Correct design, in the balance of policies for pricing, capital plant acquisition, and sales force, can often make the difference between growth burdened by debt and growth out of earnings. Correct design can help avoid the adoption of policies offering short-term advantage at the expense of long-term degradation. Correct design can help prevent expenditure of managerial time in debating policies that are inherently of low leverage and therefore unimportant. Correct design can help identify the very small number of high-leverage policies capable of yielding desirable change.

Future training in enterprise design will include study of a library of generic management situations combining descriptive case studies with dynamic computer models, each of which have wide applicability in business. I estimate that about 20 such general, transferrable, computerized cases would cover perhaps 90 percent of the situations that managers ordinarily encounter. Several powerful examples already exist. They include a model of stability and fluctuation in a distribution system (Forrester, 1961, Chapters 2, 15, and 16), a model of sales budget and capital investment as they often restrict growth (Forrester, 1968), a model of promotion chains and evolution into a top-heavy distribution of management personnel when growth slows, and a model dealing with imbalances between design, production, marketing, and service as these influence market share. Each such model manifests many different modes of behavior ranging from troublesome to successful depending on the policies employed within it.

In management and in government, there is a tendency to identify a weakness, then try to find ways to relieve the symptoms. But it would be more fundamental to insist on understanding why the objectives are not already being met. What is it in the design of a corporation or a government that is inhibiting success? A frontal assault on the symptoms, while the underlying causes remain in place, almost always fails. Success will follow when the designs of social systems give greater emphasis to removing the causes of problems rather than to trying to counteract the symptoms. I see the solution of many corporate and social problems coming in time from a new profession of system designers.

MODELING FOR WHAT PURPOSE?

System dynamics does not impose models on people for the first time. Models are already present in everything we do. One does not have a family or corporation or city or country in one's head. Instead, one has observations and assumptions about those systems. Such observations and assumptions constitute models. Such mental models are then used as a basis for action (Senge, 1990).[2]

The ultimate success of a system dynamics investigation depends on a clear initial identification of an important purpose and objective. Presumably a system dynamics model will organize, clarify, and unify knowledge. The model should give people a more effective understanding about an important system that has previously exhibited puzzling or controversial behavior. In general, influential system dynamics projects are those that change the way people think about a system. Mere confirmation that current beliefs and policies are correct may be satisfying but hardly necessary, unless there are differences of opinion to be resolved. Changing and unifying viewpoints means that the relevant mental models are being altered. But whose mental models are to be influenced? If a model is to to have impact, it must couple to the concerns of a target audience. Successful modeling should start by identifying the target audience for the model.

2 "*The Fifth Discipline*" by Senge discusses systems thinking and mental models. It provides a transition from non-systemic thought processes to the field of system dynamics. The necessary further step, after becoming aware of systems, leads into system dynamics and introduces computerized simulation models to provide the discipline needed to help the unaided thought processes from arriving at fallacious conclusions about dynamic behavior.

Unifying Knowledge

Complex systems defy intuitive solutions. Even a third-order, linear differential equation is unsolvable by inspection. Important situations in management, economics, medicine, and social behavior usually lose reality if simplified to less than fifth-order nonlinear dynamic systems. Often the model representation must be twentieth-order or higher. (The "order" of a system is the same as the number of integrations, or accumulations, or states that the system contains—such as inventories, number of employees, bank balances, product quality, reputation, level of pollution, trust, debt, and all other variables that accumulate as a result of rates of flow that increase or decrease the variables.)

Attempts to deal with nonlinear dynamic systems, using ordinary processes of description and debate, lead to internal inconsistencies. Underlying assumptions may have been left unclear and contradictory. Mental models are often logically incomplete. Assumed resulting behavior is likely to be contrary to that implied by the assumptions being made about underlying system structure and governing policies.

System dynamics modeling can be effective because it builds on the reliable part of our knowledge about systems while compensating for the unreliable part. The system dynamics procedure untangles several threads that cause confusion in ordinary debate. The modeling process separates consideration of underlying assumptions (structure, policies, and parameters) from the implied behavior. By considering assumptions independently from resulting behavior, there is less inclination for people to differ on assumptions, with which they actually can agree, merely because they initially disagree with the dynamic conclusions that might follow.

Figure 3 divides knowledge of systems into three categories to illustrate wherein lie the strengths and weaknesses of mental models and simulation models.

The top of the figure represents knowledge about structure and policies, that is, about the elementary parts of a system. This is local non-dynamic knowledge. It describes information available at each

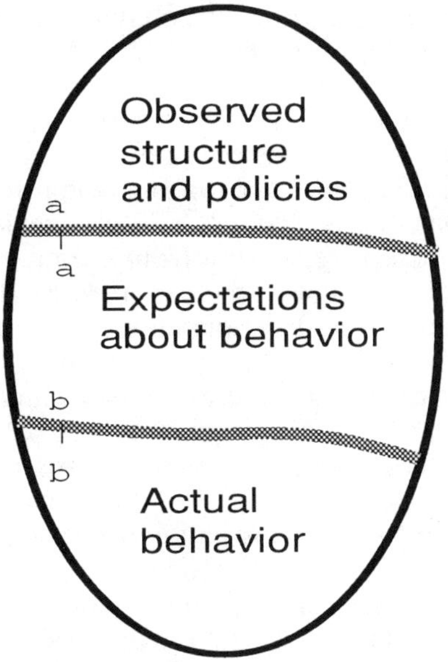

Figure 3. Three categories of information in the mental data base.

decision-making point. It identifies who controls each part of a system. It reveals how pressures and crises influence decisions. In general, information about structure and policies is far more reliable, and is more often seen in the same way by different people, than is generally assumed. It is only necessary to dig out the information, guided by knowing how structure is related to dynamics, that is, by using system dynamics insights about how to organize structural information to address a particular set of dynamic issues.

The middle of the figure represents assumptions about how the system will behave, based on the observed structure and policies in the top section. This middle body of beliefs are, in effect, the assumed intuitive solutions to the dynamic equations described by the structure and policies in the top section of the diagram. The center section of the diagram represents the solutions, arrived at by introspection and debate and compromise, to the high-order nonlinear system described in the top part of the figure. In the middle lie the presumptions that lead managers to change policies or lead governments to change laws.

Based on assumptions about how behavior is expected to change, policies and laws in the top section are altered in an effort to achieve assumed improved behavior in the middle section.

The bottom of the figure represents the actual system behavior as it is observed in real life. Very often, actual behavior differs substantially from expected behavior. Discrepancies exist across the boundary b-b. The surprise, arising from the fact that observed structure and policies do not lead to the expected behavior, is usually explained by assuming that information about structure and policies must have been incorrect. Unjustifiably blaming inadequate knowledge about parts of the system has resulted in devoting uncounted millions of man-hours to data gathering, questionnaires, and interviews that have failed to significantly improve the understanding of systems.

A system dynamics investigation usually shows that the important discrepancy is not across the boundary b-b, but across the boundary a-a. When a computer simulation model is built from the observed and agreed upon structure and policies, the model usually exhibits the actual behavior of the real system. The existing knowledge about the parts of the system is shown to explain the actual behavior. The dissidence in the diagram arises because the intuitively expected behavior in the middle section is inconsistent with the known structure and policies in the top section.

The discrepancies of Figure 3 can be found repeatedly in the corporate world. A frequently recurring example in which known corporate policies cause loss of market share and instability of employment arises from the way delivery delay affects sales and expansion of capacity. Rising backlog and the accompanying increase in delivery delay discourage incoming orders for product even while management favors larger backlogs as a safety buffer against business downturns. As management waits for still higher backlogs before expanding capacity, orders are driven down by unfavorable delivery until orders equal capacity. The awaited signal for expansion of capacity never comes because capacity is controlling sales rather than potential demand controlling capacity. When sales fail to rise because of long delivery delays, without management perceiving the true cause, management then lowers prices in an attempt to stimulate more sales. Sales increase briefly but only long enough to build up sufficient additional backlog and delivery delay to compensate for the lower prices. Price reductions lower profit margins until there is no longer economic justification for expansion. In such a situation, adequate information about individual relationships in the system is always available for successful modeling, but managers are not aware of how the different activities of the company are influencing one another. Lack of capacity may exist in manufacturing, product service, skilled salesmen, or even in prompt answering of telephones (Airlines cut fares to attract passengers. But, how often, because of inadequate telephone capacity, are potential customers put on "hold" until they hang up in favor of another airline?)

In a similar way at the national level, the System Dynamics National Model shows that puzzling and controversial economic behavior arises directly from known structure and managerial policies (Forrester, 1979). By building production sectors of the National Model using managerial policies derived from 20 years of corporate modeling, we find that most economic behavior arises from the private sector. Governmental taxation and monetary policies have less effect than usually assumed and lack the expected leverage for controlling economic behavior. The Great Depression of the 1930s has been blamed both on restrictive monetary policy and on protective tariffs, but we find that depressions arise at 45-to-60-year intervals as a result of the economic long wave, or Kondratieff cycle, which is driven primarily by major shifts in private-sector incentives for investing in capital plant, borrowing, and saving (Forrester, 1977; Sterman, 1986).

Debate about the economic long wave illustrates the situation depicted in Figure 3 (Freeman, 1983; Kondratieff, 1984; van Duijn, 1983). There is little acceptance by economists of the idea that structures could exist capable of producing a major economic fluctuation with some 50 years between peaks. Yet much of the theory for such a long economic wave already is established in the mainstream of economic thought.

In teaching macroeconomics, the classic multiplier-accelerator process is often used to explain short-term business cycles having 3 to 10 years between peaks. The multiplier (rising consumer income causing increased demand) and the accelerator (rising demand causing increased capital investment, wages, and consumer income) represent widely accepted and fundamentally correct assumptions about structure and policies belonging in the top category in Figure 3. However, the belief that the multiplier and accelerator interact to cause short-term business cycles arises from an assumed dynamic solution to the equations describing the structure. The assumed dynamic solution belongs to the middle category in Figure 3 where beliefs are often incorrect.

While investigating cyclic economic behavior, several system dynamics investigators have shown that the multiplier and accelerator are not significant in creating short-term business cycles but are powerful contributors to generating much longer cycles having several decades between peaks (Forrester, 1982, Nathan B.; Low, 1980; Mass, 1975).

Even though the economic long wave has been broadly rejected in economics, the accepted multiplier-accelerator relationships go far in explaining long-wave behavior. Here we see a common situation. Both sides in a debate can usually agree on underlying assumptions. But there is disagreement about the dynamic consequences. Building those accepted assumptions into a dynamic model begins to resolve differences arising from incorrect intuitive solutions to complex systems.

Enhancing Mental Models

Because of errors of dynamic interpretation in mental models, policy changes have often led to ineffective results, or worse, to the opposite of the intended results. A policy giving opposite of the intended result was identified in *Urban Dynamics* (Forrester, 1969), a book that was more fully developed by others (Alfeld and Graham, 1976; Mass, 1974; Schroeder, et al., 1975). Economic distress in declining American cities in the 1960s generated symptoms of high unemployment and deteriorating housing. It appeared natural enough to combat such symptoms by government intervention to build low-cost housing. But the modeling study showed, as events have since confirmed, that such urban areas already have more low-cost housing than the economy of the city can sustain. Public policy to build more such housing merely occupies land that could instead have been used for job-creating businesses, while at the same time the housing attracts people who need jobs. A low-cost housing program introduces a powerful double force for increasing unemployment, both by reducing employment while at the same time attracting people seeking work. Low-cost housing programs in inner cities become a social trap. The policy of building low-cost housing actually creates poor and unemployed people, rather than alleviating personal hardship. The lesson here is to avoid attacking symptoms of difficulty until the causes of those symptoms have been identified, and a high-leverage policy has been found that will cause the system itself to correct the problem.

Very often, a comprehensive dynamic analysis will show that strongly supported policies are actually causing the problems that the policies were intended to correct. Questioning such policies can elicit intense defensive responses. *Urban Dynamics* was the first of my modeling work that produced strong, emotional reactions. The book suggested that all of the major urban policies that the United States was following lay somewhere between neutral and highly detrimental, from the viewpoint either of the city as an institution, or from the viewpoint of the low-income, unemployed residents. Furthermore, the most damaging policy was to build low-cost housing, which at that time was believed to be essential to reviving the inner cities. The conclusions of our work were not easily accepted. I recall one professor of social science at MIT coming to me and saying, "I don't care whether you're right or wrong, the results are unacceptable." So much for academic objectivity! Others, probably feeling the same way, put it more acceptably as, "It doesn't make any difference whether you're right or wrong, urban officials and the residents of the inner city will never accept those ideas." (Forrester, 1989).[3]

As it turned out, officials and inner city residents were the two groups we could count on for support if they became sufficiently involved to understand the *Urban Dynamics* message. That is a very big "if"—if they came close enough to understand. Three to five hours were required to come to an understanding of what urban dynamics was about. Urban officials and members of the black community in the inner city would become more and more negative and more and more emotional during those three to five hours. If they were not a captive audience, they would walk out before they understood and accepted the way in which low-cost housing was a double-edged sword for making urban conditions worse. Such housing used up space where jobs could be created, while drawing in people who needed jobs. Constructing low-cost housing was a process for creating poverty.

---

3 Several paragraphs from that paper are incorporated here.

My first experience with reactions to *Urban Dynamics* came soon after the book was published. For a management program for urban executives, I was asked to take a Monday afternoon and a Wednesday morning to present results of our study. I have never had a lecture go as badly as that Monday afternoon. In the group was a man from the black Harlem community in New York and a member of the city government. He was intelligent, articulate, not buying a thing I was saying, and carrying the group with him. At one point he said, "This is just another way to trample on the rights of the poor people and it's immoral." At another point he said, "You're not dealing with the black versus white problem, and if you're not dealing with the black versus white problem, you're not dealing with the urban problem." And when I said decay and poverty in Harlem in New York or Roxbury in Boston was made worse by too much low-cost housing, not too little, he looked at me and said, "I come from Harlem and there's certainly not too much housing in Harlem." That is a sample of the afternoon. On Tuesday evening, a dinner was held for the group. I could not go; but several of our students attended. After dinner one student called me at home to report what was already obvious—that the group was very hostile. On that bit of encouragement, I started Wednesday morning.

An hour into Wednesday morning, the New Yorker's comments began to change character. He was no longer tearing down what was being said. His questions began to elicit information. Two hours into the morning, he said, "We can't leave the subject here at the end of this morning. We must have another session." I agreed to meet them again if he could find a time and place in the program. He went to the administration and scheduled another session. Later he came to my office to ask that I talk to a group he would invite in New York—his colleagues on his home turf. Relaxed in my office he said, "You know, it's not a race problem in New York at all, it's an economic problem," after telling me four days earlier that I was not even addressing the urban problem if I was not dealing with the black versus white issue. He gave me a report out of his brief case documenting the amount of empty housing in every borough of New York, including Harlem, and the rate at which it was being abandoned. My point had been that too much housing meant that there was too much for the economy of the area to support. He had all the proof in his brief case. He simply had not realized what his knowledge meant until it was all put together in a new way.

Two years later a journalist asked me what people thought in the aftermath of *Urban Dynamics*. I suggested that he talk to others, and especially with the man in New York with whom I had not been in recent contacted. After the interview, the journalist called me to report that he had been told "they don't just have a solution to the urban problem up there at MIT, they have the only solution." The lesson about urban behavior had stayed clear and alive for two years even back in his home environment. The five hours of exposure to *Urban Dynamics* had made a lasting change in his view of urban evolution. The same kind of conversion of viewpoint has been seen on several other occasions. But we have not solved the challenge of how to bring enough people across the barrier separating their usual, simple, static viewpoint from a more comprehensive understanding of dynamic complexity.

System dynamics models have little impact unless they change the way people perceive a situation. A model must help to organize information in a more understandable way. A model should link the past to the present by showing how present conditions arose, and extend the present into persuasive alternative futures under a variety of scenarios determined by policy alternatives. In other words, a system dynamics model, if it is to be effective, must communicate with and modify the prior mental models. Only people's beliefs, that is, their mental models, will determine action. Computer models must relate to and improve mental models if the computer models are to fill an effective role.

Small Models versus Large Models

What kind of system dynamics model interacts best with mental models? Clearly, a small model has advantages over a large model. A recent trend in system dynamics has been toward small models to be used for enhancing insight. Often, such models have been built directly from mental models. The process is one of discussing with a small group their concerns, assumptions, and expectations. While the conversation is in progress, a system dynamics model can be created on a desktop computer. Recent software advances, especially the user-friendly STELLA and ITHINK (High Performance Systems, 1990; Richmond, 1985, pages 706-718), facilitate the interaction between mental models and computer models (Vescuso, 1985, pages 964-974). Simple models used as interactive games, such as one demonstrating the economic long wave, or Kondratieff cycle (Sterman and Meadows, 1985), can also create a dramatic impact as they reveal unexpected implications of existing mental models.

If small models align best with mental models, and thereby have the greatest effect, what is the role for large models? The answer must depend on the circumstances. First, the size of a model that can interact with mental models depends on the amount of time and effort that will be devoted to making connections between the mental and the computer simulation models. If the available time is a half day, clearly the computer model can have no more than a few variables. On the other hand, if the computer model is for research purposes and months or even years are available to explore its implications, then the model can be of far wider scope. Even with more time available, there must be a clear justification for a large model. Special system dynamics software is also available for professional work with larger models (Eberlein, 1991; Pugh, 1986).

The System Dynamics National Model serves to put large and small models into perspective. The National Model is large, with more than two thousand equations. However, it is much smaller than was originally projected. As we have come to understand the Model better, and to relate its behavior to actual economic behavior, it has become apparent that the originally envisioned far larger model was not necessary. The proper balance between size and clarity suggested simplification. Many planned production sectors have now been aggregated into just two—capital plant and equipment, and consumer goods. Within sectors, there has been simplification especially in labor mobility and banking.

Research with the National Model focuses on four distinct modes of economic behavior—growth, money inflation, business cycles of 3 to 7 years between peaks, and the economic long wave with peaks 45 to 60 years apart. The intermediate Kuznets cycle of some 25 years between peaks also appears but is of minor importance. Simple models have been created for demonstrating most of these modes separately (Sterman, 1985). Such simple models are possible because the separate modes arise from different structures within an economy. Simple models are far easier to understand than the full National Model and for many purposes are more effective. However, simple models alone do not answer certain important questions.

There are many interactions between the four basic modes of economic behavior (short-term business cycles, economic long waves, money inflation, and growth) that do not reveal themselves in separate simple models of individual modes. An example is the way in which the economic long wave, having some 45 to 60 years between peaks, modulates the amplitude of short-term 3-to-10-year business cycles. During a long-wave expansion, as in the 1950s and 1960s, excess demand and limited output, caused by shortages of both capital and labor, suppress business cycles. Near and after the long-wave peak, as in the 1970s and 1980s, the amplitude of business cycles becomes larger because the oversupply of capital and labor allows business-cycle expansions to be more aggressive, to overbuild inventories, and then to induce sharper cutbacks to rebalance supply and demand. This interpretation of economic behavior during the last several decades is different from that which has been conventionally accepted. After World War II, mild business cycles were attributed to Keynesian economics and fine tuning of monetary policy, but such beliefs were shattered in the 1970s when business cycles again became more severe. Our work shows the shifting nature of business-cycle behavior as arising from interactions among major dynamic modes in the private economy, rather than from governmental policies.

THE SYSTEM DYNAMICS PARADIGM

System dynamics adheres to viewpoints and practices that set it apart from other fields dealing with the behavior of systems (Randers, 1980). But even so, the unique character of system dynamics has never been adequately set forth. Each aspect of system dynamics is accepted by some other professional group at least to a degree. System dynamics is distinguished not only by the particular cluster of beliefs that guide the work but also by the degree to which those characteristics are indeed practiced.

System dynamics as a way of looking at the world. It is a perspective for interpreting and understanding people, social systems, and natural processes. It is a body of insights about simple dynamic structures that reappear time and again as a person moves from one situation to another. It is a way to bring rigorous analysis and synthesis to otherwise poorly defined and unstructured debate.

However, there are no descriptions of system dynamics and its foundations that are fully meaningful to those outside the field. We have not found a way to quickly convey the significance of dynamic behavior to those immersed in static thinking. With regard to foundations and theoretical structure, only partial and rather disorganized concepts exist. Some pioneering work is underway to clarify and justify the concepts that support the field. Fragmentary glimpses of the underlying rationale for system dynamics lie in scattered locations. But there is as yet no organized body of material that sets forth, explains, and defends the concepts supporting the field.

There is more to system dynamics than just building models. Many system dynamics models lack important ties to the real world being represented. Often the models lack a clear coupling to the mental models already being used. System dynamics models will not be effective unless they relate to and ultimately change the mental models that are already controlling actual social systems. The modeling process is not complete until it relates smoothly to mental models of practicing managers and political leaders so that they take a model and its behavior and policy implications into their own subjective thinking.

Much remains to be done in how to educate people for a dynamic understanding of systems. How should education be tailored to different audiences? System dynamics as a foundation for junior and senior high school education is showing dramatic promise in some school experiments. What is the best age at which to start learning about systems? Some still think instruction in dynamics is best deferred until graduate school, while others argue for starting in the fourth grade at age eight or nine. What is the appropriate systems education for nonprofessionals, that is, what is the proper role in a liberal arts education? For all audiences, how are lectures, book study, and laboratory practice best combined? What kinds of people are most receptive to understanding dynamic behavior? Results from high schools suggest that there is no correlation between students previously classified as fast or slow learners and how well they do in system dynamics thinking and modeling. I see the educational process in system dynamics as frontier territory in which the big issues and opportunities are as yet little understood.

Endogenous Behavior

The best system dynamics practice puts rather extreme demands on a model for generating within itself the behavior modes of interest. That is, the model boundary is to be established so that the causal mechanisms lie inside the boundary. This expectation of finding endogenous causes of behavior is in sharp contrast to the view often found elsewhere (Richardson, 1991).[4] People are far more comfortable blaming their troubles on uncontrollable external causes rather than looking to their own policies as the central cause. Business managers attribute product and corporate failures to competitors, bankers, and government rather than to their own handling of resource allocations, pricing, and interpretation of customer needs. Governments blame balance of trade difficulties on other countries rather than recognizing the cause in domestic deficits, tax policies, and monetary actions.

In contrast to the endogenous viewpoint, economists often imply that the economic system is almost in equilibrium almost all the time with important behavior arising only from unexpected exogenous forces. The exogenous viewpoint common in economics leads to seeing the monetary authority as a free-will arm of government policy for unilaterally controlling economic behavior, whereas, in the National Model, we represent the monetary authority as an integral, interacting part of the economic system and as being responsive to forces such as unemployment, liquidity, and interest rates. Economists have explained business cycles in terms of exogenous actions of government, whereas, we find that business cycles arise out of internal oscillatory tendencies in production, employment, and inventories excited by those continuous streams of small random variations existing in all decision processes.

The system dynamics emphasis on endogenous behavior is more like that of an engineer in designing an oil refinery. The engineer looks at the individual working characteristics of the chemical reactors, evaporators, and distillation towers; considers how they are interconnected and controlled; and evaluates the dynamic behavior im-

4 Richardson discusses how system dynamics relates to other traditions of thinking about closed-loop behavior in the social sciences—cybernetics, servomechanisms, TOTE units (Test-Operate-Test-Exit), etc.

plied by their feedback loops. The engineer does not attempt to improve a refinery by using only information about the feed stocks that go in and the products that come out. He does not assume that the refinery exists in a state of equilibrium that is affected only by exogenous events that impact the plant from outside its surrounding fence.

System dynamics models build from the inside to determine and to modify the processes that cause desirable and undesirable behavior.

Sources of Information

Effectiveness of a model depends on how it uses the wide range of information arising from the system being represented (Forrester, 1980). In creating a system dynamics model, information is used in a substantially different way from that in other branches of the social sciences. The differences arise from the system dynamics focus on policy statements as the basic building blocks of a model and from a broader range of information sources used for creating a model.

Information is available from many sources. Figure 4 suggests three classifications of information—the mental data base, the written data base, and the numerical data base. Although "data" is a term that is often used to mean only numerical information, the dictionary meaning is far broader. Data is "something that is given from being experientially encountered" and "material serving as a basis for discussion, inference, or determination of policy" and "detailed information of any kind" (Webster's Third, Unabridged, 1961).

Human affairs are conducted primarily from the mental data base. Anyone who doubts the dominance of remembered information should imagine what would happen to an industrial society if it were deprived of all knowledge in people's heads and if action could be guided only by written policies and numerical information. There is no written description adequate for building an automobile, or managing a family, or governing a country. People absorb operating information from apprenticeship and experience. The dominant significance of information from the mental data base is not adequately appreciated in the social sciences.

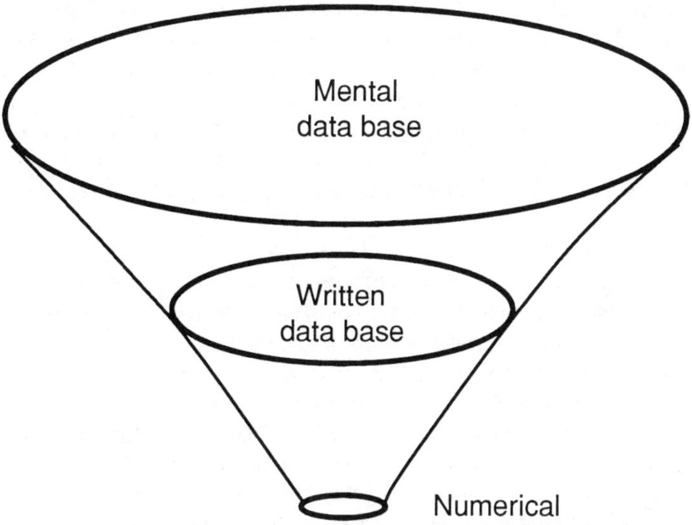

Figure 4. Decreasing information content in moving from mental to written to numerical data bases.

The mental data base contains vastly more information than the written data base, which, in turn, contains far more information than the numerical data base. Furthermore, the character of the information differs in the three categories. As one moves down the diagram, each category of information contains a smaller fraction devoted to structure and to description of policies. That is, the written and numerical data bases contain not only less information, but progressively smaller proportions of the information needed for constructing a dynamic model.

If the mental data base is so important to the conduct of human systems, then a model of such a system should include relevant knowledge from all available sources, including that which resides only in the mental data base. The mental data base is rich in structural detail; in it is knowledge of what information is available at various decision-making points, where people and goods move, and what decisions are made. The mental data base is especially concerned with policy, that is, why people respond as they do, what each decision-making center is trying to accomplish, what are the perceived penalties and rewards, and where self-interest clashes with institutional objectives.

In general, the mental data base relating to policy and structure is reliable. Of course, it must be cross-checked with all other available information. Exaggerations and oversimplifications exist and must be corrected. Interviewees must be pressed beyond quick first responses. Interrogation must be guided by a system dynamics knowledge of what different structures imply for behavior. But from the mental data base, a consensus usually emerges that is useful and sufficiently correct.

The written data base contributes to a dynamic model at several stages. Published material makes information more widely available than if it is only exchanged between mental data bases. In terms of usefulness for modeling of business and economic systems, the daily and weekly public and business press is frequently more useful than the professional press or historical accounts that adopt a longer time horizon. The current press reports the pressures of the moment that surround decisions. The temporal nature of a decision sharply restricts the kind of literature in which operating policy will be revealed. Policies govern decisions and decisions control action. Decisions are fleeting. There is only a single instant in time when one can act. That time is now. Action must take place in the present moment that separates history from the future.

The ever-advancing present moment is the business person's and politician's world of action. It is the world of placing orders, hiring people, buying equipment, borrowing money, bargaining with unions, and extending credit. As a consequence of the short life of a decision, it is primarily in the literature of the present that decisions are discussed in terms of goals, threats, limited information, and restraints on action. The multifaceted conflicting pressures of real decision making are almost absent from economics textbooks and professional journals. The professional literature emphasizes how decisions should be made rather than how they actually are made, how equilibrium is determined rather than how dynamic behavior arises, and how macroeconomic theory might apply rather than how the microstructure creates the macrobehavior.

The numerical data base is of narrower scope than either the written or mental data bases. Missing from numerical data is direct evidence of the structure and policies that created the data. The numerical data do not reveal the cause and effect directions among variables. In complex nonlinear feedback systems, statistical analysis of historical data should be used cautiously (Graham, 1980; Senge, 1978). Even so, numerical data can contribute to system dynamics model building in three ways. First, numerical information is available on some parameter values. For example, average delivery delays for filling orders, typical ratios of factory inventories to production, normal bank balances, and usual inventory coverages can be determined from business records. Second, numerical data have been collected by many authors in the professional literature summarizing characteristics of economic behavior such as average periodicity of business cycles and phase relationships between variables. Third, the numerical data base contains time-series information that in system dynamics is often best used for comparison with model output rather than for determining model parameters.

With regard to the use of data, system dynamics operates more like the engineering and medical professions, and less like practices in economics. All information is admissible to the process of model building. Information from the mental data base is recognized as a rich source of knowledge about structure and the policies governing decisions. Parameter values are drawn from all available sources, not merely from statistical analysis of time series. The mental and written data bases are the only sources of information about limiting conditions that have not occurred in practice, but which are important in determining the nonlinear relationships that govern even normal behavior.

LEARNING FROM MODELS

Model building should be a circular process of creating a model structure, testing behavior of the model, comparing that behavior with knowledge about the real world being represented, and reconsidering structure (Forrester, 1975). During the process of modeling, the system dynamicist should always be alert to new discoveries about behavior. The new discoveries may relate either to the particular system being studied or to the general nature of systems.

Surprise Discoveries

Only if there is a standard against which the model is being compared—existing knowledge of the real system—can one be prepared for surprises from the model. Surprising behavior means behavior that was not expected in terms of what was known about behavior of the actual system. Surprising behavior will usually point to model defects. But the modeler must be always alert to the possibility that unexpected behavior of a model is revealing new insights about the real system.

Our work on the economic long wave, or Kondratieff cycle, in the System Dynamics National Model Project arose as a surprise discovery. When sectors for consumer goods and capital equipment were first connected, a large fluctuation arose in the demand for capital equipment with peaks some 50 years apart.

In the National Model, after study, the 50-year rise and fall of economic activity seemed plausible. Turning to historical economic behavior, we found extensive and diverse evidence of behavior like that which the model was generating. As other sectors of the model were added, additional model variables became involved in the long-wave process and repeatedly the expanded model behavior was found to have a real-life counterpart. For example, we found that real interest rate (nominal interest minus inflation) in the model is low or negative before the long-wave peak, just as it was in the 1970s, and that real interest rate in the model moves quickly positive after the peak, as it did in the early 1930s and as it did again in the 1980s. Such behavior of real interest arises mostly from the private sector rather than from government policy as is commonly assumed.

The economic long wave is controversial, both as to its cause and even to its existence. Those who believe in its existence see it as a great rise and fall of economic activity with peaks and valleys some 45 to 60 years apart. It is considered the cause of the great depressions of the 1830s, 1890s, and 1930s. The nature of the economic long wave remained unclear because there has in the past been no theory for how it could be generated. Instead, depressions have often been attributed to accidents or mismanagement. For example, the Great Depression of the 1930s has been blamed on mistaken policies of the Federal Reserve. Most economic theory has been based on equilibrium, or steady-state, conditions that leave little room for even imagining an economy that can persist in major long-term fluctuations between booms and depressions.

But now, a theory exists for the causal mechanisms of the long wave. A computer simulation model is a theory of the behavior that it creates. The structure of the model and the decision-making policies within it cause the resulting behavior. When a model containing structure and policies representing the actual system generates behavior like that observed in real life, the model can be used to understand how production, investment, savings, construction, and credit can interact to produce, over a period of decades, great waves of excessive economic expansion and contraction.

Government and private-sector policies from 1960 to 1990 have led to severe economic imbalances that will almost certainly be corrected with considerable trauma in a major long-wave downturn during the 1990s. After examining a broad range of evidence from the last three decades, from the 1920s and 1930s, and to a lesser extent from the 1890s, one can conclude that the peak of the recent long-wave expansion was in about 1980 (by the same criteria that put the previous peak at 1920). The low point in the long-term downturn probably will occur in the mid-1990s. This means a decade or more of economic crosscurrents and readjustments as the great imbalances in the world's economies are corrected.

General Characteristics of Systems

Even more important than finding unexpected behavior of a specific system is the discovery of general characteristics that are applicable to a broad class of systems, or even to nearly all systems. In complex nonlinear systems, such generalizing must be interpreted with caution, but, even so, rules of thumb can be identified that are usually valid and give a useful basis for thinking about systems.

In such generalizing, one should make ties to history, myths, fables, and lessons from the great religions. The lessons that come to us from such traditional sources contain powerful threads of truth that are being ignored in modern attitudes dominated by short-run considerations. Several general characteristics of systems were identified in *Urban Dynamics* (Forrester, 1969, pages 107-114). Two examples will illustrate.

First, a characteristic like the long-term versus short-term trade off applies to most decisions. But the inherent conflict between immediate and ultimate consequences is not given its proper weight in management and political decisions. On the other hand, the recognition of the trade off goes back at least as far as the ancient Greeks. Aesop's fable of the grasshopper and the ant contrasts the short-term advantage of playing in the summer with the long-term penalty of freezing in the winter. In building a public understanding of systems, we should seek general insights and make connections to where the same themes have already appeared.

Second, another inadequately appreciated general characteristic of systems lies in high resistance to policy changes. Perhaps as many as 98 percent of the policies in a system have little effect on behavior because of the ability of the system to compensate for changes in the majority of policies. Most managerial and political debate focuses on such low-leverage policies that matter very little. Low-leverage policies are defeated by feedback loops that warp just enough to compensate for the effect of a policy that in linear thinking appears promising. One author criticized the *Urban Dynamics* book on the basis that it contained a very bad model because the critic had been unable to find any policy in the model that substantially changed the behavior of the model. But, have not mayors of cities also discovered most of their policies to have been without effect? Such insensitivity is not a defect in the model, it is the nature of the cities being represented by the model. Governments of American cities have expended billions of dollars over several decades without substantially altering the social problems with which they started. The waste and frustration are a consequence of attempting corrective action through policies having inherently low influence.

In a similar way, national governments have debated monetary policy, have tried all variations and theories, and are still left with worsening economic problems. In our work with the National Model, we find that monetary policy often has low leverage over economic conditions.

SYSTEM DYNAMICS AND PUBLIC RESPONSES

System dynamics models have the potential for raising the quality of managerial and political debate. The books, *World Dynamics* (Forrester, 1971), *The Limits to Growth* (Meadows, et al., 1972, Donella), *Toward Global Equilibrium* (Meadows and Meadows, 1973, Dennis), *Dynamics of Growth in a Finite World* (Meadows, et al., 1974, Dennis), and the recent *Beyond the Limits* (Meadows, et al.,

1992) led to intense world-wide debate even though their subject had been treated in many preceding publications on population growth and environmental degradation. Why? I believe there are two reasons.

The first reason for intense public response to the books arose because of the way the books illuminated long-run issues. It is commonplace to assert that people take only a short-run view of life, but that is only partially true. In fact, most people live in a world of split personalities in which business and political actions are dominated by short-run objectives while at the same time personal goals remain long-term. Individuals hope for the future well-being of their children and grandchildren even while responding unknowingly to short-run pressures in ways that jeopardize that future. The four books offered a way to understand the past and present that could assist in seeing into the future. Good system dynamics modeling contributes to relating the legacy of the past to decisions of the moment, and actions of the present to their implications for years to come.

In *World Dynamics* and the successor books, readers found an embodiment of their concerns. The models contained assumptions that had everyday meaning, and behavior that corresponded to what people were seeing in the world around them. In the intervening twenty years since the books appeared, newspaper headlines have been revealing the reality of the limits to growth theme. Every week one sees accounts of polluted wells, acid rain damage to forests, falling water tables, atomic-waste disposal uncertainties, hunger in many parts of the world, and social pressures from crowding such as terrorism and illegal immigration.

The second reason for intense public involvement arose because a presentation based on a system dynamics model can have an internal consistency that is beyond the reach of the usual discussion processes. Such consistency commands attention and yields persuasiveness. By contrast, the usual writing and debate about a complex social system contains internal contradictions. Those contradictions usually occur in going from the structural assumptions to the implied dynamic consequences. In the step from assumptions to behavior, a writer tries to solve intuitively the high-order nonlinear equations of the system; such is rarely done correctly. But a model simulation provides certainty in going from the assumptions about structure and policies to the resulting behavior.

A presentation based on a model can have complete internal consistency. One knows the assumptions in the model. Simulation gives the behavior implied by those assumptions. Policy changes can be made and the resulting changes in behavior can be determined beyond doubt within the context of the model. Within the modeling process, there need be no contradictions.

But internal consistency is not enough. An argument can be internally consistent and still erroneous in comparison with the real world. But, the persuasiveness of the system dynamics process reaches its full power when the listener or reader finds agreement wherever his independent knowledge matches the presentation of assumptions, behavior, or policy implications.

A NEW BASIS FOR PRE-COLLEGE EDUCATION

The greatest impact of system dynamics on public understanding can be expected from pioneering projects now starting for introducing an understanding of systems into high school and undergraduate studies (Forrester, 1976; Forrester, 1991; Roberts, 1978, Nancy; Roberts, et al., 1983, Nancy; Roberts, et al., 1987, Nancy). Traditional educational methods have tended to discourage synthesis and use of the knowledge that a student has already acquired. Too much emphasis has been put on the written data base and not enough on the mental data base. Education has taught static facts rather than dynamics of natural and social change.

System dynamics offers a basis for a new kind of education that leads to a better understanding of change in social and environmental conditions. But the dynamic viewpoint takes time to absorb. Several years are needed to organize a person's thinking to a dynamic frame of reference. By starting in the first year of junior high school and weaving a dynamic thread through high school and college, we can hope for a society that is better able to cope with growing social complexity.

Only when dynamic considerations are introduced throughout the educational process, will students have time to develop improved mental models to guide personal and public action. Just as understanding of the natural world rests on science studies woven into all educational levels, so will a comparable understanding of dynamic systems in society and nature need to be made a part of the entire educational sequence.

Education in the United States is generally recognized as serving less and less well in meeting modern needs. Failures in education appear in the form of corporate executives who cannot cope with the complexities of growth and competition, government leaders who are at a loss to understand economic and political change, and a public that supports inappropriate responses to immigration pressures, changing international conditions, rising unemployment, the drug culture, governmental reform, and inadequacies in education.

The weakness in education arises not so much from poor teachers as from the inappropriateness of the material that is being taught. Students are stuffed with facts but without having a frame of reference for making those facts relevant to the complexities of life. Responses to educational deficiencies are apt to result in demands for still more of what is already not working—for more science, humanities, and social studies in an already overcrowded curriculum—rather than moving toward a common foundation that pulls all fields of study into a unity that reinforces all subjects and is far easier to teach and to understand.

Education is fragmented. Social studies, physical science, biology, and other subjects are taught as if they were inherently different from one another even though dynamic behavior in each rests on the same underlying concepts. For example as in Figure 5, the dynamic structure that causes a pendulum to swing is identically the same as the core structure that causes employment and inventories to fluctuate in a product-distribution system and in economic business cycles. Humanities are taught without relating the dynamic sweep of history to similar behaviors on a shorter time scale that the student can experience in a week or a year.

Missing from most education is a direct treatment of the time dimension. What causes change from the past to the present and the present to the future? How do present decision-making policies determine the future toward which we are moving? How are the lessons of history to be interpreted to the present? Why are so many corporate, national and personal decisions ineffective in achieving their intended objectives?

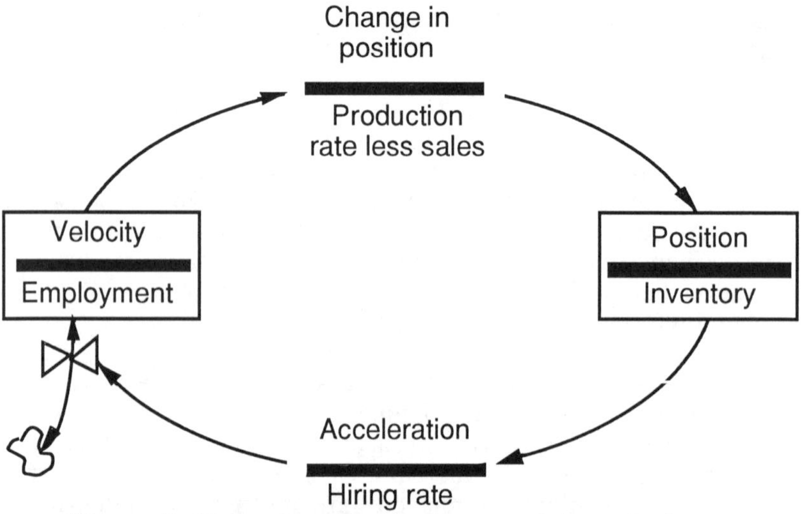

Figure 5. Upper legends are for a pendulum, the lower for the core structure relating inventory and employment. In this simple two-level feedback loop, only sustained oscillation (or equilibrium) can exist. Parameters in the loop affect only the periodicity, not the growth or damping of oscillation. Additional cross loops must be inserted to cause an oscillation to die out. Parameter values for a pendulum yield an oscillation measured in seconds. Reasonable parameters for production and employment yield an oscillation measured in months or years.

Two mutually reinforcing developments now promise a learning process that can enhance breadth, depth, and insight in education. These two threads are system dynamics and learner-directed learning.

System dynamics can provide a dynamic framework to give meaning to detailed facts. Such a dynamic framework provides a common foundation beneath mathematics, physical science, social studies, biology, history, and even literature.

From early authors, several introductory books on system dynamics became available (Forrester, 1961; Goodman, 1974; Richardson and Pugh, 1981; Roberts, et al., 1983, Nancy). None were ideally organized for classroom use. However teaching materials are now under development in many places in the United States and abroad.

In his penetrating discussion of the learning process, Jerome Bruner states, "the most basic thing that can be said about human memory... is that unless detail is placed into a structured pattern, it is rapidly forgotten" (Bruner, 1963, page 24). For most purposes, such a structure is inadequate if it is only a static framework. The structure should show the dynamic significance of the detail—how the details are connected, how they influence one another, and how past behavior and future outcomes are influenced by decision-making policies and their interconnections. System dynamics can provide such a dynamic framework.

"Learner-directed learning," refers to a way of organizing a school so that students work together in teams of two or three to cooperate in meaningful projects for which they must do research and creative thinking. Learner directed learning shifts the role of a teacher from being a dispenser of knowledge to being a guide and resource person. Students are no longer merely passive receptors of what the teacher says. Instead the students work together to help one another and to explore issues that are new to both them and the teacher.

There are now several dozen high schools and junior high schools making substantial progress in combining system dynamics and learner-directed learning. In several hundred schools some activity is under way.

The most advanced experiment in the United States in bringing system dynamics and learner-directed learning together into a more powerful educational environment appears to be in the Catalina Foothills School District of Tucson, Arizona. In that community the necessary building blocks for successful educational innovation are coming together. The process combines a fundamental new concept of education, a receptive community, talented teachers who are willing to try unfamiliar ideas, teachers who are at ease in the nonauthoritarian environment of learner-directed learning, a supportive school administration, and a "citizen champion" operating outside the school system. Without a personal vested interest except a desire to facilitate improved education, the citizen champion inspires teachers, finds funding, arranges for computers, and facilitates convergence of the political processes in the community.

To quote an eighth-grade biology teacher who has been a key player in this dramatic experiment in moving toward a fundamentally new kind of education:

> "We now see students come early to class (even early to school), stay after the bell rings, work through lunch and work at home voluntarily (with no assignment given). When we work on a systems project—even when the students are working on the book research leading up to system work—there are essentially no motivation or discipline problems in our classrooms." (Draper, 1989)

The results were so persuasive in this junior high school that in 1990 the district voted a $30 million bond issue to build a high school to carry on in the educational pattern that has been established in the junior high school.

It is time to explore a new frontier. We have been through the frontier of science and technology. The next frontier is to achieve a broadly based understanding of social systems that can provide a foundation for effectively dealing with economic and social stresses through the development of improved policies.

REFERENCES

Alfeld, Louis Edward, and Alan K. Graham, 1976. *Introduction to Urban Dynamics*, Cambridge, MA: Productivity Press. 333 pp.

Bruner, Jerome S., 1963. *The Process of Education*, New York: Vintage Books.

Draper, Frank, 1989. *Letter to Jay Forrester*. Personal communication, Orange Grove Junior High School, 1911 E. Orange Grove Rd., Tucson, AZ 85718. May 2.

Eberlein, Robert, 1991. *Vensim Reference Manual*. IBM PC, DEC VMS, and UNIX workstations. 72 Grafton St., Arlington, MA 02174: Ventana Systems, Inc.

Forrester, Jay W., 1958. "Industrial Dynamics--A Major Breakthrough for Decision Makers." *Harvard Business Review*, 36(4), 37-66. Also appears in revised form as Chapter 2 in the author's *Industrial Dynamics* 1961 and Chapter 1, pp. 1-29, of *Collected Papers* 1975; also as Chapter 2, pp. 37-65 in Edward B. Roberts (ed.), *Managerial Applications of System Dynamics*, 1978, all from Cambridge, MA: Productivity Press.

Forrester, Jay W., 1959. "Advertising: A Problem in Industrial Dynamics." *Harvard Business Review*, Vol. 37, No. 2, pp. 100-110. Also appears in revised form as Chapter 16 of the author's *Industrial Dynamics* 1961; and as Chapter 2, pp. 31-44, in *Collected Papers,* 1975; and as Chapter 11, pp. 191-204, in Edward B. Roberts, ed., *Managerial Applications of System Dynamics*, 1978, all from Cambridge, MA: Productivity Press.

Forrester, Jay W., 1961. *Industrial Dynamics*, Cambridge, MA: Productivity Press. 464 pp.

Forrester, Jay W., 1964. "Common Foundations Underlying Engineering and Management." *IEEE Spectrum*, 1(9), 66-77. Also appears as Chapter 4, pages 61-80, in the author's *Collected Papers,* 1975, Cambridge MA: Productivity Press.

Forrester, Jay W., 1968. "Market Growth as Influenced by Capital Investment." *Industrial Management Review (MIT)*, Vol. 9, No. 2, pp. 83-105. Also appears as Chapter 7, pages 111-132, in the author's *Collected Papers* 1975; and as Chapter 12, pp. 205-226, in Edward B. Roberts, ed., *Managerial Applications of System Dynamics*, 1978, both from Cambridge, MA: Productivity Press.

Forrester, Jay W., 1969. *Urban Dynamics*, Cambridge, MA: Productivity Press. 285 pp.

Forrester, Jay W., 1971. *World Dynamics*, (1973 second ed.). Cambridge, MA: Productivity Press. 144 pp. Second edition has an added chapter on physical vs. social limits.

Forrester, Jay W., 1975. *Collected Papers of Jay W. Forrester*, Cambridge, MA: Productivity Press. 284 pp.

Forrester, Jay W., 1976. "Moving into the 21st Century: Dilemmas and Strategies for American Higher Education." *Liberal Education*, Vol. LXII, No. 2, pp. 158-176.

Forrester, Jay W., 1977. "Growth Cycles." *De Economist*, Vol. 125, No. 4, pp. 525-543. H. E. Stenfert Kroese B. V., Leiden, The Netherlands.

Forrester, Jay W. 1979. "An Alternative Approach to Economic Policy: Macrobehavior from Microstructure." In Nake M. Kamrany and Richard H. Day (ed.), *Economic Issues of the Eighties*, pp. 80-108. Baltimore: Johns Hopkins University Press.

Forrester, Jay W., 1980. "Information Sources for Modeling the National Economy." *Journal of the American Statistical Association*, Vol. 75, No. 371, pp. 555-574.

Forrester, Jay W., 1989. *The Beginning of System Dynamics*, 49 Bedford Road, Lincoln, MA, USA: System Dynamics Society. 16 pp. An address at the International Meeting of the System Dynamics Society, Stuttgart, Germany, July 13, 1989, edited and printed by Peter M. Milling and Erich O.K. Zahn, conference organizers. Also available from the System Dynamics Group, Sloan School, MIT, Cambridge, MA, as report D-4165.

Forrester, Jay W. 1991. "System Dynamics—Adding Structure and Relevance to Pre-College Education." In Kenneth R. Manning (ed.), *Shaping the Future*, pp. 118-131. Cambridge, MA: MIT Press. Also available as memo D-4227-1, System Dynamics Group, MIT Sloan School of Management, Massachusetts Institute of Technology, Cambridge MA 02139, USA.

Forrester, Nathan B., 1982. *A Dynamic Synthesis of Basic Macroeconomic Theory: Implications for Stabilization Policy Analysis.* Ph.D. thesis, Cambridge, MA: Sloan School, Massachusetts Institute of Technology. 268 pp.

Freeman, Christopher, ed., 1983. *Long Waves in the World Economy*, London and Boston, MA: Butterworths, 245 pp.

Goodman, Michael R., 1974. *Study Notes in System Dynamics*, Cambridge, MA: Productivity Press. 388 pp.

Graham, Alan K. 1980. "Parameter Estimation in System Dynamics Modeling." In Jørgen Randers (ed.), *Elements of the System Dynamics Method*, pp. 143-161. Cambridge, MA: Productivity Press.

Harvard Business School, 1983. *People Express* (483-103). Case study. Boston, MA: Harvard Business School Case Services. 22 pp.

High Performance Systems, 1990. *STELLA II Users Guide*. Macintosh. 45 Lyme Road, Hanover, NH: High Performance Systems.

Kondratieff, Nikolai, 1984. *The Long Wave Cycle*, New York: Richardson and Snyder, available through E. P. Dutton. 138 pp. A translation into English of papers written in Russian in the 1920s.

Low, Gilbert W. 1980. "The Multiplier-Accelerator Model of Business Cycles Interpreted from a System Dynamics Perspective." In Jørgen Randers (ed.), *Elements of the System Dynamics Method*, pp. 76-94. Cambridge, MA: Productivity Press.

Mass, Nathaniel J., ed., 1974. *Readings in Urban Dynamics: Volume I*, Cambridge, MA: Productivity Press, 303 pp.

Mass, Nathaniel J., 1975. *Economic Cycles: An Analysis of Underlying Causes*, Cambridge, MA: Productivity Press. 185 pp.

Meadows, Donella H., Dennis L. Meadows, and Jørgen Randers, 1992. *Beyond The Limits*, Post Mills, VT: Chelsea Green Publishing Co. 299 pp.

Meadows, Donella H., Dennis L. Meadows, Jørgen Randers, and William W. Behrens III, 1972. *The Limits to Growth*, New York: Universe Books. 205 pp.

Meadows, Dennis L., William W. Behrens III, Donella H. Meadows, Roger F. Naill, Jørgen Randers, and Erich K. O. Zahn, 1974. *Dynamics of Growth in a Finite World*, Cambridge, MA: Productivity Press. 637 pp.

Meadows, Dennis L., and Donella H. Meadows, ed., 1973. *Toward Global Equilibrium: Collected Papers*, Cambridge, MA: Productivity Press, 358 pp.

Pugh, Alexander L., III, 1986. *Professional DYNAMO Plus Reference Manual.* IBM PC computers. 5 Lee St, Cambridge, MA: Pugh-Roberts Associates.

Randers, Jørgen, ed., 1980. *Elements of the System Dynamics Method*, Cambridge, MA: Productivity Press, 488 pp.

Richardson, George P., 1991. *Feedback Thought in Social Science and Systems Theory*, Philadelphia, PA: University of Pennsylvania Press. 374 pp.

Richardson, George P., and Alexander L. Pugh III, 1981. *Introduction to System Dynamics Modeling with DYNAMO*, Cambridge, MA: Productivity Press. 413 pp.

Richmond, Barry M., 1985. "STELLA: Software for Bringing System Dynamics to the Other 98%." In *Proceedings of the 1985 International Conference of the System Dynamics Society: 1985 International System Dynamics Conference*, pp. 706-718, 49 Bedford Road, Lincoln, MA 01773 USA: System Dynamics Society.

Roberts, Edward B., 1978. *Managerial Applications of System Dynamics*, Cambridge, MA: Productivity Press. 562 pp.

Roberts, Nancy, 1978. "Teaching Dynamic Feedback Systems Thinking: an Elementary View." *Management Science*, Vol. 24, No. 8, pp. 836-43.

Roberts, Nancy, David Andersen, Ralph Deal, Michael Garet, and William Shaffer, 1983. *Introduction to Computer Simulation: A System Dynamics Modeling Approach*, Reading, MA: Addison-Wesley. 562 pp.

Roberts, Nancy, Tim Barclay, and David Kreutzer, 1987. "Modeling Tools for Problem Solving." In *Proceedings of the 1987 International Conference of the System Dynamics Society, Shanghai, China: 1987 international Conference of the System Dynamics Society*, pp. 888-895, 49 Bedford Road, Lincoln, MA 01773 USA: System Dynamics Society.

Schroeder, Walter W., III, Robert E. Sweeney, and Louis Edward Alfeld, ed., 1975. *Readings in Urban Dynamics: Volume 2*, Cambridge, MA: Productivity Press, 305 pp.

Senge, Peter M., 1978. The System Dynamics National Model Investment Function: A Comparison to the Neoclassical Investment Function. Ph.D thesis, Cambridge, MA: Sloan School, Massachusetts Institute of Technology.

Senge, Peter M., 1990. *The Fifth Discipline*, New York: Doubleday. 424 pp.

Skinner, B. F., 1971. *Beyond Freedom and Dignity*, New York: Bantam Books. 215 pp.

Sterman, John D., 1985. "A Behavioral Model of the Economic Long Wave." *Journal of Economic Behavior and Organization*, 6, 17-53.

Sterman, John D., 1986. "The Economic Long Wave: Theory and Evidence." *System Dynamics Review*, Vol. 2, No. 2, pp. 87-125.

Sterman, John D., 1989. "Modeling Managerial Behavior: Misperceptions of Feedback in a Dynamic Decision Making Experiment." *Management Science*, Vol. 35, No. 3, March, pp. 321-339.

Sterman, John D., and Dennis Meadows, 1985. "STRATEGEM-2: A Microcomputer Simulation Game of the Kondratiev Cycle." *Simulation and Games*, Vol. 16, No. 2, pp. 174-202.

van Duijn, J. J., 1983. *The Long Wave in Economic Life*, London: George Allen & Unwin. 239 pp.

Vescuso, Peter, 1985. "Using STELLA to Create Learning Laboratories: An Example from Physics." In *Proceeding of the 1985 International Conference of the System Dynamics Society:* pp. 964-974, 49 Bedford Road, Lincoln, MA 01773 USA: System Dynamics Society.

CHAPTER EIGHT

WHOLE-SYSTEM CONCEPTS IN SOCIETAL TRANSFORMATION

WILLIS HARMAN

Taking a whole-system view is far more often talked about than done. The central theme of this book is that everything is connected to everything else, the interconnections become ever more evident, and policymaking must increasingly be based on a whole-system view.

The merit of dealing with the society, or global society, as a whole, organic system would seem to be obvious. Despite this inescapable connectedness, we persist in attempting to solve the environmental degradation problem, or the problem of the homeless, or the unemployment problem, as though it could be dealt with in isolation. The global warming problem is perhaps the first case where even the politicians have had to recognize that it involves choices of energy sources, economic issues, life styles, equity issues—all aspects of the total system.

Previous chapters have discussed such topics as whole-system evolution, and paradigm shifts in thinking. In this chapter this discussion continues, with special emphasis on the basic premises and values which characterize modern society and which, in the decade of the 1990s, appear to be undergoing epochal change. The whole-system view is made especially imperative by this historical uniqueness of the times: The 1990s is a decade of fundamental transition, and if we fail to take this into account we cannot possibly make appropriate policy decisions.

SIGNS OF A FUNDAMENTAL TRANSFORMATION

Those who, like myself, perceive modern society to be passing through a fundamental transformation tend to emphasize three kinds of evidence. One is the *global dilemmas* that are proving to be unresolvable through the usual approaches—technological "fixes," new management practices, and legislative programs—and are "pushing" the system toward needed fundamental change. The second is shifting *value emphases* that are "pulling" toward an envisioned positive future. The third is indications of a fundamental shift in the *worldview* underlying the whole structure of modern society. Let us explore each of these three briefly.

The Global Dilemmas

The Brundtland Commission Report Our Common Future (World Commission on Environment and Development, 1987) summarizes today's challenging global-scale dilemmas, and poses as the single most critical task of the future achieving "sustainable development" not only for individual countries, but for the global system itself. There have been a number of such reports, beginning with the U.S. Government's Global 2000 (The Council on Environmental Quality and the Department of State, 1980), which have essentially the same message: The world is beset with a complex of global problems of environmental degradation, resource depletion, toxic chemical concentrations, man-made climate change, chronic hunger and poverty, species extinction, the ability of modern military powers to obliterate whole populations in the name of "national security," etc. These problems are becoming progressively more severe; they are highly interactive with one another and with industrialization and population concentration. Without major changes in the present trends the problems would become intolerably grievous by early in the next century.

Alvin Toffler (1980, p. 17) writes of "growing millions who recognize that the most urgent problems of the world—food, energy, arms control, population, poverty, resources, ecology, climate, the problems of the aged, the breakdown of urban community, the need for productive, rewarding work—can no longer be resolved within the framework of the industrial order." Because of the interconnections, the problems are not solvable one at a time; they will become solvable only through fundamental change throughout the entire social-economic-political system. Therein lies the dilemma. The price of bringing about the necessary system change to resolve these problems and support "sustainable development" seems high, but the cost will apparently be even higher if the change does not take place.

Since the events of late 1989, one often hears in the United States and parts of Europe the verdict that the Communist system has clearly demonstrated its non-viability in the long term, and that Capitalism has clearly "won" over socialist philosophy. The celebration may be premature. It overlooks the fact that socialist thought arose in the first place because of perceived flaws in materialist Capitalism. Those flaws are still present: they include lack of intrinsic guiding values and tendency to foster cynicism; proneness to generating alienation and anomie; economic incentives acting *against* taking care of the earth; inherent tendency for wealth and power to accumulate; proclivity toward exploitation of an underclass; inherent tendency toward the world economy's becoming one vast gambling casino such that the clever profit at the expense of the moral and industrious. On a world level the present system produces marginal people and marginal cultures. For all these reasons it is not at all clear that the Capitalist world economy, in anything like its present form, can survive in the long term.

Christopher Lasch (1991) masterfully documents the rise of the modern idea of progress, based on belief in unlimited technological progress, unlimited conquest of scarcity, unlimited economic growth, and a predominant value of acquisitive individualism. It "pins its hopes on the eventual triumph of critical intelligence over superstition, cosmopolitanism over provincialism, man over nature, abundance over scarcity. . . . [This] belief in progress . . . transcends commitment to any particular system of production" (p. 528). But "progressive optimism rests, at bottom, on a denial of the natural limits on human power and freedom, and *it cannot survive very long in a world in which an awareness of those limits has become inescapable. . .*" (p. 530, italics added). "Economic equality cannot be achieved under an advanced system of capitalist production. . . . The exhaustion of the progressive tradition . . . betrays itself in its inability to confront these fundamental questions of modern politics or the equally urgent question of how the living standards of the rich can be extended to the poor, on a global scale, without putting an unbearable burden on the earth's natural resources" (p. 532).

In short, since both socialism and capitalism are simply variations on the Western industrial paradigm, the traditional tension between the two has become a political irrelevancy. The problem is with the "modern" paradigm itself, and its characteristic belief in material progress. It becomes ever clearer that we must seek a "third way" through some deeper level change.

Changing Values

There is a change of mind taking place in the modern world which is only partially reflected in survey data. It is seen more clearly in the Green political movements in various countries; in the ecological, peace, and alternative lifestyles movements; in the new-values emphasis in the women's movement; in the suddenly rising interest in "12-step" programs for dealing with addictions and "co-dependency." It is characterized by a repudiation of the competition, exploitation, materialism and consumerism of modern society, and an increased emphasis on alternative values: quality of relationships; cooperation, caring, and nurturing; oneness of humanity; social justice; humane, ecological and spiritual values; respect and caring for the other creatures on the planet. It is further characterized by an emphasis on the power of the individual to take charge of his/her own life, and the power of people to change whole societies through challenging the legitimacy of the prevailing order. It is represented in only a minority of the population, but such a minority exists in practically every developed country, and the force is strengthening.

Self-Liberation Thrust. One of the most interesting of current social phenomena is the rapid spreading of the group-help approach to problems of addiction, which has moved from being the most effective way to deal with alcoholism (Alcoholics Anonymous) to being a broad social movement in its own right.

The 12-step program is a key component of the Alcoholics Anonymous approach to supporting the recovering alcoholic. These steps involve, first, admitting powerlessness over alcohol, and recognizing that life as an alcoholic had become unmanageable; recognizing the help possible from "a Power greater than myself" to aid in the restoration to sanity; taking moral inventory; admitting problems to others; trying to make amends to those persons harmed in the past; seeking contact with God "as we understand him"; and carrying the message of spiritual awakening to other alcoholics in a mutual support system.

The 12-step approach results in profound personal change as the individual, with the aid of a small supporting group, assumes total responsibility for his/her life. It has proven so successful with alcoholism that it has been extended, over the past decade, to other forms of addiction—including all forms of chemical dependency, food abuse, gambling, sexual compulsions, and smoking. In addition, recognition of the mutual interdependence that occurs where a relationship involves addiction has led to the creation of support groups for spouses and children of alcoholics and other forms of co-dependence.

More recently, the ideas of addiction and co-dependence have been extended still further, to apply to the individual's relationships with organizations and society (Schaef, 1987). Functioning as a closed system, and presenting the individual in it with few options in terms of roles and behaviors—or even ways of thinking and perceiving—the typical bureaucratic organization manifests dozens of patterns that amount to institutionalized co-dependence. These include: vague, confused and indirect communication; written memos used to avoid face-to-face confrontations; secretive behavior; executive cover-up of real problems; avoidance of conflict and difficult issues; inability to tolerate "straight talk" and the honest expression of feelings. The same approach has been extended to include our collective "addiction" to the features of "consumer society."

The corporate sector particularly, employers and employees alike, are recognizing the undesirability of institutionalized co-dependence. The key feature of the 12-step programs is how deeply into the person's unconsciously held beliefs they aim to effect change. Once people sense the possibility of self-determination and freedom in one area of their lives, they are likely to want to extend it to the rest. Thus this spreading mutual-help activity is part of a broad and growing self-liberation movement.

The Assertion of the Importance of the Feminine. Among the many social movements which convey and denote the shifting values emphasis, one stands out as not only a majority movement, but beyond a movement. That is the assertion of the importance of the feminine—not only of women, and a feminine point of view, but of the feminine side of all humans. It involves giving attention to such values as caring, nurturing, cooperating, creating; a perception of relationships over things, of wholeness over parts; the authenticity of inner knowing. It aims at the liberation of the feminine in the male as well as the female gender. The emphasis is placed primarily on changed "consciousness"—on not "buying in to" the belief system of patriarchal society.

The full dimensions of the value shift are made even more clear when the feminist and environmental thrusts combine in the "eco-feminist" movement. On superficial examination the values espoused may seem to be rather obvious virtues that everyone believes in—taking care of the environment, appreciating quality relationships, trusting inner wisdom—but closer examination reveals that the eco-feminist change of consciousness equates with liberation from oppression by the belief system of modern industrial society.

Liberation Forces in the Developing World. A counterpart liberation movement is taking place within the "sleeping giant" known as the developing world: the giant is awakening. Those who had, for so long, accepted the role of privation, inferiority, and servility, are less and less willing to do so. There has been in recent years growing insistence on a different international economic order and exploration of alternative development paths. The cultures of the indigenous peoples around the world, almost stamped out by colonialism and modernism, are showing a resurgence; there is growing appreciation of the ecologically wholesome relationship to nature that is embodied in the spiritual traditions of these native peoples.

If these attitudes continue to strengthen, the thrust will be toward, not Western industrial monoculture around the globe, but rather, an ecology of diverse cultures, each with its own interpretation of human development, societal goals, and ultimate meaning—and toward a world system that tends to support rather than diminish, bias against, or thwart this diversity. The momentum and influence of the single industrial world economy is so great, however, that any such change in the development thrust is tantamount to major change in the world's economic and financial institutions.

Revolutions in Europe. Recent developments in Eastern Europe and what was once the Soviet Union (and a somewhat more retarded movement in China) are dramatic exemplars of *the potential force for change that resides in the public's power to grant or withhold legitimacy*. There really are three contemporary revolutions in Europe, all in the end interrelated and all global in their implications. One is the revolution of liberty and democracy in Eastern Europe. The second is the revolution of unity in the European Community which, although it proceeds by fits and starts, still has come a long way in half a century. The third is the revolution of underlying values and premises which we might term "Green thinking" (a phenomenon that is far more widespread than its political manifestation in the "Green parties," especially since the 1992 "Earth Summit"). The last is the most fundamental of the three.

Shifting Basic Premises

Presently, throughout the industrialized world (particularly in the English- speaking portion and in northern Europe), there are widespread indications of a fundamental shift in worldview. In brief, this shift is characterized by two features. One is emphasis on *interconnectedness and wholeness*; though we may compete we are nonetheless each part of a unity, so that no one of us "wins" unless we all do. The other is a shifting attitude toward our inner, subjective experience—affirming *inner wisdom, authority, and resources*, and challenging the authority of "experts" in general, but in particular the scientific materialism of the earlier part of the century. This new interest is manifested both in the broader society (in involvements in various meditative disciplines and religious philosophies, and in the nature of some of the newer forms of executive development seminars) and more specifically, in the scientific and medical communities (in research and practice relating to the nature of consciousness, and to the role of the mind in the body's immune system and its self-healing and regenerative functions), and in the abovementioned 12-step programs.

Reassessing Science. Discerning persons have long been aware that there was something wrong with the scientific picture of reality. Almost by definition it fails to include human intention, conscious awareness, and that broad spectrum of "exceptional abilities" that appear to be scattered through the population, which includes miraculous healings, spirituality, and the paranormal. William James had said long ago, about the vast realms of inner experience, "No account of the universe in its totality can be final which leaves [these] disregarded."

Western culture has for a long time been trying to manage a society based on two incompatible pictures of reality—one "scientific" and the other in some sense "spiritual." The former tended to deny the spiritual, and the latter found science irrelevant to the important questions of life. This schism should have suggested to us that something was wrong somewhere.

The Danish philosopher Søren Kierkegaard opined that there are two ways of being fooled—to believe something that isn't so, and to refuse to believe something that *is* so. One of the things we have been most reluctant to believe (although it has been pointed out by many well-known figures in the past, from Plato to Ouspensky) is that each of us is hypnotized (quite literally) by our culture to experience reality in the culturally approved way. That is true for us in the modern world just as much as it is true for those cultures which we term "prescientific" or "primitive."

It is a humbling observation, but essential to the understanding of the present social change forces, that our "modern," "scientific" picture of reality is in some sense arbitrary and parochial. We are all accustomed to the idea that, as science advances, it impacts powerfully on society—through the technologies to which it contributes, through the power to manipulate our total environment which it makes possible, through the influence of its insights on our values, attitudes, and lifestyles. Less noticed is the fact that this is a two-way street. Society in turn influences science.

This is obvious if we recall that science in its present form is the product of a particular historical stage of

European society, a stage that placed high value on such attributes as the ability to predict, control, and manipulate the total environment. A different society would have created a science with significantly different emphases and priorities, different methodologies and implicit values. Society shapes science by the kinds of scientific activities it supports and the kinds it does not support, the places it bestows honor and those it does not. Science is as much a product of culture as culture is a product of science. "Nothing evokes as much hostility among intellectuals as the suggestion that social forces influence or even dictate either the scientific method or the facts and theories of science. . . . [Yet] science, in *all* its senses, is a social process that both causes and is caused by social organization" (Levins and Lewontin, 1985; p. 4).

Present society is clearly ambivalent about its science. Science is still the official knowledge validating system, the only generally accepted cognitive authority. Yet more and more people have been led to question the validity of science because the belief context simply cannot deal with the pressing problems of human experience, of mental and physical health, of sanity and the urgent need for spiritual nourishment.

The Call to Reconcile Science and Metaphysics. There has been a long-standing conflict between modern science and the human experience of spirituality—important to policymaking because the former is the source of human power to impact the world around us, while the latter is one of the chief sources of any society's guiding values. Science and religion will always be somewhat separate activities; there are important areas of science that have little to do with religion, and important activities in religion that have little to do with science. Nevertheless, there is an area of overlap, and *there the two should agree*. That they do not is a basic source of modern confusion.

This common ground comprises mainly three areas:

1. The basic phenomenon of the sorts of "*meaningful coincidences*" that appear in people's lives that are not amenable to physical explanation but do not feel like random events. Examples of "meaningful coincidences" include, for instance, apparently "telepathic" communication; seemingly clairvoyant "remote viewing;" the "coincidence" between the act of prayer and

the occurrence of the prayed-for, such as healing; the feeling of having a "guardian angel" when a person feels warned about a danger, or provided with a particularly fortuitous circumstance in life.

2. The basic phenomenon of *states of consciousness*: of "religious experiences," of "mystical" states of consciousness, of "other dimensions of reality."
3. The accounting for *the origin of the universe and the evolution/creation of humankind.*

About the first two of these areas, science has so far had very little to say. Indeed, it tends to deny their existence or importance and is intrinsically unsuited to their exploration. This fact underlies the limitations of present scientific research as its findings relate to the exploration of human spirituality. With regard to the third area of evolution and cosmology, science and religion have tended to have very different, and apparently incompatible, stories to tell. Yet these stories must eventually converge if we believe that at their best both science and religion somehow lead us toward a worldview on which we can "bet our lives" and by which we can reliably guide our societies. (See also the related discussion in Chapter 4.)

The word "metaphysics" is sometimes used, and will be so used here, to refer to accumulated knowledge about the experiential side of religion, the study of the transcendent or the supersensible. This knowledge is to be found in the "perennial wisdom" at the esoteric core of the world's spiritual traditions. As a cultural phenomenon, the apparently growing interest in metaphysics shows up in book purchases, groups joined, movie themes, and even everyday conversation, where one hears such "code phrases" as "we are all one" or "there are no coincidences" or "we create our own reality."

Modern science presents a worldview based in an assumption of separateness—separateness of observer from observed; of man from nature; of mind from matter; of "fundamental particles" from one another; separability of the parts of a system or organism to arrive at a "scientific" explanation of phenomena. Metaphysics, on the other hand, takes seriously the root mystical experience of oneness, unity, wholeness. The growing persuasion seems to be that both of these viewpoints have something to offer; that an adequate science will somehow reconcile them. More particularly, in the three areas of overlap mentioned above, metaphysics and science should ultimately come to the same conclusions about the interpretation of pertinent evidence. While the discrepancy remains, many well-informed persons are turning to metaphysics as the more reliable of the two in these overlap areas.

The Importance of the Shift in Basic Premises. The practical importance of this shift in basic premises may not be immediately apparent. Modern industrial society, like every other in history, rests of some set of, largely tacit, answers to the questions basic to any culture: What kind of beings are we? What is our relationship to one another? What kind of universe do we live in, and how do we relate to it? What is ultimately important?

The scientific materialism which so confidently held forth its answers to these questions a couple of generations ago is now a dying orthodoxy. Its basic positivistic and reductionistic premises are challenged by various forms of transcendentalist beliefs that include increased faith in reason guided by deep intuition. In other words, a respiritualization of society appears to be taking place, but one more experiential and noninstitutionalized, less fundamentalist and sacerdotal, than most of the historically familiar forms of religion. Such a change in basic assumptions must inevitably be accompanied by a lasting shift in value emphases and priorities. All institutions can be expected to go through major changes—as was the case in Western society with the last fundamental change in worldview, which we know as the scientific revolution. The institutions which will be particularly affected, because they are so thoroughly imbued with the presently dominating paradigm, are the industrial, technological, scientific, economic and financial ones. (Refer again to Chapter 1.)

Nature of the Transformation

It should be clear from the above that in using the term "transformation" to describe this transition period we do not mean simply change—even rather radical change. Persons living today have all experienced rapid, life-transforming technological change throughout their lives. We are not referring to more change of a familiar sort, but rather to *trend-breaking change of a kind that is unknown in our life experience.*

Perhaps the closest comparison is with the change that took place in Western Europe in the 17th century. This period was essentially the transition from medieval to modern times; basically it amounted to *a shift in the basic perception of reality*. The world perceived by a person living in Western Europe in the year 1600 was still the world of the Middle Ages. By 1700 the educated person would perceive—literally—a different reality, one which gives rise to faith in human progress and guiding values not too different from those characterizing modern society. This fundamental shift eventually transformed every institution and way of life throughout most of the world.

The defining characteristics of modern society—centrality of the economic and technology-centered institutions, steadily rising demand for democracy and equity, unbridled national sovereignty, and supremacy of the "reality" definition developed through empirical science—were more or less set in place by the end of the 17th century. By now these characteristics have influenced practically all other societies around the globe. They account for the present dominance of Western culture. *They are also the ultimate source of the global dilemmas we now face.*

Confidence in the "reality" of empirical science grew steadily. It was astoundingly successful at providing the power to predict and control in the physical world, as modern technological prowess attests. Yet by the latter part of the 20th century there were unmistakable signs that, however useful science might be for some purposes, its arrogant materialism had a serious negative effect on our understanding of values. This effect was to undermine the common religious base of values and to replace it with a kind of moral relativism. Into the vacuum came, as a sort of pseudovalues, economic and technical values—material progress, efficiency, productivity, etc. Decisions that would affect the lives of people around the globe, and generations to come, were decided on the basis of short-term economic considerations. The "technological imperative" to develop and apply any technology that could turn a profit or destroy an enemy, in the end endangered both the life-support systems of the planet and human civilization.

Thus the "new heresy" came to be the spreading belief that reality is not found in the reductionistic scientific worldview any more than it was to be found in dogmatic theology. As the modern perception of reality differed from the medieval, so a growing band of individuals today are betting their lives on a different picture of reality than that of conventional science. It is not just that some "New Age" values are loose in the land; rather, a competing picture of reality infuses holistic health care approaches, new concepts of business management, and people seeking to replace the lost meaning in their lives. As the "scientific heresy" of the 17th century affected, in the end, every institution in society—so the "new heresy" will do the same, but far more rapidly.

POLICIES FOR WHOLE-SYSTEM TRANSFORMATION

Whatever the policy issue under consideration, it is part of the whole-system evolution. Even recognizing this, there are two possible approaches which contrast sharply. The first is to treat the transforming environment as context, and to judge the decision by conventional criteria. The second is to consider one's organization, its activities and policies, as part of the transformation, and to judge the decision by how it fits into the transforming whole.

The distinction will be clearer from an example. Take, for instance, some major policy decision in the area of national security. Adopting the former approach, one might consider the fundamental changes taking place in the global environment in defining and comparing alternative policy options, and judge the policy decision on the basis of how each alternative would affect the country's military strength and its ability to deter aggression and achieve its international goals. The latter approach, however, would put at the center of the analysis the likelihood that part of the transformation involves a redefinition of "national and global security" in some broader way involving an expanded definition of basic human rights, and would recognize the new reality that with modern weapons of mass annihilation, no nation can achieve the goal of security in any meaningful sense through military strength. There is no security for anyone until there is security for all, and that premise dictates a totally different set of policy options.

This example displays what is true in general, that the second approach makes the policy analyst and policymaker part of the transformation. They—their internalized assumptions, and hence they themselves—are being transformed, along with the world they are examining. It is this situation we shall now explore in more detail.

The Whole-System Nature of Global Problems

In the presently dominant paradigm we seek quantifiable models of the system elements and their interconnections, in the form of causality links. The aim of the analysis is increasing ability to predict and control. Goals and guiding values are assumed to come from outside the system—from tradition, culture, and individuals' preferences. Policy is decided through compromise and balancing tradeoffs, in the political process.

The new paradigm involves an initial assumption of the interconnectedness of everything, including interconnectedness where there is no obvious "mechanism." Thus there is a whole system, made up of intrinsically interconnected quasi-autonomous or self-motivated units, in which—as policy pundit Harlan Cleveland expresses it—"there is nobody in control, and that's the way we want it." Goals and guiding values are not extraneously imposed, but part of the system. There is not cause and effect, input and response, but rather a whole system evolving. The societal structure, dominant worldview, aims and aspirations, and approach to societal and global problems are all parts of this evolution. They may move together, or they may be to some degree in dissynchronization. In this view the chief aim of policymaking is to aid "wholesome" evolution and to avoid impeding it (much as the proper aim of medicine is to aid wholesome growth and development—with the meaning of "wholesome" defined by the nature of the organism).

To more fully appreciate the interconnectedness of the global problems, it may help to try to imagine that you are visiting "modern" society with fresh eyes, as if from some other civilization. You note that among the problems which loom large in the future are deteriorating environmental quality, air and water pollution, deforestation, desertification, toxic chemical concentrations, resource depletion, man-made climate change—and that all of these are strongly correlated with the amount of economic product. On the other hand, nations approvingly point to the amount of economic product and economic growth as though these were appropriate measures of societal well-being, and the more the better.

You observe that the era of cheap energy from fossil fuels is destined to come to an end within a period that is measured in decades. Yet policy choices continue to be made which keep prices down, thus encouraging high levels of consumption and hastening the day when a major restructuring will have to take place because cheap energy is no longer available. (You also observe a widespread millenarian belief that scientists will come up with a totally new source of cheap energy just in time to save the day.)

You find that past and present policies are based on the assumption that "progress" involves increasing participation in the one master economy—in terms of goods and services, much of which in earlier times was in the informal and household economies, and also in terms of jobs as the chief way in which individuals find a social role. This assumption drives the necessity for more economic growth to create jobs—hence more induced consumption, more makework, more environmental degradation, more resource depletion. In spite of this rather obvious contradiction, you find little inclination to challenge the basic assumption.

Clearly not everyone in the world needs to have jobs in the mainstream economy in order for it to be able to produce all the goods and services that are really needed for a good quality of life, or even all the economic product the planet can stand. The obvious question before modern society is: What meaningful and constructive activities can now be engaged in because economic production does not require the full efforts of all? But you observe very little tendency to ask this question seriously; rather, society engages in frenzied activities to increase economic consumption.

One of the more unique characteristics of modern society, as contrasted with any before it, is the extent to which economic rationality and economic values affect major policy decisions. Individual politicians and business leaders may bring in personal values here and there, but by and large the bureaucratic way is to defend policy decisions with cost-benefit or similar economic analyses. Major investment decisions are based on anticipated rate of return and little else. Because of discounting the future at a rate depending on current interest rates, the well-being of future generations is formally ignored in economic logic. The older tradition, for example of the farmer's leaving the land for the next generation better than he found it, is a sentimentality that is "contraindicated" by "sound" financial management which puts the emphasis on the near-term financial bottom line.

The modern agricultural industry is highly efficient in terms of product per acre or per person-hour. It accomplishes this with generous inputs of tractor fuel, synthetic fertilizers, and other energy-intensive inputs. The net result is that (taking into account energy used in long-distance hauling, packaging, freezing, thawing, and cooking) food on the table represents many times more fossil fuel energy than solar energy. We are literally eating petroleum. From an ecological standpoint that doesn't make sense, particularly in the long term; but it seems to make economic sense.

The industrialized world, having lost any consensus on ultimate meanings and values, steers itself mainly by economic and financial signals serving as pseudo-values. Part of the developing world is scrambling to catch up with the industrialized countries; other parts are seeking some attractive alternative to scrapping their own cultural roots and adopting the alien culture of "Western civilization." Many of the developing countries are spending more for sophisticated arms than on health, education, and social services combined. Many countries of the South stagger under debt loads such that oppressive interest payments impair their ability to improve their lot, and the net transfer of wealth is out of these countries to the North.

The global environmental condition worsens steadily, notwithstanding temporary or localized reversals of the general deterioration. Much of this degradation is associated with the economic activities of the industrialized countries—air and water pollution, acid rain, toxic chemical disposal, solid waste disposal, fertilizer and pesticide runoff, resource extraction, and so on. Much of it also is a consequence of the desperate quest of the poor for food, firewood, grazing land,

and shelter; poverty is one of the planet's greatest polluters. Almost all of the policies and actions aimed at environmental problems are in the nature of patch-up, and the efforts are inescapably losing ground. Because these problems are so intimately related to economic conditions, the need to reverse present environmental trends is equivalent to a need for fundamental changes in the world economic structure. But who is willing to talk about that?

Policies in the highly industrialized countries such as the U.S. assume the future to be one of continued economic growth; most development policy assumes a similar pattern for the rest of the world. Throughout the world, economic, social, and international policy is being implicitly based on concepts of global development that seem inexorably to lead toward continuing widespread misery and conflict—because of the gross and increasing disparities between the rich and the poor; because of policies pushing peasants off the land and into urban slums; because of the wrenching apart of community by the temptations and imperatives of modernization; because of deforestation and water pollution and soil destruction and desertification and a hundred other kinds of environmental spoliation; because of sudden wealth stemming from natural resources that are coveted by the industrialized world; because of irreversible changes in plant and animal species, soil composition, and climate. *Present economic, development, social and international policies are implicitly based on assumptions which are built in to present institutions and practices, and hence have great inertia—and which are totally incompatible with a viable global future.* It is not surprising, in view of that sober fact, that bad policy choices continue to be made.

Characteristics of the Evolving System

How may the global system be evolving toward an ultimate resolution of this dilemma—a resolution which may be aided or impeded by policy choices? To try to see at least a partial answer to that question, we need to look at one of the more significant characteristics of organic systems.

Self-Healing Forces in Society. Living organisms are self-healing to considerable extent. The same can undoubtedly be said of societies even though the mechanisms are less studied. The Gaia hypothesis (Lovelock, 1988), which has recently been attracting much favorable attention, suggests that the planet, as a living system, may be self-healing as well. Of course, there is no assurance that the healing process of the planet guarantees the continued existence of human civilization; humans will have to see to that themselves.

Recognizing this capacity for self-healing, we see several things in a new light. The question is not so much how the problems arose, nor even how their effects can be ameliorated; the key question is: What went wrong with society's self-healing system such that it failed to handle the pathogenic challenges as they came along? What might be done to help the restoration of the societal self-healing processes?

Identifying the Evolutionary Thrust. Adopting the optimistic hypothesis that much of people's innovative activities these days can be interpreted as society-healing impulses, partially unconsciously guided, we find it easy to recognize many indications of spontaneous creative response. These indications include the various social movements mentioned above, as well as a host of innovative experiments in nonprofit organizations, intentional communities, alternative economies, alternative health-care programs, new forms of business entrepreneurship, citizen approaches to assisting new enterprise and community development in Third World countries, and many others.

From these various indicators we can identify an overall thrust in the direction of:

- An awareness of the inextricable interdependence of all human communities and dependence on the planetary life-support systems, and awareness that resolution of the world's dilemmas will come only from a whole-system view.
- An affirmation of the brotherhood of man/woman and of a non-exploitative stewardship relationship to the planet.
- A stand that social institutions (the economy, the military) are legitimate only when they are guided by the highest values and principles.
- A growing global awareness, and a willingness to have rights that have long been guaranteed within the nation

(e.g. Rule of Law, freedom of association, minimal satisfaction of basic needs) apply globally.

- An insistence that the legitimacy of war and of preparation for mass destruction as instruments of national policy is obsolete.
- A rejection of extreme positivism and reductionism in science, and a reassertion of the value of the inner search.
- An affirmation of the existence of consciousness in the universe, and of the essential spirituality of the individual.
- Sensitivity to the need to counterbalance "masculine" competitive, aggressive, exploitative values with "feminine" nurturing, cherishing, conserving, cooperative values and appreciation of the feminine perspective.
- Conviction that achievement of high quality of life will require decentralization of much social and political activity, rebuilding of community, and an appropriate technology (i.e. human scale, ecologically compatible, "kind" to human beings and to the planet).
- Conviction that the richness of cultural diversity is a planetary resource just as much as the richness of diversity of plant and animal life, so that the tendency of the world industrial economy to stamp out competing cultures must be curbed.
- Commitment to global change through a spreading awakening, the political manifestations of which must use nonviolent means.
- An emphasis not on goals but on process—on people becoming empowered to take responsibility for their own lives and for changing society as necessary: the process is an evolutionary one, and the goals are emergent.

Dynamics of societal transformation. Identification of the thrust for change leaves open the question of how the transformation actually comes about. It is reasonable to question whether a fundamental transfiguration of the modern world could possibly happen over a short period of a few decades. However, the great convincer here is the peaceful revolution of Eastern Europe during the final months of 1989, followed by the fall of Communism in the Soviet Union a year and a half later. It would not have seemed realistic, in the mid-1980s, to anticipate the profound changes that had actually taken place in Eastern Europe by the end of the decade. When masses of people change from perceiving the locus of *authority* to be external, and make it internal; when masses of people recognize that they can challenge the *legitimacy* of whatever oppresses them—then the balance of power between institutions and people shifts. Change that would otherwise seem quite improbable suddenly becomes possible.

We in the modern world are just becoming aware that forces for radical change have been mounting and may be close to a critical level. This is not the classical revolutionary force of the oppressed. It is, rather, the revolutionary awareness that *we have been oppressing ourselves with a belief system that we bought into, a belief system on which our whole techno-economic structure rests, which is not compatible with a viable future for human society on the planet.* Even many of those who have been most successful at rising to positions of affluence and power in the present system have begun to question whether that is ultimately satisfying in terms of the way they invest their lives.

Identifying Appropriate Policy Choices

The above general characteristics fail to address the question of what kinds of policies are going to be both appropriate to the goals and politically feasible. Interventions are possible at various levels of the system, and the policy package may be multi-leveled. An analogy may illustrate.

A Health-Care Analogy. Imagine a business executive (you have probably known someone more or less fitting the description) whose life has become rather frantic. The strains at the office are almost too much for a mere human to deal with. Time at the office is experienced as one crisis or impossible deadline following the last. Frequent feelings of anxiety are incapacitating. Health problems intrude: There have been attacks of ulcers; colds and minor illnesses are frequent; and cardiovascular disease lurks constantly in the background. Also, there has been a tendency recently to drink more alcohol than appears wise. Nor is the situation alleviated by going home in the evening. The marriage has been fractious lately; there are financial worries; and the children seem alienated. When these many problems are encountered one on top of the other, there seems almost nowhere to turn.

Yet these varied problems may also be seen as a single pattern, a gestalt, within which they all have common roots in deeply held beliefs—beliefs which result in stress-creating attitudes to work, home, and family. Now a very different remedial action is suggested. If these fundamental beliefs and attitudes can be changed, this puts all the problems in a different perspective and reduces stress. That in turn strengthens the functioning of the body's immune and healing system, and health improves. By seeing a different interpretive pattern, a different and more constructive response is called forth. This response may amount to a rather revolutionary transformation of the person's reality perceptions, major value commitments, and overall life pattern.

Yet we all know how resistant such a person can be to re-perceiving his life situation. The analogy I am suggesting is obvious. If we come to see the societal and global situation in a different interpretive pattern, that may suggest different expectations and a different response.

In the health care analogy we may take aspirin for the headache, antibiotic for the disease, bland diet for the ulcers, and these may all bring temporary relief and thus supplement affirming a new worldview to deal with the stress. However, by themselves without that new worldview, they fail to get at the root of the problem.

Similarly, if we recognize that the serious global problems are interconnected and can be viewed at various levels, there is no contradiction involved if the policy package consists of some "bandaid" or "patch-up" approaches; some at a subsystem level; and some measures to facilitate smooth whole-system change.

Symptom Amelioration and System Change. Thus we arrive at one of the most important observations about policymaking in a transforming world. Policy interventions are appropriate at various levels in the system, toward various ends.

For both humanitarian and political reasons, policies to ameliorate symptoms will no doubt be called for. Their goals will be fairly specific and apparent to all. But if this is all that is done (as has become rather typical in politics) the actions will not only fail to solve the deep-rooted problems, but are likely to have unanticipated undesirable consequences as well.

Policy choices aimed at the deeper level change have quite different ends. Their purpose is not to bring about the change, because fundamental societal change does not come from top-down policy. Their objective, recognizing the presence of a growing force for transformation, is to help the change be relatively smooth and non-disruptive. History is replete with examples of fundamental societal change characterized by disruption, violence, and human misery; that is much more the norm.

The Example of Policy Relating to Work

Let us try to make this discussion a little more specific by using as an example policymaking relating to the issues of work and unemployment.

Politicians continue to promise that they will adopt economic policies to create jobs for all. There was actually an attempt, following World War II, to outlaw unemployment. One faction in the debates that led to the Employment Act of 1946 argued that it should be a "Full Employment" act. In short, they proposed to enact legislation such that the Federal government would take whatever measures were necessary to insure full employment, essentially defined as having jobs available for everybody who wanted one. Even now it is not generally recognized that this is a totally infeasible goal.

We spoke above of the negative environmental and social effects of the unrelenting pressure to create jobs in the mainstream economy. This pressure comes from a basic dilemma of the modern world. On the one hand, if a country does not continually increase labor productivity, the industry of that country tends to become noncompetitive in the international market. On the other hand, if productivity does increase then, by definition, to maintain the same number of jobs the economic product must increase. Thus as various constraints—resource, environmental, political, and social—tend to limit economic growth, chronic unemployment becomes an intrinsic characteristic of the future. Since that is an unpleasant thought to contemplate, countries and individuals have tended to use tortuous logic and subtly evasive actions to avoid confronting it.

In a few countries demographic trends are obscuring this unemployment dilemma for the short term, but the tendency is inexorable in the longer term. Some of the obfuscation is more contrived, as is illustrated by the progressive redefinition, in the United States, of the official unemployment rate corresponding to "full employment." This rate, which some claim represents the unemployment level necessary to maintain low inflation rates, has risen steadily from around 2.5% in the 1940s to over 7.5% by the early 1990s. A nation can, as did the Soviet Union for many years, insure that there is no unemployment by giving every adult a job whether or not his or her labors are really needed. But this expedient of "makework" creates a serious morale problem, because no one likes to feel unneeded.

Why is modern society unable to arrange things so that practically every citizen has ample opportunities for meaningful, fulfilling work? That is the fundamental question, underlying all the more obvious aspects of work-related problems from unemployment and underemployment, through chronic inner-city poverty and homeless people, to environmental impacts of economic growth. "Meaningful work" is not necessarily work that is exciting and challenging at every moment; it is enough that it be part of a larger endeavor which is infused with meaning. It is mainly this "larger endeavor" for our modern, mass-consumption society which is lacking.

Agriculture is an example. Once one has experienced the tractor with air-conditioned cab, one is unlikely to be content following a horse with a plow; such work no longer has the same meaning. One can hardly wish to abandon all labor-saving technology. At the same time, there is a profound truth in the observation that hand labor is not something to be avoided because it is wearisome or beneath one's dignity. The pleasantness or unpleasantness of hand labor depends on such factors as the cultural value put on relationship to nature, the context of meaning created by the farming community, the satisfaction from seeing the results of one's own work, having good tools, and ownership. Some of those who currently leave the cities behind and seek a new meaning in a more pastoral existence, working close to the land, feel themselves to be part of a larger endeavor—namely that of bringing about a new and viable order in which human activities are viewed in the perspective of the entire global ecological system.

We tend to overlook how basic the work/jobs dilemma really is. Contemporary concepts of business and labor, of employment and welfare theory, of liberal and Marxist analysis, are all based in *production-focused* society. *It is this central concept that is obsolete.* Throughout much of the industrial period it may have made sense for society to be oriented around material progress, just as during the Middle Ages society centered around building great cathedrals for the glory of God. However, in the 1990s centering society around mere economic growth and technological advance, unguided by some overarching values, seems decreasingly suitable; the need is strongly sensed to reorient around some new focus.

All of history supports the observation that the desire to create is a fundamental urge in humankind. *Fundamentally, we work to create, and only incidentally do we work to eat.* That creativity may be in relationships, communication, service, art, or useful products. It comes close to being the central meaning of our lives. It is not without significance to note that the world has never found a more congenial environment for creative work than small free enterprise (either profit-making or non-profit).

In a technologically advanced society where production of sufficient goods and services can be handled with ease, one seems forced to the conclusion that *work and employment are primarily for self-development, and are only secondarily concerned with the production of goods and services.* "Self-development" is meant here to imply all that enriches the self, including not only personal and professional development, but quality relationships and meaningful service to others.

Not all such creative work needs to be accommodated in the master economy; in fact, it probably never can be. In the long term, many people and activities have to move *out* of the master economy. There are many alternative ways to handle distribution of the overall societal income; jobs in the master economy is only one.

When the economy goes into recession or depression, emergency policies to alleviate the attendant human misery are perfectly in order. Policies to alleviate the problems of unemployment in the somewhat longer term might include measures such as those of the 1930s to create federally- and state-funded work that brings social benefit in a broad sense. Programs something like the "local initiatives programs" initiated by Canada in the 1970s might be appropriate. These programs were intended to be a substitute for welfare handouts; from many standpoints they worked very well. The basic concept was that a small group of persons could initiate a project to benefit the environment or community, and then ask the federal government to provide money for salaries and supplies. The work tended to bolster self pride, much as any self-employment does; the same money distributed to the same persons in the form of welfare payments would seem demeaning.

In the long term, assuming that the system evolution is more or less in the direction identified above, appropriate policy would facilitate a partial disintegration of the master economy and a proliferation of local, informal economies. An example of the latter is the "local employment trading system" concept of a local economy with local currency ("green dollars," which are really just electronic signals on a computer), superimposed on the master "real dollar" economy. Because "green dollars" can be borrowed from the system on an essentially interest-free basis, they tend to circulate with high velocity; thus the local economy can be flourishing even when the main economy is in depression. Furthermore, the kinds of goods that get produced, and the services provided, in the local economy tend to be environmentally benign and resource-conserving, compared with those of the master, highly industrialized, international economy. The two can exist side by side, and the main economy can produce the VCRs and the fiber-optics communication systems while the local economy concentrates on quality of individual and community life.

But with all that said, there is a fundamental tension of which the policymaker is all too aware. In the long term there must be a restructuring of the economy, with elimination of the drive to have ever more economic product, because the environmental and social costs of that drive are too great to bear. However, in the short term the realities include strong political pressure to increase economic product and create jobs.

The overall point is that by analyzing the direction of evolution of the overall system (which may include some devolution as well), it is possible to discern what policies would help to facilitate that development. At the same time, some of the symptom-alleviation and "illness control" policies will be necessary as well.

CONCLUDING OBSERVATIONS

Thus we are able to make a few general observations. If we really take the whole-system view, desirable policies are those which aid the system to evolve in a "wholesome" direction, with a minimum of disruption, dissynchronization, and human misery. Timing is extremely important. The right policy, if an attempt is made to institute it too early, may be a disaster. If it is adopted too late, it may be ineffective.

Arriving at Appropriate Policies

The entire policy package will typically include some policies to alleviate immediate symptoms, some to ease the strain of the transition period, and some to facilitate the longer-term evolutionary goals. It would not be politically realistic to ignore the first two, and it would not be responsible to ignore the third. The temptation on the part of some "Green" enthusiasts has been to try to bring about by legislative fiat deep-level changes which, if they are to be accomplished in a way compatible with preservation of liberty and justice, can only come about through slower evolution involving major cultural change.

One of the chief obstacles to this kind of "sapiently adaptive" policymaking is interest-group politics. The United States of America became a great and prosperous country because there were some among its leaders who rose far above the narrow interests of particular groups, and behaved as true statespersons. In this time of need, we may hope that such will appear again. The likelihood will be greatly increased if the people are wise enough to demand vision, foresight, and whole-system thinking.

Because so much depends upon public support of wise policies, probably the single most needed and most fruitful action is talk. Although "just talk" is often disparaged, serious explorative talk among concerned citizens and peers is one of the better ways of coming to new understandings and diminishing the fear that tends to be generated by the prospect of profound change.

The Role of Business. We think of governments as having the chief responsibility for policy. But by the latter part of this century, business had become the dominant institution on the planet. The policies set in the corporate boardroom and the CEO's office are also components of the aggregate pattern that in the end influences the evolution of the whole system.

Business leaders have a natural penchant towards action. But if the action is to be pertinent to the needs, there must be clarity regarding those needs. These change with time. Right now, the most important task may be to promote broader understanding, and maybe some enthusiasm about, a change in paradigm that could well be good for ourselves, good for relationships, good for the environment, good for the planet. But that awareness raising has already been going on for quite a while—not just by a few people in business, but by great numbers of people throughout all of the modern world, for at least a quarter of a century.

If we are indeed approaching the point of "critical mass," where people suddenly realize (as in late-1989 Eastern Europe) that in some fundamental way the legitimacy of the present techno-economic system, and the values and beliefs that underlie it, has to be challenged, then the task shifts. If the transformation starts to happen with such rapidity that it generates a good deal of fear—fear of instability, of economic collapse, of mass unemployment, of an uncertain future, of the "crazies" in our midst—then the one thing people will crave most is stability. The overriding concern will be how the transition can be managed; how balance can be maintained; how we can keep the machine on the tracks. At that point the critical task changes (as the reconstruction task in Eastern Europe in the 1990s is completely different from the awareness-raising task of the 1980s).

That is the time the *really* critical role of business comes in. Whatever business can do now is preparation for the truly crucial time in history when somebody has to reassure the fearful that the needed transformation can be accomplished without widespread social disruption and attendant human misery. The experience of business leaders can be critical at that point—assuming that they are sensitive and aware enough to play a constructive rather than reactionary role.

The dominant institution in any society needs to take a responsibility for the whole—as the Church did in the days of the Holy Roman Empire. But business has not had such a tradition. This is a new role, not yet well understood and accepted. Built in to the concept of Capitalism and free enterprise, from the beginning, was the assumption that the actions of many units of individual enterprise, responding to market forces and guided by the "invisible hand" of Adam Smith, would somehow add up to desirable outcomes. But in the last decade of the 20th century it has become clear that the "invisible hand" is faltering. It depended upon a consensus of overarching meanings and values which is no longer present. So *business has to adopt a new tradition which it has never had throughout the entire history of Capitalism. That is, as the most powerful institution on the planet, to take responsibility for the whole.* Every policy decision that is made, every action that is taken, has to be viewed in the light of, in the context of, that kind of responsibility.

Worldviews and Scenarios. We all carry around "in our heads" pictures of the world—how it is, how it got that way, where it seems to be heading. For each of us that internalized picture influences how we interpret our experiences, and how we choose to act. One of the most important kinds of learning involves re-examination of that picture; in fact Pierre Wack, onetime chief planner for Royal Dutch Shell, claimed that his chief job was to help the top executives change their internal pictures.

One of the more popular tools for policymaking has come to be the construction of *alternative scenarios*. Basically the central idea is to deal with context uncertainty by creating a number of plausible stories about how the total environment might change over the planning period, selected to bracket the major uncertainties. For instance, one scenario might portray severe political unrest and economic disruption; another might assume more benign conditions. Policies can then be tested against these various scenarios: Typically it will be found that policies that will continue to look good if future developments are more or less as portrayed in one scenario, appear far less desirable in some other scenario. This then gives information not only with regard to what policies to choose, but also what warning signals to watch out for, lest events go in a direction unfavorable to those policies (See Meadows, 1992). (Refer to the system dynamics approach in Chapter 7.)

One fact that often seems to receive little attention is that the group of scenarios that will be selected for examination is critically dependent on the accepted *interpretation of the present*. What is the meaning of the pattern of present events and trends? If the interpretation explored above is accepted, the policy-guiding scenarios will be a totally different set from those that would be selected if the transformational interpretation were to be ignored or denied.

Trust. One last observation relates to fear and trust. When the whole-system view is really internalized, when one knows from experience and repeated affirmation that the nature of the universe is oneness, unity, wholeness—then it is natural to trust. After all, the only things we fear are what we perceive to be fearful; if the perception of oneness does not include anything fearful, trust is the most practical state. No experience is to be feared; all experience is feedback to be learned from.

Management consultant Roger Harrison once expressed it this way (1983):

> "Seen from a global viewpoint, the organization exists only as part of a larger reality, supported and nurtured by the larger system on which it depends: the nation, its culture, the world economic and political system, the physical and biological planet itself. . . . Were we to approach strategy from the point of view of endeavoring to discover the place of the organization in the larger [evolving] systems of which it is a part and on which it depends we would do far better [than with the usual strategic planning efforts]. From such a viewpoint, organizational purpose is not simply decided by its members, but is in large part 'given' by its membership in the larger system. . . .
>
> "We can take the view that our organization has an appropriate place in the larger system, and that our task as managers and leaders is to attune our organization to its

environment in order to discover what our part is and play it. The difficulties we experience are interpreted as signs and signals from the environment that we are somehow out of resonance with our true role. . . . According to this point of view it should not be difficult for an organization to survive and thrive, any more than an organ in a healthy body has to work especially hard to survive. When it plays its part, it receives the nourishment it needs. From a system point of view, then, strategic thinking is a search for meaning, rather than a search for advantage."

One can hardly imagine a better example of a policy stance that makes sense in the transforming-system view, but may sound like dangerous nonsense from the conventional standpoint.

REFERENCES AND SUGGESTED READING

Bateson, Gregory, 1972. *Steps to an Ecology of Mind.* New York: Ballentine.

Berman, Morris, 1981. *The Reënchantment of the World.* Ithaca, NY: Cornell University Press.

Berry, Thomas, 1988. *The Dream of the Earth.* San Francisco: Sierra Club Books

Berry, Wendell, 1978. *The Unsettling of America: Culture and Agriculture.* New York: Avon/Sierra Club Books.

Boulding, Kenneth, 1978. *Ecodynamics: A New Theory of Societal Evolution.* London: Sage.

Brown, Lester R., 1981. *Building a Sustainable Society.* New York: Norton.

Brown, Lester R., 1990. *State of the World, 1990: A Worldwatch Institute Report on Progress Toward a Sustainable Society.* New York: Norton. (Similar publications every year since 1984.)

Capra, Fritjof, 1982. *The Turning Point: Science, Society, and the Rising Culture.* New York: Bantam.

Commoner, Barry, 1971. *The Closing Circle: Nature, Man, and Technology.* New York: Knopf.

Daly, Herman, 1977. *Steady-State Economics: The Economics of Biophysical Equilibrium and Moral Growth.* San Francisco: Freeman.

Daly, Herman, and John B. Cobb, 1989. *For the Common Good: Redirecting the Economy Toward Community, the Environment, and a Sustainable Future.* Boston: Beacon Press.

Davis, W. Jackson, 1979. *The Seventh Year: Industrial Civilization in Transition.* New York: Norton.

Devall, Bill, and George Sessions, 1985. *Deep Ecology: Living as if Nature Mattered.* Salt Lake City: Peregrine Smith Books.

Eisler, Riane, 1987. *The Chalice and the Blade: Our History, Our Future.* New York: Harper and Row.

Elgin, Duane, 1981. *Voluntary Simplicity: Toward a Way of Life That is Outwardly Simple, Inwardly Rich.* New York: William Morrow.

Ellul, Jacques, 1964. *The Technological Society.* (Original French edition 1954) New York: Vintage/Random House.

Ferguson, Marilyn, 1980. *The Aquarian Conspiracy: Personal and Social Transformation in the 1980s.* Los Angeles: Jeremy Tarcher.

Harman, Willis, 1988a. *Global Mind Change.* Indianapolis: Knowledge Systems.

Harman, Willis, 1988b. "The Need for a Restructuring of Science." *ReVision,* 11(2) (Fall), pp 13-21.

Harman, Willis and John Hormann, 1990. *Creative Work: The Constructive Role of Business in a Transforming Society.* Indianapolis: Knowledge Systems.

Harrison, Roger, 1983. "Strategies for a New Age." *Human Resource Management* 22(3) (Fall), 209-235.

Henderson, Hazel, 1992. *Paradigms in Progress: Life Beyond Economics.* Indianapolis: Knowledge Systems.

Johnston, Charles M., 1988. *The Creative Imperative: Human Growth and Planetary Evolution.* Berkeley: Celestial Arts.

Kohr, Leopold, 1978. *The Overdeveloped Nations: The Diseconomies of Scale.* (Original edition in 1962, in Spanish and German.) New York: Schocken Books.

Lasch, Christopher, 1991, *The True and Only Heaven: Progress and Its Critics.* New York: Norton.

Levins, Richard, and Richard Lewontin, 1985. *The Dialectical Biologist.* Cambridge: Harvard University Press.

Lovelock, James E.,1988. *The Ages of Gaia: A Biography of Our Living Earth.* New York: Norton.

Lutz, Mark A. and Kenneth Lux, 1979. *The Challenge of Humanistic Economics* Menlo Park, California: Benjamin/Cummings.

Meadows, Donnella H., Dennis L. Meadows, and Jørgen Randers, 1992. *Beyond the Limits: Confronting Global Collapse, Envisioning a Sustainable Future.* Post Mills, Vermont: Chelsea Green.

Merchant, Carolyn, 1980. *The Death of Nature: Women, Ecology, and the Scientific Revolution.* New York: Harper and Row.

Milbrath, Lester W., 1989. *Envisioning a Sustainable Future: Learning Our Way Out.* Buffalo: State University of New York Press.

Muller, Robert, 1982. *New Genesis: Shaping a Global Spirituality.* New York: Doubleday.

Ornstein, Robert, and Paul Ehrlich, 1989. *New World, New Mind.* New York: Doubleday.

Robertson, James, 1979. *The Sane Alternative: A Choice of Futures.* St. Paul, Minnesota: River Basin Publishing.

Robertson, James, 1990. *Future Wealth: A New Economics for the 21st Century.* England: Cassell Publishers.

Roszak, Theodore, 1978. *Person/Planet: The Creative Disintegration of Industrial Society.* Garden City, NY: Doubleday.

Schaef, Anne Wilson, 1987, *When Society Becomes an Addict.* New York: Harper and Row.

Schumacher, E. F., 1973. *Small is Beautiful: Economics as if People Mattered.* New York: Harper and Row.

Sorokin, Pitirim, 1941. Social and Cultural Dynamics. New Brunswick, New Jersey: Transaction Publishing.

Sperry, Roger, 1987. "Structure and Significance of the Consciousness Revolution in Science." *Journal of Mind and Behavior* 8 (1), 37-66.

Spretnak, Charlene, Ed., 1982. *The Politics of Women's Spirituality.* New York: Anchor/Doubleday.

Spretnak, Charlene, and Fritjof Capra, 1986. *Green Politics.* Santa Fe, New Mexico: Bear and Co.

The Council on Environmental Quality and the Department of State, 1980. *The Global 2000 Report to the President: Entering the Twenty-First Century.* Washington, D.C.: U.S. Government Printing Office.

Theobald, Robert, 1987. *The Rapids of Change: Social Entrepreneurship in Turbulent Times.* Indianapolis: Knowledge Systems.

Toffler, Alvin, 1980. *The Third Wave.* New York: William Morrow.

Wald, George, 1988. "The Cosmology of Life and Mind." In Ravi V. Gomatam and T. D. Thoudam D. Singh, Eds., *Synthesis of Science and Religion: Critical Essays and Dialogues.* San Francisco: Bhaktivedanta Institute.

Winner, Langdon, 1977. *Autonomous Technology.* Cambridge, MA: The MIT Press.

World Commission on Environment and Development (The Brundtland Commission), 1987. *Our Common Future.* New York: Oxford University Press.

Chapter Nine

BREAKING OUT OF THE SYSTEMS QUANDARY

Harold A. Linstone [1]

THE SYSTEMS QUANDARY

The more our knowledge grows, the more complex we perceive the world to be. And the exploding population and technology further compound this complexity. We are hurtling on a wild, uncharted river into the 21st century, sensing a desperate need of some control in navigating our way past the rapids to attain "a just society for future generations". And the systems community has not been able to keep up with the need. In recognizing this quandary, I must look at both theory and praxis (or craft). What follows is unabashedly a personal perspective and reflects my four decades of experience, about evenly split between industry and academe.

On Systems Theory

It is hardly surprising that *systems science* engenders confusion. Some identify it with cybernetics, others with computer programming, with preliminary design in engineering, or with policy analysis. There is von Bertalanffy's *General System Theory.* He includes information theory, set theory, graph theory, game theory, decision theory, automata theory, as well as cybernetics. His purpose is to bring forth principles which apply to all systems, whether they are physical, biological, or sociological. As an example, he holds the exponential growth curve to be "a law of nature" (Von Bertalanffy, 1968, pp. 32-33, 62-63). We immediately notice confusion between *system* and *model.* The ordinary textbook definition of a system comprising a set of elements with relations between them and their attributes really specifies a *perception*, one way of looking at an entity, in other words, a model. (Let us keep in mind that it may be impossible to distinguish a perception from an illusion – a dilemma not uncommon in policymaking.)

---

[1] An earlier version of this chapter was originally presented as the Von Bertalanffy Lecture, which the author was invited to give at the meeting of the International Society for Systems Sciences (ISSS) at Portland State University, Portland, Oregon, July 9, 1990.

An exponential growth curve serves as a model of certain phases of growth in living systems. Similarly, Jay Forrester's *system dynamics* offers a model for the study of certain electrical engineering problems. Like Von Bertalanffy, he sees this as a generalizing principle and applies it to the corporation (*Industrial Dynamics*), the city (*Urban Dynamics*), and the world (*World Dynamics*). He writes:

> All systems that change through time can be represented by using only levels and rates. The two kinds of variables are necessary, but at the same time sufficient, for representing any system.
>
> (Forrester, 1971, p. 18)

James Miller (1978) perceives all living entities – from cell to supra-national system – as systems comprising twenty subsystems, eight processing matter and energy, ten processing information, and two processing all three. These generalizing efforts are remarkable *tours de force.*

Categorization, ordering, and structuring define the most intrinsic characteristic of the systems approach. It can thus serve as a mediator for learning about complexity and for gaining valuable insights to facilitate decision and action. But we must also be aware of its limits. Categorization and imposition of a structure act as constraints. They easily fixate our thinking along a certain direction – an impediment to creativity. For example, I cannot conceive of systems analysis generating a novel mathematical concept such as the use of imaginary numbers.

The systems approach abstracts and idealizes, replacing the real world entity with a much simpler one. It is a technique that has worked beautifully for centuries in science, yielding very elegant and useful models, exemplified by the gas law and the laws of optics. Unfortunately, our prime concern today is with entities which are *seen* as so complex, that they defy this abstraction and idealization process. Exponential growth is severely constricted in a world of limited space and resources. Living systems – individual human beings, businesses, societies – do not confine themselves to the "rational" behavior prescribed by the paradigms of science. Like Buridan's ass, which found itself between two bales of hay and starved to death while unable to decide which to go to, the analyst is in danger of perishing between two options: addressing unsolvable, exceedingly complex, nonlinear problems or reducing them to solvable, but irrelevant, linear ones.

Systems Craft

I came to systems by a different path, applied mathematics rather than general systems theory. In the 1940s, mathematics was applied to military operations as it has been on many occasions over the centuries. For example, in the 16th century Niccolo Fontana, also known as Tartaglia, applied mathematics to problems in artillery operations (Newman, 1956, p. 119). In World War II, the defense of Britain against German bombers and rockets, as well as the defense of ship convoys in the Atlantic, challenged applied mathematicians and led to the evolution of operations research. The connection is very evident in the first text on the subject by Morse and

Kimball (1951). The war proved to be a decisive turning point in military technology – the beginning of the high-tech era. R&D budgets went from millions of dollars to billions. The nuclear bomb, jet aircraft, missiles, radar, and computers were examples of the new age. Planning for weapon systems on the basis of military officers' experience was now no longer possible.

It was a recognition of this state of affairs by the newly formed US Air Force, that led to the creation of the RAND Corporation as a think tank to support military planning. Operations research and policy analysis were the main tools in strategic defense system studies. Economics joined mathematics and engineering as a basic discipline. The initial tasks concerned nuclear weapon strategy; they were perceived to be well-structured problems and quickly resulted in great success and prestige for RAND. Game theory, linear programming, inventory theory, cost-effectiveness analysis and the planning-programming-budgeting concept were developed there in the 1950s. The term *systems analysis* came to be used as a label for RAND's activities.

There was a second source for this term, however. In the late 1940s MIT's Lincoln Laboratory and the Bell Telephone Laboratories were developing complex technological hardware – radar and telephone networks. These were labelled *systems* and preliminary design activities were soon termed *systems analysis*. The military electronics industry took over the label when key Bell people, such as Simon Ramo and Dean Wooldridge, joined this exciting new high-tech world. In 1961 Robert McNamara, the whiz kid president of Ford, was appointed Secretary of Defense and the *systems approach* became an official activity in Washington. To the dismay of the career military officers, the position of Assistant Secretary of Defense/Systems Analysis was created. New weapon procurement proposals had to pass the hurdle of cost-effectiveness analysis and the planning-programming-budgeting concept was formalized for the defense sector.

Alain Enthoven (1967, pp. 72-73), the director of the new office, explained to a Congressional Subcommittee:

> Systems analysis is a reasoned approach to highly complicated problems of choice characterized by much uncertainty...[It] is an effort to define the issues and alternatives clearly...every decision should be viewed in some meaningful context. In most cases, decisions deal with elements that are parts of a larger system. The word analysis is used to emphasize the need to analyze complex systems, that is, to reduce them to their component parts. Systems analysis is not a substitute for judgment; it is an aid to judgment.

He opined that, like medicine by the middle of the nineteenth century, systems analysis was now doing more good than harm. The same year, Max Ways (1967) wrote in *Fortune*'s "The Road to 1977":

> The further advance of this new style [systems analysis] is the most significant prediction that can be made about the next ten years. By 1977

> this new way of dealing with the future will be recognized at home and abroad as a salient American characteristic.

Success emboldened the practitioners to apply their tools to modeling an ever widening range of problems, in particular, ill-structured sociotechnical systems such as transportation, criminal justice, and health care.

In 1970 the International Institute for Applied Systems Analysis (IIASA) was formed through the efforts of America's McGeorge Bundy and Howard Raiffa, Italy's Aurelio Peccei, and others. Today fifteen nations are partners in this think tank, including the United States, Russia, and Germany. Its work has focused on transnational problems – energy initially, environment and climate change today.

The state of the systems craft is well described in the recent *Handbook of Systems Analysis* edited by Miser and Quade (1988).

Critiques

In the 1970s the pendulum began to swing the other way. Criticism of systems theory and praxis began to be heard.

Ida Hoos, in her book *Systems Analysis in Public Policy: A Critique*, objects to von Bertalanffy on three grounds (1984, p. 40):

• The so-called isomorphisms are nothing but tired truisms about the universality of mathematics, i.e., 2 + 2 = 4 prevails whether we consider soap, chickens, or missiles.

• Superficial analogies may camouflage crucial differences and lead to erroneous conclusions.

• Adherence to an alleged irreducibility doctrine renders the approach philosophically and methodologically unsound because it can impede analytic advances. For Von Bertalanffy this is a dubious loss because the isomorphisms have effected the reduction of chemistry to physical principles and life phenomena to molecular biology.

David Berlinski wrote his brilliant, but scathing, book *On Systems Analysis: An Essay Concerning the Limitations of Some Mathematical Methods in the Social, Political, and Biological Sciences*. His basic argument:

> The craving for generality is, of course, an immemorial intellectual inclination; without it, mathematics would be sterile and physics rather uninteresting. What separates systems analysis from such serious studies is simply the gap that inexorably opens between the conception and execution of a set of intellectual ambitions. Mathematical systems theory is...a perfectly serious mathematical discipline. It is only when these subjects are appropriated for analyses executed in the political or social or biological

> sciences that one has systems analysis, with results that are generally, although not inevitably, disappointing.
>
> (Berlinski, 1976, p. 180)

William Thompson saw systems analysis as a Faustian bargain, the evil in *Evil and World Order*:

> The tongue cannot taste itself, the mind cannot know itself, and the system cannot model itself.
>
> (W. I. Thompson, 1976, p. 109)

There was disillusionment in the military application sector as systems experts could not repeat the success of their nuclear superpower confrontation analysis in the messy Vietnam conflict – measures such as "body count" in village security analysis had been dismal failures. There was also disillusionment with system dynamics models – they were seen as little more than cargo cult models, that is, primitive belief systems based on the products of modern technology. By 1980 global modelling was receding as a lively subject of interest for systems analysts and many corporations stopped buying econometric models.

These models were criticized for fatally oversimplifying the real world, giving the false impression that they replicate its behavior. They do not deal with structural change which societal systems experience when under strain. Data-based models rest on an understanding of the past, since all data are past information. This is appropriate if the trends and patterns continue, as in much of basic physics, demographics, and medical diagnosis. It is a fatal flaw in regimes of instability and turmoil or in periods of creativity and innovation. These points have been emphasized elsewhere in this book.

Tversky and Kahneman (1974, 1981) have shown that utility theory and subjective probability theory fail to the extent they assume scientific rationality. They have demonstrated that the framing of a problem changes preference responses, that there is insensitivity to prior probabilities, that recent data outweigh old data, that memorable events are confused with the larger picture, that discouraging odds are ignored, that chance occurrences are underestimated, and that there is inconsistent tolerance for risk. Limitations to the assumption of rationality were discussed in considerable detail earlier in this book.

Probability theory has also been discredited when used for risk analyses of industrial accidents involving the combination of very low probability and very severe consequence. The familiar concept of the "expected value", the product of likelihood of occurrence and outcome of the occurrence, becomes inappropriate when considering, say, an accident probability of 10^{-5} and a consequence of 10^6 fatalities. The calculated product is 10 fatalities. This obviously makes no sense in the real human world. The consequence, 1 million fatalities, is decisive, not the probability or the expected value. No one can give assurance that the catastrophe will not occur next year, even acknowledging its very low likelihood. Such a happening would in no way invalidate the calculated probability 10^{-5}. The decision becomes a question of societal or community acceptance of the consequence rather than a technical question.

Expert systems also prove to be oversimplifications in many applications. The model is only as good as the knowledge engineer's understanding of the superstructure of the experts' thinking – which may be inaccessible. The experts may not be able to convey their expertise in simple rules. Human experts work with an entire gestalt in social and sociotechnical areas. They may apply rules that are not preexisting and the combination of factors is always novel to a certain degree (Ascher, 1989). For exceptionally intuitive experts, even important links of deduction may remain unknown. Classical mathematicians like Cardan, Poincaré, and Hadamard, as well as modern physicists like Penrose, are in awe of the power and mystery of intuition.

Another criticism is the wide chasm between theory and craft. General systems theory is seen as pretentious, yet bereft of content. In pursuit of the craft there is a similar chasm between analysis and implementation. The analysts work hard on their analysis, present it to the client, and walk away. The output is a study with little concern for problems of implementation.

Summing Up

In a world of explosive growth in population and technology, we must increasingly address real-world systems which exceed our grasp. The decisionmaker badly needs help and systems theory and praxis fall far short. What to do?

There are at least four options:

1. Focus on general systems theory (GST). As with all basic research, the ultimate usefulness is impossible to fathom. Who could have forecast the value of the algebra of Galois or Boole? It is conceivable that, with its unique approach, GST may make a significant contribution to the great problems of science – the origin of life, quantum gravity, brain theory, sociobiology, artificial intelligence, or language acquisition. On the other hand, major advances are coming from researchers outside the confines of GST, as evidenced by catastrophe theory and chaos theory. Although GST is intellectually stimulating, its survivability in an environment of institutional crises and complex real-world challenges may be precarious.

2. Tackle applications to messy real-world problems by the old system/model thinking, yielding deceptively imposing, yet woefully simplistic, models. As with *trompe l'oeil* paintings, there is much less there than meets the eye – a sure path to demise.

3. In the area of ill-structured societal systems, rely on the self-organizing nature of such systems. Focus instead on applications to well-structured problems using well-known tools and improving them – a modest aim to assure survival of the practitioners. There are plenty of opportunities; for example, in operations research the Department of Labor forecasts needs that imply an increase in professionals of

55% by 2000 – from 55,000 to 85,000. As for total reliance on self-organization of ill-structured systems in a competitive world environment, the risk is obvious: as with all aspects of evolution there are winners and losers.

4. Focus on new thinking relevant to messy real-world problems in both theory and praxis – also a risky path, but less passive and more appealing to an optimist like myself.

It is this latter option 4 to which I shall now turn.

BREAKING OUT

On Theory: Develop Deeper Insights About Complex System Behavior

The bad news is that in the real world almost all systems are nonlinear. They can be stable, converging to an equilibrium, oscillate stably, diverge unstably, or exhibit persistent chaotic behavior within predictable bounds. We labor under much ignorance in handling nonlinear relationships. Nothing illustrates this better than our astonishment only recently (Gordon and Greenspan, 1988) in discovering the strange, chaotic behavior of a simple recursive equation such as

$$x_{next} = x\,k\,(1 - x),$$

where k is a constant and each subsequent value of x (i.e., x_{next}) is calculated from the preceding one (x). The surprising behavior of this simple nonlinear equation can in no way be predicted by the usual examination of the equation. More generally, chaos theory shows that systems once thought to behave in an orderly fashion may simply be operating in a regime that later will shift to chaotic or unstable behavior. Apparent patterns may, in fact, be random and the information that these systems appear to hold really may be illusory. Furthermore, in the real world almost everything interacts with everything. Neither the elements nor the couplings nor the boundaries are uniquely definable.

The good news is that we are learning rapidly. The recent surge in chaos theory also tells us that systems that appear disorderly may, nevertheless, have an underlying order that yields some predictability. We are beginning to rethink the concept of system evolution. Living systems may be viewed as combining chaos with feedback, a kind of learning process. Chaos and feedback appear to underlie system adaptability and diversity, in part shaping the dynamics of social systems (Hansson, 1991). Thus randomness plays a vital role in successful complex system development. And the principle of "success breeds failure" or "stability leads to instability" is now seen as central to progress generally.

The Lotka-Volterra predator-prey equations are another excellent example of the appropriate use of models, that is, to provide insight for understanding complex system behavior. Two species x and y prey upon each other. If a, b, p, and q are constants, the equations

$$dy/dt = (a - bx)\,y \quad \text{and} \quad dx/dt = (py - q)\,x$$

describe the population change behavior of the two species. As Berlinski points out, the equations are simple, but the analysis quite subtle.

> This suggests a maxim for the mathematical modeler: start simply and use to the fullest the resources of theory. It is this prescription taken neatly in reverse that characterizes [Forrester's system dynamics]; there the prevailing pedagogical maxim has been: pile up an imposingly complex system of equations and then subject them to an analysis of ineffable innocence...[It] is not always necessary to subject an analytically intractable system to simulation in order to understand it qualitatively; correspondingly, qualitative insights are at greater depth than partially quantitative results....[While] the predator-prey equations offer an appraisal of reality that is at once severely idealized and incomplete, what insights they offer are at some depth. In [Forrester's system dynamics] the reverse holds true: the scope is unbounded but the insights are slight. The moral: look to systems for which a qualitative analysis is possible.
>
> (Berlinski, 1976, pp. 83-85)

C. S. Holling (1977) showed how the introduction of a time lag into the Lotka-Volterra equations produces instability. Limit cycles and stability boundaries appear. In a totally known world, we can readily determine a *fail-safe* policy once we know the stability domain. In a partially known or unknown world, we are forced to a *safe-fail* policy. We can minimize the cost of failure, not the probability of failure. It is a vital distinction often missed, particularly by engineers – surprising because, as all other human beings, they are themselves examples of safe-fail systems.

In the two areas discussed here, chaos theory and stability domain analysis, the vital role of the computer as laboratory to experiment with complex system behavior is apparent. Chaos theory has helped us to understand better that there are regions of irreducible unpredictability. Coming from a very different direction, French computer expert J. C. Simon (1979) arrives at a similar conclusion about the limitations of computable models of reality. Using the computer as a laboratory tool, complexity is reflected in large computer programs through the number of errors remaining in the steady state after debugging. It is almost impossible to eliminate all semantic errors. In large programs the minimum number of errors rises exponentially somewhere between 10^4 and 10^5 instructions. Beyond a certain level of complexity it is impossible to completely predict the behavior of a deterministic algorithm in the program: a small variation in coding may have extraordinary consequences. Furthermore, there does not exist a program able to correct the errors of any program. This is actually a consequence of Gödel's theorem, one form of which may be stated as follows (Casti, 1990, p. 381):

> There is no universal algorithm U able to decide if any algorithm A will stop computing by examining only the data and the algorithm A.

Or in another form,

> For every consistent formalization of arithmetic, there exist arithmetic truths that are not provable within that formal system.

This also has a bearing on the question of what is simple and what is complex. Complexity depends in part on language. A representation may be simple in one computer language and complex in another.

There are intriguing implications for evolving systems in the next century. Complexity has advantages and disadvantages. An advantage: partial unpredictability facilitates human freedom of choice. A disadvantage: complexity increases errors; although these may be minimized by system decentralization or decomposition into subsets with considerable local autonomy, diffusion of control can also lead to system paralysis. We see in the United States disturbing signs of the tyranny of paralyzing political power exercised at the lowest levels. One-issue voting and insistence on individual rights over responsibilities can prevent action and foil the common good (Garrett Hardin's "the tragedy of the commons").

Indeed, information technology in all its aspects, from communication to computers, from neural networks to biochips, makes possible remarkable institutional change in both directions simultaneously - globalization and localization, concentration and distribution of control, centralization and decentralization, hierarchical (vertical) and flat (horizontal) organization. What is a desirable balance? How fluid should a structure be? How dangerous is the increasingly high speed of information movement for good decisionmaking? These questions are of profound interest for governments and corporations fearing impending obsolescence in the information age of the 21st century (Linstone, 1989b). National boundaries are losing their importance as we see signs of diffusion of power in the Soviet Union, and separatist pressures in Canada, the United Kingdom, Italy, and Spain, while simultaneously transnational integration is occurring in Europe. Corporations are gearing for the new era with smaller, flatter core organizations as well as new regional and global linkages with suppliers, markets, and venture groups. The "national corporation" is becoming passé.

More insight about the behavior of complex real-world systems has been provided by Perrow (1984) with regard to the physical risk of catastrophe. We have already noted the weakness of probability theory when confronted with very low likelihood/very severe consequence situations. Understanding the nature of such system coupling (tight or loose) and the interactions (expected/uncomplicated or unexpected/intricate) proves useful. Tight coupling denotes a lack of buffer or slack between system elements; loose coupling correspondingly indicates a lack of tight control. Nuclear plants, dams, and some drug processing plants are tightly coupled; universities, most manufacturing, and post offices are loosely coupled. An example of a simple interaction is a linear production assembly line. Intricate/unexpected interactions are found in cases of many common-mode connections, multiple feedback loops, and interacting controls. Dams and most manufacturing plants involve simple interactions; nuclear plants and universities intricate interactions. In general, tight coupling calls for system centralization; intricate interactions are easier to manage with decentralization. Consider the following combinations and their control implications:

- simple interactions/tight coupling (dams) — centralization
- simple interactions/loose coupling (manufacturing) — either centralization or decentralization
- intricate interactions/tight coupling (nuclear plant) — incompatible demands!
- intricate interactions/loose coupling (universities) — decentralization

The dilemma posed by tightly coupled and intricate interaction systems such as the Three Mile Island and Chernobyl nuclear power plants, and the Bhopal chemical plant thus becomes evident. Advancing technology and exploding population will present us with an ever growing number of such catastrophic system accidents.

In one sense the behavior of complex systems is un-American. We are culturally true believers in the maxim "more leads to more" or "bigger is better" – more defense means more security, more money means a better quality of life, for example, better education and more health. Mitroff (1989) has pointed out the paradoxes which appear to accompany system complexity in both our national defense and economy. We are reluctant to accept complex system behavior when more may lead to less and less may mean more. Some of us are very uncomfortable with the insistence of the true systems thinker that "everything interacts with everything".

On Praxis: Rethink the Task of the Systems Craft

I see a shift in emphasis to two tasks:

1. *systems analysis as a pedagogical or learning/training concept for planners and decisionmakers* (Michael, 1989)

Models can serve two distinct roles:

(a) a representation of a complex system for *prediction*, replicating it in a mathematical format, for example, simulation using a large-scale, computerized model;

(b) an abstract thinking aid aimed at *explanation*, revealing or illuminating some aspect of system behavior in a simple way, unlocking an insight, answering important questions of a prescriptive nature about the system.

We yearn for the first, indeed, we are in the practical world frequently more interested in a good forecast than in a good explanation. But we are often disappointed. As Nobel Laureate Herbert Simon (1988) has observed:

> The rapid rise in the last decade of chaos theory...has shown the fundamental reasons why such prediction may be impossible, now and forever. These are linked to the complexity of many systems of interest. Nature is capable of building, on a scale of microcosms or macrocosms or any scale in between, systems whose complexity lies far beyond the reach of our computers and supercomputers – present or prospective.

But we are increasingly using models to good effect in the second role, as explanatory tools or thinking aids.

Proposed steps for improvement include:

- Provision of training to decision makers (a) to overcome the common tendencies of reductionism, trivialization, self-delusion, issue avoidance, and nearsightedness; (b) to reinforce receptivity to change and sensitivity to new insights.

- More effective use in system design of *Gedankenexperimente* (thought experiments) and insights drawn from the models developed in systems theory, for example, fractals, cusps, and stability domains.

- More effective use of historical case studies to observe complex system behavior, for example, imperial overreach leading to dissipation of resources (England), the rich pursuing their own ends and ignoring the widening gap between them and the poor (Roman Empire), institutional obsolescence (General Motors – U.S.), and corporate myopia (U.S. consumer electronics industry).

- Appreciation of the "fuzzy gambling" nature of decisionmaking.

> Far-reaching changes are required in systems-analysis praxis in order to handle hard uncertainties correctly and provide help with fuzzy decision gambles. (Dror, 1988, p. 264).

- Use many "what-if" scenarios rather than a single one to train personnel to respond to the unpredictable behavior of complex systems.

To illustrate, let us expand on the last point. Complex self-organizing systems are not plannable in the traditional business sense of planning. We have a clue when we compare business planners with military planners. The former look for the most likely scenario in determining a desirable strategy; the latter do not. They seek robustness instead, a strategy that can be effective in an environment subject to all kinds of surprises. The training of officers using numerous war games and the development of many contingency plans reflect this approach. Monitoring the unpredictable environment and maintaining strategic flexibility become essential. A possible analogous path for business and government planners might be the use of many (quasi-random) "runs" of a currently stable system subject to external perturbations to determine unstable future states and identify desirable long-term outcomes. This would provide feedback analogous to that which we see in successfully evolving biological systems. Thus chaos theory may open up entirely new ways to deal with complex systems, to use randomness constructively.

2. *systems analysis as an evaluative mechanism rather than a problem-solving framework, with output focus on the process, not merely the product* (Thompson and Warburton, 1985)

Several steps for improvement can be mentioned here also:

• Recognition that, with complex systems, the problem frequently is to know what the problem is; each party involved with a given system may define the problem quite differently.

For example, energy development is usually seen in economic terms by industry and in physical hazard terms by environmentalists.

• Recognition that the most meaningful "facts" may be the institutions, not the data.

As Thompson and Warburton (1985) learned in their environmental study of the Himalayas, there are often highly unreliable quantitative data, whereas the institutions and their concerns are quite clear.

• Recognition that uncertainty is generated by institutions for institutions to maintain their viability.

• Development of a process orientation by (a) focusing on the bottom-up (tinkering), not just the top-down (grand design), approach, and (b) addressing the conflicting planning horizons (or discount rates) of the involved parties.

• Development of robust and resilient action strategies, including identification of points of leverage.

• Use of multiple perspectives rather than a single perspective.

It should be stressed that creative leadership is essential for these steps to be effectively implemented and this places additional importance on task 1.

I shall expand briefly on just two of these aspects, uncertainty and multiple perspectives.

a. *Treatment of uncertainty.* Textbook exercises do not prepare us for the world of complex systems. The spectrum between certainty and complete ignorance is wide, and it has been poorly served by systems analysis. For example, decision making under uncertainty is conventionally addressed by introducing probability, either frequency or subjective. We have already pointed out the unreliability of such probability calculus in important situations. That is only the beginning and we need to probe much further.

As we have just seen, there are situations where certainties prove to be false or contradictory. There are situations where the alternatives are known but not their probabilities (quantitative uncertainty). There are situations where the alternatives are not even known (qualitative uncertainty). An example is the nature of the stable societal state following temporary system instability (revolution). As Michael has noted, human activities have an emergent quality – the whole is unpredictably greater than the sum of its parts (Michael, 1989, p. 80).

There are also the surprise events, like the low probability/severe consequence industrial occurrences noted earlier. There are cases where the uncertainties may be reduced (soft uncertainties) and cases where they cannot be resolved (hard uncertainties). Finally, there is the case where we do not even know whether the uncertainty is hard or soft, that is, complete ignorance (Dror, 1988).

Uncertainty is normally seen as undesirable, to be avoided or minimized. As uncertainty increases when we move into the future, it has been the common reponse to shrink our planning horizon or discount the future. It seems paradoxical to think of uncertainty in a positive way. But we recall that the Heisenberg principle recognizing irreducible uncertainty had a vital positive effect on the development of quantum mechanics – perhaps one of the two most important modern contributions to understanding our physical universe (relativity theory being the other). Thompson and Warburton (1985, p. 9) have noted that a complex system draws in multiple institutions. They all have their own perceptions, problem definitions, expectations, and rationalities. These institutions may well generate and impose uncertainty to protect their legitimacy. For the diverse institutions to survive, sufficient uncertainty is generated to swamp all the contradictions that are inherent in the pluralized positions they hold. Thus uncertainty fills a vital functional role in facilitating institutional flexibility and adaptability – and survival.

b. *Use of multiple perspectives.* In discussing the management of complex systems, Churchman (1977, p. 90) has drawn on Hegel to emphasize that

> the mature individual is the individual who can hold conflicting world views (*Weltanschauungen*) together at the same time, and act, and live, and that his or her life is enriched by that capability – not weakened by it....That is complexity, that is really complexity! To be able to see the world globally...and to see it as a world of unique individuals...

He found a way out of the systems dilemma by using the ideas of Edgar Singer, Jr. (Churchman, 1982). A Singerian mode of inquiry

• is a meta-inquiring mode – it "sweeps in" data-based, model-based, dialectic, and other modes as needed;

• is pragmatic, that is, the truth content is relative to the overall goals and aims of the inquiry;

• constantly attempts to sweep in new aspects, is, in fact, non-terminating;

• recognizes that the system designer and analyst are themselves fundamental parts of the system; their psychology and sociology are inseparable from the system's physical representation.

We at once notice that models, as "objective" representations of reality, are now only one facet of the systems approach – reflecting one kind of perspective. We have a clear break with the von Bertalanffy – Miller – Forrester view of the systems field as virtually synonymous with such modelling.

I set off along this path from the policy application by Harvard's Graham Allison (1971). Three types of perspective are now used: technical (T), organizational/institutional (O), and personal/ individual (P) (Linstone, 1984, 1989a). It is vital to recognize that each type uses distinct paradigms. The first of these (T) corresponds to the traditional systems approach with its problem-solution focus, reductionism[2], model- and data-based analysis, optimization, cause and effect, quantification, and objectivity. However, these do not carry over into O and P. Table 9.1 compares the characteristics of the three types.

For example, T focuses on problem solving, O on process and action. Cause and effect are associated with T, challenge and response with P. The use of uncertainty discussed under the preceding heading is characteristic of the O perspective.

The concept has now been applied successfully over a wide spectrum of complex system problems, ranging from chemical to oil industry disasters, from technology diffusion within a U.S. corporation to Third World development, from California's health care delivery to its medfly crisis (see, for example, Bowonder and Linstone 1987, Linstone et al. 1987, Linstone 1990, Lorraine 1991, Tarr 1991, Umbdenstock 1981).

---

[2] A reason for the popularity of reductionsim: chopping a problem into pieces ultimately leads to Heinz Von Foerster's First Law: "the more complex the problem which is being ignored, the greater are the chances for fame and success" (Von Foerster, 1972, p.1). Success is obviously viewed here in academic terms, as the publication of papers, even if they are pedantic explorations of minute details of trivial significance.

Table 9.1 – Paradigms Associated with the Three Perspective Types

	Technical (T)	Organizational (O)	Personal (P)
World view	science-technology	view unique to group or institution	individuation, the self
Goal	problem solving, product or output	action, stability, process	power, influence, prestige
Mode of inquiry	sense-data, modeling, analysis	consensual and adversary	intuition, learning, experience
Ethical basis	logic, rationality	justice, fairness	morality
Planning horizon	far	intermediate	short for most
Communication	technical report briefing	insider language, out- siders misunderstood	personality and charisma desirable
Other characteristics	cause and effect, problem simplified, idealized	agenda (problem of the moment), problem delegated and factored	challenge and response, hierarchy of individual needs
	need for validation, replicability	political sensitivity, loyalties	compassion, altruism, greed, love, hate
	claim of objectivity	reasonableness	copes with only few alternatives at a time
	optimization (seeks best solution)	satisficing (seeks first acceptable solution)	filters out inconsistent ideas
	quantification	incremental change	fear of change
	trade-offs, cost-benefit studies	standard operating procedures	leaders and followers
	use of averages, probabilities	compromise and bargaining	creativity and vision by the few
	uncertainties noted (on one hand...)	uncertainties used for self-preservation	need for certainties, beliefs, illusions

The crucial lessons learned:

• Each perspective provides insights not obtainable with the others;[3] the limitations of the conventional perspective of systems analysis in dealing with complex systems involving human beings becomes crystal clear.

• The perspectives may reinforce or conflict with each other; their cross-cuing and integration present us with a Singerian inquiring mode.

• There is no "correct" or "comprehensive" set of perspectives and intraparadigmatic concepts cannot be assumed to be interparadigmatic[4].

• The introduction of the organizational and personal perspectives bridges the gap between analysis and action, between the modeler and the real world.

It becomes quite obvious that the concept leads us from a quasi-one-dimensional to a quasi-three-dimensional appreciation of complex systems. At this point it is interesting to recall the concerns raised by Operations Research Society past-president William Pierskalla at the 1984 Society meeting and by Tomlinson and Kiss in their 1984 book: *Rethinking the Process of Operational Research and Systems Analysis*.

Pierskalla (1984) asked: Have we reached the limits of the operations research paradigm? And suggested that we may need to coopt other paradigms including those of human behavior. Tomlinson and Kiss (1984, p. xi) saw the weaknesses in systems analysis as follows:

> One... assumption was that the 'hard' part of a problem – which could be expressed in mathematical terms – could usefully be isolated from the human and organizational elements which could thus be eliminated from the analysis. Another false assumption was that the implementation was an entirely separate activity from the analysis itself.

Whereas the traditional analysis focuses on T, implementation must focus on O and P. This explains the common failure of communication between the systems practitioner and the client or user of his or her output.

---

[3] Analogies: in physics - the simultaneous use of both wave and particle theories of light; in biology - the information perspective and the dynamic system perspective on living systems.

[4] An analogy in mathematics: theorems in Euclidean geometry do not carry over into Riemannian or Lobachevskian geometry. One cannot prove the theorems in one geometry within the context of another. The replicability and validation concepts standard with the T perspective are not appropriate to O or P.

The use of multiple perspectives also helps to avoid misunderstanding when systems cross cultural boundaries (Bowonder and Linstone, 1987, Linstone, 1987). Far too many systems studies (for example, technology transfer process) have been assumed to be culture-independent. With the fading of old national boundaries and the rise of networks with greater local autonomy as well as transnational linkages, successful corporations and governments will be those best able to communicate across cultural boundaries. An interesting illustration of a valuable communication technique unfamiliar to Westerners is the Japanese *nemawashi* (a word that defies translation) to solve conflicts and promote group harmony (Saito, 1982, 1990).

Systems and Communication

Besides the interface between systems analyst and client, the inherent nature of the subject calls for some additional comments.

a. *Systems theory/praxis and the academic.* We began our discussion of systems theory with a recognition of the confusion engendered by general systems theory and systems analysis. In the university there is skepticism about the soundness of interdisciplinary programs generally and systems programs specifically. As noted in the review of critiques, there is considerable basis for concern. Consequently, these programs have a difficult time defining themselves to the academic community as well as to funding agencies. Institutes that focus on the craft and generate useful contributions, such as RAND and IIASA, have a somewhat easier time. But even they have to struggle; for example, the U.S. government recently suggested that financial support for IIASA would be easier if it changed its name to the International Institute for Environmental Studies.

b. *Complex systems and the public.* The irony today is that information technology is advancing our information handling capabilities by leaps and bounds, yet communication of complex ideas remains a very serious stumbling block. In a media-oriented society, rapid visual communication dominates. Most people obtain their news in short, easily digestible spurts from television. Symbolism displaces substance. Easily understood issues elbow out presentation of complex ones. Trillion dollar national debts and the enormous savings and loan industry losses have surprisingly little impact because they cannot be grasped visually. On the other hand, flag burning, oil spills, abortion protests, gun owner rights, and terrorist/hostage events are effectively communicated. Complex issues are avoided or converted into false, simple ones that are more easily communicated – exploiting a strength of the P perspective.

Systems and Ethics

Churchman defines wisdom as thought combined with a concern for ethics, which he translates into an "unbounded" systems approach that draws in the ethical aspect (Churchman, 1982; Tomlinson and Kiss, 1984, pp. 70-71). Using multiple perspectives, I associate the ethical basis for the T perspective with rationality, for the O perspective with justice or fairness, and for the P perspective with morality (Linstone, 1984, 1989a). Integration of the perspectives itself presents an ethical challenge. The weighting of perspectives in integrating them and reaching a decision clearly implies ethical considerations. How should individual and societal needs be balanced? How should near-term and long-term needs be balanced?

Discounting the future for the present (Linstone, 1973; 1984, pp. 20-24), organizational rigidity, educational failures, pandering to the basest human instincts – these all raise ethical questions that have a place in the "unbounded" systems approach. How can we even discuss "a just society for future generations" when America's young people 18 to 24 are poorly educated and indifferent to societal concerns. They strongly discount in both space and time dimensions, focusing only on the here and now. According to the most recent surveys:

> The typical young person doesn't want to hear about it "unless it's knocking on my door"...[a 22-year old entering law school: "people my age group are] only concerned about issues that affect them. When the drinking age went up, quite a few people were upset." (*New York Times* June 28, 1990, pp. A-1, A-12)

The fact that most care only about their rights, not their responsibilities, is the culmination of a trend that seems to have begun after World War II, a shift of responsibility from the individual to "the system" – government, companies, schools, etc. – abetted by the lawyerization of American society.

This leads us directly to what, in my view, is one of the most intriguing ethical aspect of systems, the question of control or the power diffusion/concentration balance which I already mentioned earlier. Individualism versus societal cohesion, autonomy versus heteronomy, democracy versus authoritarianism, self-management versus outside management, increasing degrees of freedom versus decreasing degrees of freedom – these all require crucial ethical choices in system management (Von Foerster, 1977).

FINAL WORDS

We considered four options:

1. Focus on general systems theory with little perceived connection to the real-world challenges requiring action.

2. Tackle applications to messy real-world problems by the old system/model thinking.

3. Focus on applications to well-structured problems using well-known tools and improving them.

4. Focus on new thinking in both theory and praxis relevant to messy real-world problems.

I personally see the last as the only promising path of escape from the systems quandary and it was the subject of the preceding discussion. I pointed to some of the promising possibilities in both theory and praxis. And I touched on the communication and ethical aspects that demand attention.

In a world of bursting population and technology, complexity appears to be growing inexorably. The evolving "knowledge society" rests on much *terra incognita*. Planners and decision makers will need stronger support and it is my conviction that they have a right to expect the systems community to "grasp the nettle". But this will mean changing the training in the university, a frustrating task in an obsolete educational bureaucracy.

The university is compartmented in a way real-life problems are not. Too many professors teach old thinking in courses devoted to well-structured systems year after year and have never themselves had to deal with messy ill-structured systems. Texts are used that are filled with cookbook techniques, neatly defined problems and solutions. Training which combines theory and praxis, as in medicine, suggests one path to improvement. Bringing successful practitioners and researchers into closer contact is another. At lower school levels development and cultivation of the students' long-term perspective should be a high priority goal, together with the strengthening of vital basic skills. Breaking out of the present quandary cannot be done by moving into the 21st century with 20th century thinking and 19th century institutions. That is where we must start.

REFERENCES

Allison, Graham T., 1971. *Essence of Decision: Explaining the Cuban Missile Crisis.* Boston, Mass.: Little, Brown & Co.

Ascher, William, 1989. "Limits of Expert Systems for Political-Economic Forecasting". *Technological Forecasting and Social Change* Vol. 36 (2), 137-152.

Berlinski, David, 1976. *On Systems Analysis: An Essay Concerning the Limitations of Some Mathematical Methods in the Social, Political, and Biological Sciences.* Cambridge, MA: The MIT Press.

Bowonder, B. and Harold A. Linstone, 1987. "Notes on the Bhopal Accident: Risk Analysis and Multiple Perspectives", *Technological Forecasting and Social Change*, vol. 32 (2), 183-202.

Casti, John L., 1990. *Searching for Certainty.* New York: William Morrow & Co.

Churchman, C. West, 1977. "A Philosophy for Complexity". In Harold A. Linstone and W. H. Clive Simmonds (Eds.), *Futures Research: New Directions.* Reading, MA: Addison-Wesley, 82-90.

Churchman, C. West, 1982. *Thought and Wisdom.* Seaside, CA: Intersystems Publications, Chapter 10.

Dror, Yehezkel, 1988. "Uncertainty: Coping With It and With Political Feasibility". In Hugh J. Miser and Edward S. Quade (Eds.), *Handbook of Systems Analysis.* New York: North-Holland, 247-281.

Enthoven, Alan, 1967. Hearings before the Subcommittee on National Security and International Operations of the Committee on Government Operations, U.S. Senate, Part 2, September 27 and October 18, 1967. Washington, D.C.: US Government Printing Office.

Forrester, Jay W., 1971. *World Dynamics.* Cambridge, MA: Wright-Allen Press.

Gordon, Theodore J., and David Greenspan, 1988. "Chaos and Fractals: Tools for Technological and Social Forecasting", *Technological Forecasting and Social Change,* Vol. 34 (1), 1-25.

Hansson, P. A., 1991. "Chaos: Implications for Forecasting", *Futures,* Vol. 23 (1), (Jan./Feb. 1991), 50-58.

Holling, C. S., 1977. "The Curious Behavior of Complex Systems: Lessons from Ecology". In Harold A. Linstone and W. H. Clive Simmonds (eds.), *Futures Research: New Directions.* Reading, MA: Addison-Wesley, 114-129.

Hoos, Ida R., 1972, 1984. *Systems Analysis in Public Policy: A Critique.* Berkeley, CA: University of California Press, 1972, revised ed. 1984.

Linstone, Harold A., 1973. "On Discounting the Future", *Technological Forecasting and Social Change,* Vol. 4 (4), 335-338.

Linstone, Harold A., 1984. *Multiple Perspectives for Decision Making.* New York: North-Holland.

Linstone, Harold A., 1989a. "Multiple Perspectives: Concept, Applications, and User Guidelines", *Systems Practice,* vol. 2 (3), 307-331.

Linstone, Harold A., 1989b. "Mediacracy, Mediocracy, or New Democracy: Where Are the Information Age Jeffersons and Madisons When We Need Them?", *Technological Forecasting and Social Change* vol. 36 (1-2), 153-169.

Linstone, Harold A., 1990. *Multiple Perspectives on the Alaska Oil Spill.* Portland State University, Portland, OR. April 1990.

Linstone, Harold A., Jacob Fried, Wang Yinling, and Shu Hui, 1987. *Multiple Perspectives in Cross-Cultural Systems Analysis: The China Case.* Report to the National Science Foundation. Portland State University, Portland, Oregon. October 1987.

Lorraine, Hilary, 1991. "The California 1980 Medfly Eradication Program: An Analysis of Decision Making Under Non-Routine Conditions", *Technological Forecasting and Social Change,* Vol. 40 (1), 1-32.

Michael, Donald, 1989, "Forecasting and Planning in an Incoherent Context", *Technological Forecasting and Social Change,* vol. 36, (1), 79-87.

Miller, James G., 1978. *Living Systems.* New York: McGraw-Hill.

Miser, Hugh and Edward Quade (eds.), 1988. *Handbook of Systems Analysis*. New York: North-Holland.

Mitroff, Ian I., 1989. "Slip, Sliding Away: Are We Any Closer to Understanding the Stuff of Which Social Reality Is Made?", *Technological Forecasting and Social Change*, vol. 36 (1), 69-78.

Morse, Philip M., and George E. Kimball, 1951. *Methods of Operations Research.* MIT Press and John Wiley.

Newman, James R., 1956. *The World of Mathematics*. New York: Simon & Schuster.

Perrow, Charles, 1984. *Normal Accidents*. New York: Basic Books.

Pierskalla, William P., 1984. Speech at the National TIMS/ORSA Meeting, San Francisco, Cal., May 1984.

Saito, M., 1982. "Nemawashi : A Japanese Form of Interpersonal Communication", *et cetera*, vol. 39 (3), Fall 1982, 205-214. Also "Observations on the Style of Japanese Negotiations", talk presented at World Futures Studies Federation Meeting, Budapest, May 1990.

Simon, Jean C., 1979. "Complexity Concepts and the Limitations of Computable Models", *Technological Forecasting and Social Change*, vol. 13 (1), 1-12.

Simon, Herbert, 1988. Speech at IIASA Conference on Perspectives and Futures, June 14-15, 1988, Laxenburg, Austria.

Tarr, Steven, 1990. The Knowledge Systems Transfer Project: A Multiple Perspective Investigation into the Integration of a New Technology within a Business Unit. Doctoral Dissertation. Portland, OR.: Portland State University.

Thompson, William I., 1976. *Evil and World Order*. New York: Harper & Row.

Thompson, Michael and Michael Warburton, 1985. "Decision Making Under Contradictory Certainties: How to Save the Himalayas When You Can't Find Out What's Wrong With Them", *Journal of Applied Systems Analysis*, vol. 12 (1), 3-34.

Tomlinson, Rolfe. and Istvan Kiss, 1984. *Rethinking the Process of Operational Research and Systems Analysis*. Oxford: Pergamon Press.

Tversky, Amos and Daniel Kahneman, 1974. "Judgment Under Uncertainty: Heuristics and Biases", *Science*, Vol. 185 (Sept. 27, 1974), 1124-1131.

Tversky, Amos, and Daniel Kahneman, 1981. "The Framing of Decisions and the Psychology of Choice", *Science*, Vol. 211 (30 January 1981), 453-458.

Umbdenstock, Linda, 1981. The Prenatal Regionalization Project: A Study in Form and Development. Doctoral Dissertation. Portland, OR: Portland State University.

Von Bertalanffy, Ludwig, 1968. *General System Theory*. New York: G. Braziller.

Von Foerster, Heinz, 1972. "Responsibilities of Competence". *Journal of Cybernetics*, Vol. 2 (2), 1-6.

Von Foerster, Heinz, 1977. "The Curious Behavior of Complex Systems: Lessons from Biology", in Harold A. Linstone and W. H. Clive Simmonds (eds.), *Futures Research: New Directions*. Reading, MA: Addison-Wesley.

Ways, Max, 1967. "The Road to 1977", *Fortune*. Vol. 75, 93-95, 194-195.

Chapter Ten

DECISION MANAGEMENT, THEN AND NOW, 1790, 1890 AND 1990: SELECTED SCENARIO MODELS

Jacques G. Richardson*

INTRODUCTION

Thus far we have examined in this book numerous facets of the nature of the systems underlying the processes of policymaking and decisionmaking. In the two chapters immediately preceding, Willis Harman reviewed some holistic approaches to societal transformation, whereas Harold Linstone sketched out some modalities for breaking away from what he calls the systems quandary. As the author of the final contribution to this volume, I wish to add some temporal, geographic, historical and complexity dimensions to the models and cases presented throughout the volume. Unlike several earlier chapters, this chapter presents little theory.

Rather, we shall examine here three cases of the evolution of systemic demand and design, together with the decision processes leading to their acceptance. The illustrations selected progress from the relatively concrete and simple to the somewhat less concrete but more complex; and from the nearly benign, through the neither benign nor malign, to the threateningly malign.

The examples or "social experiments" analyzed in varying degrees of depth are:

1. - The selection of the locale and configuration for the capital at Washington (United States);
2. - The choice of the kind of police system for adoption at the national level (Japan);
3. - The options open or closed to both sides in the Arabian or Persian Gulf Crisis/War of the 1990s (Iraq and part of the rest of the world).

The first two examples are presented as little more than vignettes. They should be useful, nevertheless, in appreciating how decision rises in increasing complexity as time passes and experience grows.

---

* The author expresses his gratitude to Mircea Malitza, Romanian mathematician, system analyst and diplomat, for his contribution to the early conception of this chapter. Ambassador Malitza is currently president of the Futures Studies Commission of the Academy of Sciences in Bucarest. The author is also grateful to Julian Behrstock, Nahum Joel, Urs Hammer, D.W. Doyle and David Spurgeon for the careful reading of various drafts and their valuable suggestions.

I cite complexity - a factor weighing more and more heavily as we proceed through the illustrations that follow - in the double sense that Warren Weaver proposed in the late 1940s:

> ... organized complexity and disorganized complexity. The latter is due only to a very large number of variables and is statistically predictable. Organized complexity has to do with patterns that do not submit to the rules of statistics. This means that in organized complexity balance is not [attributable] to statistics but to how parts interact. Complexity measures this interaction. (Hansson, 1991)

The examples presented here subsume, besides organized complexity:

1. -Growth in knowledge of how to coordinate and manage affairs, including the organization of citizens and their daily activities.
2. - The accumulation of experience of the body politic as a function of the passage of time and the transmittal of received ideas from one generation to the next, at the discrete levels of the local constituency, the nation, and the international community.

Societal effects and long-range implications, together with the symbolism of the illustrations used, are indicated too. All these are especially germane in terms of experience to be considered - and then selected, adapted or rejected - when similar circumstances recur.

SOME DEFINITIONS AND LIMITATIONS

Definitions of Terms

For the purposes of the present chapter, the reader may consider the expressions "example," "experiment," "paradigm" and "model" as virtually interchangeable. *Example* is usually a random illustration. *Experiment* connotes a trial intended to reveal something hitherto not well understood. *Paradigm* means an example with built-in comparisons and contrasts. A *scenario*, by extension, is an outline or plot connoting the drama of interrelated sequential events.

The experiments proffered here (all of which are based on historical fact and largely on known quantities) contrast sharply with each other and differ somewhat from others of their times. Although the reader may demand more examples in each of the three categories for purposes of comparison and contrast, space forces us to limit ourselves to discussion of one of each type.

Each paradigm described is separated by a century of time and the correspondingly exponential growth and structural change of societal experience and human knowledge during the hundred years' span. Such growth and change, experience tells us, are less likely to be linearly continuous and smooth; rather, they bear a resemblance to the notion of "punctuated equilibrium" in evolutionary biology introduced by Stephen Jay Gould and his associates (Gould, 1983). These concepts are allied to the discussions of evolution and structural change in Chapters 1,2 and 4 of this book.

Changes accruing with time and experience underline, indeed, the rise in the number of factors to be considered, then interpreted, by those formulating policy and making the decisions that will take effect in proximal or distant futures.

Limitations to Knowledge

Hegel, Coleridge, Carlyle and Santayana are among those who have expressed their puzzlement about the inability of our species consistently to learn from past experience. Yet we know that "the tongue of experience," according to the proverb in Arabic, "has the most truth."

Kenyon De Greene (1990, 1991) often poses the underlying question: Is humanity a learning system? In partial response, my colleagues and I, writing in *Windows on Creativity and Invention* (Richardson, 1988) justified our view that *Homo sapiens* is involved in an incremental (if not often tediously heuristic) learning process which bridges and binds the generations, the centuries, and the ages. Mankind often learns, indeed, to learn again.

Discovery and revelation, the historical record shows fairly clearly, seldom come to us in bursts of brilliant inspiration or sagacious breakthrough or felicitous serendipity. Sequential accomplishment and achievement are therefore essential elements in the learning process, to be sure, but so is error.

Our models (1) and (2) can also be termed accomplished for better or worse, whereas experiment (3) should be characterized as under way (or, at least, historically unresolved) and therefore to some extent inferential.

A BASIC DECISION-AND-COMMAND MANAGEMENT MODEL

Decision-and-Command Management

Command is the prerogative of control by *de facto* authority (whether invested or usurped), devolving from a decision process, and exercising the direction of subordinate entities and leaders forming a societal structure or institution amenable to such management, i.e., the application of decisions.

The approach I have elected in order to profile the effects of command, what historian William H. McNeill calls "the management of armed force" (McNeill, 1982), is what seems most appropriate here. I base this choice on my experience in business, in the military and civil services, with universities, and with those institutions most closely emulating the military - (a) national diplomatic establishments and (b) intergovernmental organizations.

What I have done is borrow in part from methods familiar to the statistician and operational researcher, derived from the propositional calculus of basic logic (after the work done in Germany by Schröder a century ago and Löwenheim and Hilbert early in our century). These I have combined with elements of game theory, as proposed by Morgenstern and von Neumann in the 1940s (Makower and

Williamson, 1967; Raiffa, 1962; von Neumann and Morgenstern, 1944).

We could then say that our examples constitute a decision-and-command management model, a system of choice based upon a five-step process of collating and evaluating information and then implementing one of several courses of the propositions for action indicated by this process. This model also bears resemblance to the long used, six-part staff study originally developed by the general staff of the United States Army. There is similarity, too, between the Army's staff study and a four-step model for the development of options devised by the Harvard Negotiation Project (Department of the Army, 1984; Fisher and Ury, 1981; see also Kremenyuk, 1991).

The model of this combined process is shown in Figure 10.1 (see following page). The logic scheme used can be applied not only to the first two cases which we shall discuss, but also to at least the Coalition side of the third case - and to many other citable examples in which future action depends on *progressio ad optimum.* The method seems particularly applicable to options surrounded by considerations of interdisciplinarity, complex intersectoral factors, conflicting goals, or any combination of these.

The literature of decisionmaking almost inevitably refers to the need and methods for assessing probabilities and holding in reserve appropriate alternatives. The assessment of the worth of an educational or training syllabus, for instance, is reasonably straightforward (i.e., via examinations). But even in the case of corporate decisions regarding business plans, the range of choice and degree of complexity are often less wide-ranging than in the cases presented here. Work on evaluation analysis done by some United Nations (UN) agencies during the 1980s, notably by the multi-mission and politically hard-pressed UN Educational, Scientific and Cultural Organization, approximates more closely the types of problems dealt with in this chapter (Unesco, 1990).

Further Definitions

The term "command," as we have seen, refers to system imposition and corresponding acceptance through a hierarchy so as to reach or implement a decision, usually within a specific frame of time. This circumstance typifies all three of the examples used.

By "decision," I refer to a hortatory or even imperative urging imposed by the managerial hierarchy within a societal entity to study and weigh alternatives and then, ulteriorly, to favor a single option or perhaps combination of options. This kind of decision is distinct and separate from the logic of preference called prohairetic (from the Greek, *proairesis*, "choosing"): which TV channel to select, what to order at the restaurant, which pair of shoes to purchase, where to spend the next holiday.

The command process thus reflects a structure in the form of its methodical analysis, one exhibiting at least internally the rigor of reasoning. The prohairetic

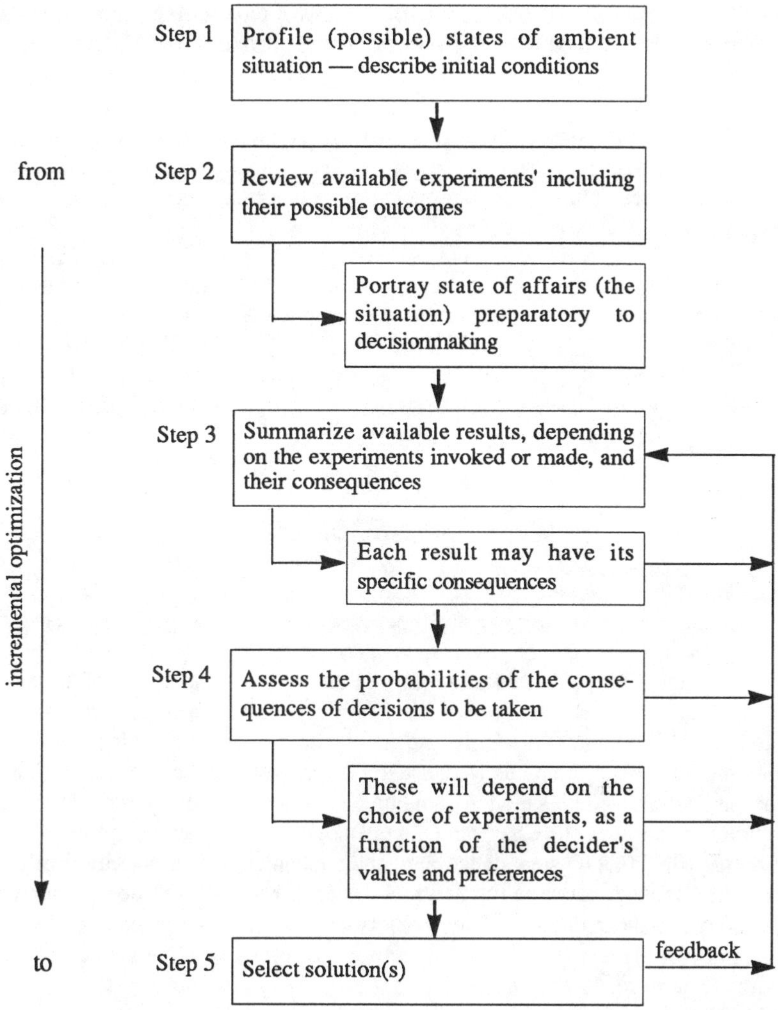

Figure10.1 Basic decision-and-command management model. The model is applicable specifically to the design of the city of Washington and the creation of Japan's national police force; it implies the appraisal of subjective and objective factors, as discussed in the text. The optimization shown at left is a largely unilateral perspective, somewhat alien however to the win-win, win-no win outcomes foreseen in game theory, previously mentioned.

mode is far more empirical and frequently affective, thus ranging in motivation from the reasonably iterated through the random to the capricious.

Mere Scenarios?

Some readers may argue, however, that the general model applied in this chapter lacks rigor in its pretension of serving as a simulacrum of choosing and deciding; that, instead of modeling a process or procedure, it is in fact no better than a scenario or option among multiple scenarios.

I prefer to compromise on the point, regarding the first two paradigms presented as reasonably specific models of decision, the third as perhaps an unfolding scenario with unseen and potentially widespread consequences - not to speak of imprecise, even confused, causes - occluded further by the very indefinitiness of the future.

The three cases are presented in narrative form, with as much historical context and circumstance as space allows.

TWO BRIEF SCENARIO MODELS

The Capital at Washington

Two centuries ago, an emerging and independent nation needed - besides a name, a constitution, a mechanism of governance, a coherent economy and a currency - a consensually agreed capital. This is the story of the selection and design of the site.

Background. Capital cities are a fairly new phenomenon in history, largely because the nation-state is a relative newcomer to the evolution of geopolitical jurisdiction. Historically, capitals were usually the largest communities of a region - whether in China's Middle Kingdom, among the seafaring or landlocked districts of the Indian subcontinent, the Chaldean Ur and Babylon near today's Baghdad, or the often rectilinear urban clusters of the indigenous peoples in Central America (e.g. the vestigial Zapotec community sites near Oaxaca, Mexico) and South America (the possibly pre-Incan citadel of Machu Picchu in Peru). "Each Pharaoh builds his own capital," in the words of Lewis Mumford; "his urban home is as exclusive as his tomb" (Mumford, 1961, p. 80).

Locales such as these and Thebes in Egypt became large or even dominating because of topography, water supply or other climatological advantage, the center of industry or the funneling of trade, or the consequences of marauding, plunder or war, pestilence or other natural catastrophe. In the new United States, Philadelphia had good historical reason to aspire to the honor. John Adams, as vice-president in 1789, wanted very much that the nearby community of Germantown be named the "Federal town." But Princeton, Trenton, Kingston (New York), Marietta in the wilderness of the Ohio territory, and Baltimore were also on the list of 40-some possible locations.

In the late 18th century, Paris was a geopolitical center at the global level, strategically situated on a commercial waterway, developed almost two millennia earlier by the Roman conquerors of Gaul. London, also situated on an important fluvial trading route, was of comparable standing and vintage. Both capitals directed empires stretching over all the seas.

The designer of a new capital. Pierre Charles L'Enfant (Paris, 1754-Maryland, 1825) studied at the Royal Academy of Painting and Sculpture (today's equally famous Beaux Arts school) in Paris during the five years before the rebellious provinces of British North America unilaterally declared their independence in July 1776. The young L'Enfant left Europe in that year to join French and American forces in the Western Hemisphere, where he was commissioned a lieutenant of engineers and eventually mustered out with the rank of major (Kennedy, 1989).

After the hostilities ended, L'Enfant remained in the brand-new United States, designing public buildings (including the New York City Hall, where President George Washington was inaugurated in 1789) and private residences, most of which were situated along the country's eastern seaboard.

When the freshly constituted Congress voted to create a national capital - and favored a quadrilateral of territory that would be ceded by the states of Maryland and Virginia - L'Enfant was awarded the commission to design the federal city-to-be. L'Enfant used land surveys made by Andrew Ellicott and and verified by astronomer Benjamin Banneker. He had already been strongly influenced by the baroque landscape design of properties such as the palace at Versailles, especially the combined geometry and vistas of André LeNôtre and J.-A. Gabriel and their relationships with the perspectives of grand, non-urban architecture.

L'Enfant thus laid out (1791-2) on squared paper an entire agglomeration suitable to serve, in time, as the political-administrative hub of what he foresaw (with astonishing lucidity, it should be added) - according to one of his biographers - as a federation ultimately grown to fifty states and a national population of several hundred million. It is curious, too, considering the nature of L'Enfant's accomplishments, that the available literature on him and his work is sparse (Hall, 1991).

Plotting the plan. The concept developed by the ex-military engineer exploited the Maryland side of the Potomac River tract's existing topography of flatness with little tree cover, broken here and there by mounds and basins, and marked irregularly by marshland behind the Potomac River's shoreline. (The Virginia grant, west of the Potomac and never used, was finally retroceded in 1846.)

L'Enfant interpreted his mandate, from both Congress and ideas he gleaned from George Washington and Thomas Jefferson, each of whom knew surveying and developed impressive estates of his own, as awesome and clearly monumental. As L'Enfant understood his instructions, the design was the thing, rather than a concern for system. Narrow, curving and discontinuous streets in the style of the

Old World were to be avoided (the first street to be straightened in Paris, the Rue de Richelieu, dated from only 1633). This would leave broad thoroughfares and grand perspectives to do honor to majestic public buildings, embellished with monumental statuary especially commissioned, and to serve at the same time the needs of security and defense.

The new capital city was named after the country's first president, and the federal tract of land on which it stood was a "territory" called the District of Columbia. The District was not a state, nor was it destined by law to become one.

L'Enfant also tried, but with notably limited success, to influence the design, construction and function of the Capitol itself. We can only surmise with the benefit of hindsight that his artistic training and experience may have limited his ability to envisage the multiplicity of workaday functions with which a city on the threshold of the 19th century must be equipped - especially a new and grandiose community expected to serve, if not forever, at least countless future generations.

To be more specific, yet for reasons that are also far from clear, L'Enfant (despite considerable coaching from the pragmatic Jefferson) demonstrated little concern for the efficacy of the future metropolis's growth in trade, industrial development, and even residential configuration and amenities. L'Enfant's vision of the total human settlement thus proved to be somewhat constrained.

Perhaps unnecessarily severely, one American historian has the following to say. Major L'Enfant "(to the subsequent confusion of motor traffic) imposed a series of avenues radiating from circles on a grid of numbered and lettered streets. Jefferson himself," emphasizes Samuel Eliot Morison, "when secretary of state, had determined the sites for the Capitol and White House after L'Enfant quit in a huff" (Morison, 1965).

Plans gone awry. This kind of discrepancy in planning (or even short-sightedness) was to be the case again a century and a half later with (1) LeCorbusier and the new city he conceived at Chandigarh, India, and (2) the innovative capital of Brasilia elaborated by city planner Lucio Costa and architect Oscar Niemeyer, both Brazilians. Kenyon De Greene theorizes that the Bauhausian characteristics of a community such as Brasilia, "large windows without drapes, wide-open spaces to suggest the socialist reality of equality," were in turn manifestations of "government attempts to shape behavior via architecture" (De Greene, 1990, 1991).

While LeCorbusier, Costa and Niemeyer were not direct disciples of Walter Gropius, founder (and, under Hitler, forced liquidator) of the Bauhaus school, it is true that there exists a certain quality of "state-inspired egalitarian design" to all of their enterprise (De Greene, 1991). L'Enfant and his generation were not equipped to foresee the environmental problems of the age of electricity, internal combustion and consumerism. Generations later, LeCorbusier and Costa-Niemeyer failed to take into account the local traditions of the informal economy (vendors' curbside stalls) and squatter living (by those disdaining municipal housing as culturally unfamiliar and costly).

For better, for worse. The nation's founding fathers added another dimension of confusion to confront residents of the new capital: law excluded them from the national franchise. Civil servants and other District of Columbia inhabitants earning their daily bread, directly and indirectly, from the business of federal governance were considered politically neutral (if not neuter) in the administration of both municipality and nation.

A congressionally appointed board of commissioners oversaw the capital city's day-to-day operations for nearly the first two centuries of the existence of the United States. Only in the last quarter of the 20th century have Washingtonians been able to cast a vote for their municipal executive.

Postlude. L'Enfant's work with the layout of Washington done, he claimed about $92,000 in professional fees from Congress - which finally paid him less than $4,000 for his efforts. His not-quite terminated model became, nevertheless, the federal city that had been sought and, in its turn, served later as the model for new cities in the United States and elsewhere.

The planner of the capital of the United States obviously had foresight, but neither Jefferson and Washington nor L'Enfant could have predicted the push westward that would characterize the century to follow their lifetimes. The Congress could not have commanded planning other than what its mandate and knowledge permitted. The scudding away from the Atlantic seaboard of much of the nation's swelling population during the 19th century and the creation, as a consequence, of geographic eccentricity in the capital's location were elements of societal evolution not readily foreseeable. Today's planners (usually) manage their designs for the future more thoroughly than ever before possible.

The conception and laying out on site of a young nation's capital community, the city of Washington - on territory where none had existed previously - may have been a historical first. (We ignore to this day the motivations creating the planned communities of the Aztec, Mayan and Incan nations.) Were the experience to be repeated in the 1990s, certain critical factors would be taken into account that escaped the urbanists of two centuries ago. Climatological setting, holistic environmental appreciation, the arrangement of satellite communities, the networking of communication and transport, long-term hydrological and sanitary engineering: all these assume roles in contemporary urban design that were virtually absent in L'Enfant's enterprise.

This case can also be examined via the multiple perspectives discussed by Linstone in Chapter 9. A rather sizable technological achievement of the time (laying out the new capital city) must be interpreted in terms of interaction with organizational (e.g., Congressional) and personal (e.g., L'Enfant, Jefferson) factors.

Deciding upon A Police Force for Japan

Here we take a look at what a fully identified nation-people probably thought as they examined their internal security needs to befit a country in transition from closed feudalism to a constitutional monarchy.

A public institution. While serving as a young officer on General Douglas MacArthur's staff during the occupation of Japan, I had ample occasion for more than two years to roam the four main islands of that country and discover many Nipponese social structures for myself - often with the aid of fellow Japanese-speaking specialists.

One of the public institutions, almost unscathed by the war, that struck us as an impressive societal regulator was Japan's national police. Throughout the economic stress, national grief and personal indignities that the vanquished Japanese experienced, especially in the years 1945-1949, the police web and its personnel remained a stabilizing force with the country's people. This swollen, hungry, postwar population included several million demobilized service personnel and an overseas contingent of civil service and commercial specialists forcibly repatriated from China and the southern seas to a Japan severely bombed out, lacking housing and food, and almost totally bereft of its industrial base.

The police system also showed itself to be an effective administrative-political-cultural link with the multi-nation occupant (besides the United States, there were also represented China, France, Great Britain and the Soviet Union). In terms of Japan's postwar destiny, the police thus found a role with their people as confidence builders.

How had this come about in a populous nation that had emerged less than 80 years before from a "closed system" of feudalistic isolation, warring clans, limited commerce, and technology mostly of the artisanal kind? The evolution of the modern police and security administration in Japan is amply documented, although mainly in Japanese. A fine synthesis of the 19th-century story in English was published a few years ago by the American social historian, D. Eleanor Westney (Westney, 1987).

Tokugawa Japan (1605-1867) had only an inchoate concept of constabulary forces and law-and-order operating at levels above that of the lordly fiefdom, vastly different from an effective police process centralized at the national level as one could find by the late 18th century in many European countries.

Historical origins. Japan had known, since at least the Nara period which had begun a thousand years earlier, refined art, religious and stately architecture and even orderly town planning. These developments were attributable in large part to the import of Buddhism from Korea during the 6th and 7th centuries. Buddhism had found its way to Korea from India via China. Public administration already existed, indeed, in both Korea and Japan; it was directly derivative of the Chinese concepts of civil service and civic management absorbed by both the Koreans and Japanese much, as they had also borrowed the written language of the great Han people of China beginning in the 5th century.

But the emergence of the European nation-states (England, France, Spain), paralleled by the outreach of the potent city-states nearby (Florence, Venice, Genoa and those of the Hanseatic League) was not to be in Japan. By the time of the opening of Japan's doors to the outside world between 1853 and 1868, the disparate leaders within Nippon were able to take stock of the many deficiencies hampering the smooth and skillful operation of a large country in the process of joining the rest of the world.

Lacking in part or in whole were, among other organizational concepts:

- a defined economic "system"
- integrated fishery, forestry and agriculture
- an interrelated industrial base
- interregional communication and transport
- health-delivery systems
- coherent education and vocational training
- scientific research and technological transfer
- a system of national defense and internal security.

Creation of a new service. After the Meiji restoration of 1867-1868, the reinstated imperial monarchy in Tokyo set about to create new, or modify existing, societal systems (education, public health, industry, postal service, the military) by selective emulation of their analogues abroad. Communication and transport were borrowed from the British. Medical schools found their inspiration in the German system. Modernized agriculture and the innovation of agricultural colleges came from North America, whereas forestry management came from Austro-Hungary and Germany. Banking was learned from the Belgians and Americans. The army followed French, then (after the Franco-Prussian War), German models. And so on.

"The selective invocation of elements of the past," comments Westney (1987, p. 33), "reinterpreted in the light of the needs of the present (particularly organizational needs for control), was an important part of the organizational development process."

To acquire a reformed and newly invigorated police apparatus, after two and a half centuries of what one Japanese called, nonetheless, "an absolute police state" (Fukuda, 1987), Japanese leadership began literally to shop around. Concurrently, samurai guards were still patrolling the new capital of Tokyo (formerly Edo) as late as 1871.

In the following year, as the samurai were further demilitarized, the centralized government sent a survey team of eight young civil servants to study the police, legal and court, and penal systems in force in Austria, Belgium, France, Germany, Italy, the Netherlands and Russia.

The effects were rapid. The immediate result of this mission - charged also with distinguishing clearly between the judicial and administrative responsibilities of a police service - was the appointment in 1873 of its leader, Kawaji Toshiyoshi, as the head of an incipient national police network. The British model had been discarded *a priori* because it responded primarily to local authority and virtually not at all to the national government in London. The French traditional system of *gendarmerie*, replicating other continental models such as Italy's *carabinieri* or the Dutch *maréchaussée* (the marshalry, or royal marshals) was deemed more consonant with the Japanese conception of central control subservient to crown authority.

The nascent *keishi-chô*, or metropolitan police agency for Tokyo, together with a newly established national police force outside the capital, underwent varying adjustments of number, structure and *modus operandi* until 1881. During the balance of the 1880s, the police authority under home minister Yamagata Aritomo (who had also been instrumental in consolidating and modernizing the army and can be considered the prime mover in our model), adopted role differentiation and specialization, modern investigational methods (including anthropometric mensuration), uniforms and military-style training, the telegraph, executive/legislative liaison with governmental authorities, and budgetary and pension schemes - all based on French forms.

It is interesting to recall that just a century earlier (7 December 1790), enabling legislation was approved by France's Constituent Assembly delineating police powers and structure for the greater Paris area during the final years of the monarchy and prior to the adoption of a new constitution in 1791. But contrary to the Japanese method based on teamwork, the French juridical approach was decided virtually by a single individual, the liberal aristocrat and mathematician, Marquis M.J.A.N. Caritat de Condorcet (Badinter and Badinter, 1988).

A salient exception made by the Japanese to the types of police methods and technologies brought from abroad, this one imported from Great Britain, was the arsenal of firearms chosen for the men (no women) wearing the Western blue-serge uniforms. In contrast to the attraction of the Japanese to British sidearms, it is interesting to note parenthetically that the British police itself remains normally unarmed to the present day.

Stability of the system. Systemic stability in Japan's police was assured by the early 1890s, two decades after the retirement of the largely illiterate samurai, when the first university graduates joined the force (1893). Except for a rise in numbers from somewhat more than 29,000 in 1890 to nearly 42,000 in 1910 (major politico-military adventures in China and with the Russian empire and in Korea had intervened in 1894-1895, 1904 and 1910), Japan's police structure and authority did not evolve significantly until the period when I first observed them after the defeat in 1945.

The organizational development of the Japanese national police system had assumed proportions, by the end of the 19th century, of at least one order of jurisdictional magnitude greater than that of the challenge confronting Pierre-Charles L'Enfant in the early 1790s when he was asked to design a capital city.

Instead of working only at the urban level, the new system of protection, investigation and security in Nippon extended to all districts of the archipelago - from the subarctic Kuril Islands to the subtropical latitudes of the southernmost islands of

Kyushu and the Ryukyu chain. Feudal guards and mercenary watchmen serving a disappearing landed aristocracy had been transformed within a single generation into an effective, country-wide institution.

Aftermath. With the economic and social modernization of Japan, especially since 1946, the nation's police have continued to serve the building of confidence in both the explicit and the unspoken relationships linking civil authority with individual citizens. The possession of personal firearms remains illegal (except for hunting), so that many types of felony are of extremely low incidence - manslaughter and armed assault included.

The neighborhood police box (*kôban*), almost as ubiquitous in Japan as fire alarms and postal drops, is a stabilizing factor in terms of the incidence of misdemeanors and traffic accidents. While the neighborhood cop may not be everyone's best friend, he/she (now there are women) is a steady and reliable authority model in a society imbued with the consensual spirit.

The decisions made in Japan regarding a police system stand in marked contrast to those possible in the United States. The fifty states have constitutional endorsement for managing and commanding their constabularies (state police, county and municipal forces, sheriffs' offices). The prerogative is historic and could be changed only by reversal of constitutional precepts.

An American super-police force that is related - both in legislative and public eyes - to the national security (the Attorney-General, the Federal Bureau of Investigation and parallel services charged with customs and drug control) is managed and commanded, by national consensus, more coherently at the federal level. Its existence and function are rationalized in terms of collective security for all fifty states.

Finally, this case shows that an institutional design may produce more stability and fewer unexpected side effects than does an apparently simpler, more straightforward design that is perceived to be essentially technological.

DECISIONMAKING IN THE GULF CRISIS/WAR: EVENTS OF 1990-1991

The coming of The Day
In the Name of Allah, the Compassionate, the Merciful
When that which is coming comes
- and no soul shall deny its coming -
some shall be abased and others exalted.
Qu'ran (The Koran), Chapter 56

In the section following, we progress to a "thickening of plot," a movement upward in complexity which - if not a quantum leap in terms of numbers - is so in terms of detail, interpretation, and impact.

Historical Background

Many countries of today's Near and Middle East are considered by their neighbors to be synthetic quantities. They are seen as territories descended from centuries of occupation, first by the Mongols and Tartars,later by Turkish, and still later by Russian, British and French regimes; and finally strongly affected after the Second World War by the partisan outlooks of the world's two superpowers - the United States and the Soviet Union - in combination with the creation of the Jewish state of Israel in 1948.

During the period since the Turks were put to rout by the politico-military shifts attributable to the First World War, large reserves of petroleum were discovered and first exploited by British and American interests. Then, concurrently with the early decolonization of most of the world and the First UN Development Decade (1960-9), oil exploitation became gradually indigenized. As a result, national regimes in Saudi Arabia, Iran, Iraq and the princely states along the Gulf's shore assumed the dominant capitalization, managerial control, and profits from the ever more indispensable natural resource that petroleum became. In the background, there festered the open wound of Arab-Israeli confrontation, today still in search of closing and healing.

Local rivalries grew among Arab entities of the region, exacerbated by two armed conflicts between several Arab countries and Israel in 1967 and 1973, the Egypto-Israeli accords signed at Camp David in 1978, the Soviet-Afghan conflict of 1979-1988, the Iran-Iraq war of 1980-1988, and protracted disagreement on oil pricing among the member-nations of the Organization of Petroleum Exporting Countries (OPEC). By 1989-1990, Iraq was struggling with neighboring Kuwait in order to gain more direct control of oil resources and their prices on the world market. (For another perception of the situation, see, further along, *A Standard View of Motivations?*)

Forms and Dimensions of Conflict

Before attempting to analyze the specific process of decision used by each side in the Gulf Crisis/War of 1990-1991, it is useful to look at how an expert in polemology - the analysis of war - interprets the forms of force available to the would-be user. James Schellenberg postulates, in *The Science of Conflict* (Schellenberg, 1982), three forms of force:

1. - Coercion with force and violence
   — physical force, backed by violence, assures coercion
2. - Force and violence without coercion
   — when it is more an end than a means, e.g., the "useless riots" of history: stadium brawls and worse
   — when force is exerted ineffectively, e.g., by a dying politico-military regime (Nicaragua, Angola and Ethiopia are recent examples)
   — when force is deliberately and successfully used for the destruction of

the adversary, leaving no one to coerce, as "when the walls of Jericho came tumblin' down"

3. - Coercion without physical force or violence
— with force, but holding violence in reserve
— "with the shadow of force [giving] indirect support, without the shadow of force being directly `suggested'"
— without either force or violence, but coercion applied nevertheless.

The Gulf episode clearly belongs within the first of these categories.

Mohammad-Reza Djalili, an Iranian political scientist, views wars in other dimensions. He proposes an eight-order typology of wars occurring in developing regions of the world (Djalili, 1991). The eight orders of "Third World conflicts," according to this author, embrace three types: global, regional (two or more countries), and national.

In the regional category, Djalili identifies three variations of modern conflict: among several nations, irredentist, and hegemonic. In the global category, conflicts are seen to occur in three modes: in the wake of decolonization, as a "South" dimension to East-West tensions, and as intervention by major powers. The Gulf War of 1991 can thus be said to have begun with the crisis caused by Iraq's annexation of Kuwait - a regional move towards establishing hegemony by one state over a second or over a region, which then became a (near-) global conflict with intervention by several major powers in order to wrest Kuwait from political integration with Iraq.

Capability, Credibility, Relevance and Legitimacy

Once the decision process points to the choice to be made among the three coercive options listed by Schellenberg, the mode selected is most likely to be one in direct relation to the user's capability. This is because, still according to Schellenberg (Schellenberg, 1982, p. 230),

> Force, to be used effectively, must be sufficient to overcome opposing forces... Speaking in military terms, we see manpower and firepower as especially important. We must not, however, neglect attention to mechanisms for supplying front-line troops or the organization of the command structure. All of these are vital factors in military capability.

While the primary judge of capability is the one who may use the force held in readiness, the measure of this capability lies with the one(s) most likely to be on the receiving end of the force applied: the perceiver of the credibility of another's threat. Schellenberg notes, too, "that force would in fact be used unless certain conditions are met which makes it most persuasive" (Schellenberg, 1982, p. 231). Force, in other words, is most effective in coercion when it is not used.

A third standard for the acceptance of the use of force is relevance. To be pertinent "in the sense that it can drastically affect, in a negative way, persons with real power," Schellenberg states that force must also help "these persons make the associations and draw the conclusions necessary to be coerced" (Schellenberg, 1982, p. 231).

A final criterion germane to the application of force is its apparent legitimacy, first in the reckoning of the one exercising physical force, second in the estimate of the one on whom force is used (or at least threatened), and third in the minds of observers. In the case of the Gulf Crisis-War, therefore, the circle of observers included all 161 member-states of the United Nations, even the oft demurring China, Cuba and Yemen.

Author Schellenberg ends his conceptual analysis of the varieties of force available with two general conclusions:

> The first is that physical force can most effectively be applied to resolve a conflict when one party has a clearly superior capability [and when this force is supported by credibility, relevance and legitimacy]. A second conclusion is that where the claim to legitimate authority can be asserted most persuasively, the use of physical force is often redundant (Schellenberg, p. 234).

Spread of Involvement to Many Countries

Involved, of course, was not a single country: neither Iraq, by invading and annexing its neighbor, Kuwait, in 1990; nor the United States, by riposting instantly and taking the lead politically and militarily to contain Iraq, urging the UN to act in virtually unanimous economic concert against Iraq via blockade and (finally) by directly coercive armed force.

In the long course of history, disputes evolving into full-blown armed conflicts have been largely bilateral affairs (McNeill, 1982, *inter alia*). It is chiefly during the past five centuries, with the advent of alliances, treaties, pacts, accords and uncounted purpose-specific protocols between nation-states, that the two sides in political struggles became multinational. (The Crusades, beginning in the 1090s, remain an exception.) In our century the principal many-country engagements were the First and Second World Wars, the Korean War, the Indochinese conflicts of 1945-1975 and the Gulf incidents of 1990-1991.

Thus, almost willynilly in the case of the Gulf episode, the entire world became involved as actors or audience (see Figure 10.2) in a complicated scenario, with 28 belligerent states in action against a twenty-ninth, Iraq.

We shall return in a moment to the subject of intermediation by the UN. The 12 resolutions adopted in 1990 by the Security Council regarding the Gulf crisis, presented in summary form in Figure 10.3, comprise a decision process that is hypercomplex when compared with those relating to the capital at Washington and a new Japanese police service.

Hypercomplexity has become increasingly a preoccupation of the management processes, especially in the contexts of multidisciplinary planning, during the past two decades. A leader in the field of developing complexity analysis has been the Organization for Economic Cooperation and Development in Paris.

RESOLUTION 660
2 August 1990

The Security Council,

Alarmed by the invasion of Kuwait on 2 August 1990 by the military forces of Iraq,

Determining that there exists a breach of international peace and security as regards the Iraqi invasion of Kuwait,

Acting under Articles 39 and 40 of the Charter of the United Nations,

1. ***Condemns*** the Iraqi invasion of Kuwait;

2. ***Demands*** that Iraq withdraw immediately and unconditionally all its forces to the positions in which they were located on 1 August 1990;

3. ***Calls upon*** Iraq and Kuwait to begin immediately intensive negotiations for the resolution of their differences and supports all efforts in this regard, and especially those of the League of Arab States;

4. ***Decides*** to meet again as necessary to consider further steps to ensure compliance with the present resolution.

RESOLUTION 678
29 November 1990

The Security Council,

Recalling and reaffirming its resolutions
660 (1990) of 2 August 1990,
661 (1990) of 6 August 1990,
662 (1990) of 9 August 1990,
664 (1990) of 18 August 1990,
665 (1990) of 25 August 1990,
666 (1990) of 13 September 1990,
667 (1990) of 16 September 1990,
669 (1990) of 24 September 1990,
670 (1990) of 25 September 1990,
674 (1990) of 29 October 1990
and 677 (1990) of 28 November 1990,

Noting that, despite all efforts by the United Nations, Iraq refuses to comply with its obligation to implement resolution 660 (1990) and the abovementioned subsequent relevant resolutions, in flagrant contempt of the Security Council,

Mindful of its duties and responsibilities under the Charter of the United Nations for the maintenance and preservation of international peace and security,

Determined to secure full compliance with its decisions,

Acting under Chapter Vll of the Charter,

1 Demands that Iraq comply fully with resolution 660 (1990) and all subsequent relevant resolutions, and decides, while maintaining all its decisions, to allow Iraq one final opportunity, as a pause of goodwill, to do so;

2 Authorizes Member States cooperating with the Government of Kuwait, unless Iraq on or before 15 January 1991 fully implements, as set forth in paragraph 1 above, the foregoing resolutions, to use all necessary means to uphold and implement resolution 660 (1990) and all subsequent relevant resolutions and to restore international peace and security in the area;

3 Requests all States to provide appropriate support for the actions undertaken in pursuance of paragraph 2 of the present resolution;

4 Requests the States concerned to keep the Security Council regularly informed on the progress of actions undertaken pursuant to paragraphs 2 and 3 of the present resolution;

5 Decides to remain seized of the matter.

Figure 10.2 Texts of UN Security Council Resolutions 660 and 678.

Iraq-Kuwait: The UN's 12 Resolutions

(660) 2 August - UN demands that Iraq withdraw its forces from Kuwait. Vote: 14-0, Yemen abstaining.

(661) 6 August - UN applies strict sanctions on all trade with Iraq with exceptions of medicine and, for humanitarian purposes, food. Vote 13-0; Cuba and Yemen abstain.

(662) 9 August - UN declares Iraq's annexation of Kuwait null and void. Vote: 15-0.

(664) 18 August - UN demands that foreign nationals be permitted to leave Iraq, that Iraq rescind its order on closure of diplomatic missions in Kuwait. Vote: 15-0.

(665) 25 August - UN authorizes member states to use limited naval force and inspect naval cargoes, ensuring compliance with UN economic sanctions. Vote: 13-0, without Cuba and Yemen.

(666) 13 September - UN approves food shipments, for humanitarian reasons, if distributed by approved international bodies. Vote: 13-2, with Cuba and Yemen opposed.

(667) 16 September - UN condemns Iraqi raids on French, other diplomatic missions remaining in Kuwait. Vote 15-0.

(669) 24 September - UN instructs Security Council's Sanctions Committee to assess requests for relief from countries suffering from trade embargo. Vote: 15-0.

(670) 25 September - UN prohibits air traffic with Iraq and occupied Kuwait except in humanitarian cases. Vote: 14-1, Cuba opposing.

(674) 29 October - UN requests member-nations to document human-rights violations, financial losses subsequent to Iraq's invasion of Iraq.

Vote: 13-0; Cuba, Yemen abstaining.

(677) 28 November - UN requests its secretary-general to assure safekeeping of smuggled pre-invasion Kuwaiti population register.

(678) 29 November - UN authorizes states the use of "all necessary means" against Iraq unless Iraq withdraws from Kuwait by 15 January. Vote: 12-2, China abstaining and Yemen, Cuba opposing.

Figure 10.3 A consolidated listing of the dozen UN resolutions on the Iraq-Kuwait crisis adopted by the Security Council between early August and late November 1990. The resolutions are, in effect, the "command" following decisions taken in near-universal concert.

On the Nature of Leadership

During the period being examined, certain of the world's leaders found themselves in extremely varied political circumstances, both internally and externally. Systemic analysis of the evolution of the Gulf episode requires brief examination of this factor.

The then incumbent Secretary-General of the United Nations (Javier Pérez de Cuéllar of Peru) had already begun to assert some of the prerogatives and authority which are appurtenances of this important post but were neglected by his immediate predecessor, Kurt Waldheim of Austria.

Most of the secretaries-general of the UN have tended, indeed, to be dutiful vassals of the majorities ruling the international roost in New York. Earliest among these majorities influencing policy and decision came the United States and its allies of the Second World War; then from the 1960s onwards came the blocs known as the Group of 77 - the developing countries together with the Soviet-Chinese amalgam - and the Non-Aligned Nations. These held sway until 1985-1986 when Mikhail S. Gorbachev, the new General Secretary of the Communist Party of the Soviet Union, began, probably unwittingly, to unbalance the world's distribution of international power.

Pérez de Cuéllar's influence as UN senior civil servant became evident in helping calm political tempers in Central America during the late 1980s, stimulating withdrawal of Soviet armed forces from Afghanistan in 1988-1989, easing tensions between the Soviet Union and the United States at the same time, and aiding thereby the political transformation beginning in Central and Eastern Europe in 1988-1989.

Some quarters see Pérez de Cuéllar's role, however, as tardy in the case of the Gulf crisis. He is reproached for having delayed in joining others to persuade the Iraqi president to leave Kuwait and, later, in helping succor the Kurdish minority (see especially *The Economist*, 1991, pp. 15-16). Yet the Secretary-General's part, while far from intrusive or interfering, remained active and effective.

George Bush, leader of the United States, was elected by popular vote in 1988 and enjoyed favorable public-opinion ratings at home and a generally positive standing abroad. Within a year after the war with Iraq, Bush's ratings plummeted as long-neglected economic and other social problems erupted domestically.

Gorbachev had been made President of the U.S.S.R. by a complicated electoral vote rather than by universal suffrage. The author of domestic reform via *glasnost* and *perestroika*, his star began to show tarnish (at least domestically) from late 1988 because of grave internal dislocations, and he was finally forced to relinquish command by December 1991.

The leader of Iraq and the Ba'ath Socialist Party, Saddam Hussein, had become dictator of his country by a *coup d'état* staged more than two decades earlier; he held totalitarian control (although he took giant strides forward to modernize his country) through a repressive internal-security apparatus whose authority was pervasive in a non-fundamentalist Islamic society.

Following the identification made by Max Weber (Weber, 1947), these four authority figures can be typed as shown in Table 10.1. The place of the Iraqi leader in this taxonomy is endorsed by the fact that he was also recognized by the rest of the world as his country's chief executive *de jure*.

Legally chosen (based on reason)	Pérez de Cuéllar; Bush; Gorbachev
Traditional (with custom prevailing)	Gorbachev (also here)
Charismatic (based on emotion)	Saddam Hussein

Table 10.1. A Weberian typology of some of the world's leadership during the late 1980s-early 1990s.

Leadership factors such as these, and their consequences - made acute in the general cognizance of the world community since the creation of the Israeli state and aggravated by the spread of terrorism since about 1970 - have helped beg the question: Is the "Middle East Crisis" not, in effect, A Third or Fourth World war, or perhaps a new Hundred Years' War, or even a renascence of the Crusades? The reply to any of these questions may well be affirmative. Readers desiring to refresh their memory on the what and the why of the Crusades are referred to Jonathan Riley-Smith's excellent recent work, *The Atlas of the Crusades* (Riley-Smith, 1991).

Further Entangling Relations

By midsummer 1990, Iraq had forcibly annexed Kuwait by direct military action, *Blitzkrieg* in the fullest sense. Expressed in terms of another dimension, Iraq forced - by way of this rapid and seemingly unexpected use of arms - a multitude of other nations, both nearby and remote, suddenly to revise their views of the foreign-relations environment affecting them in western Asia. This can be depicted by the schematic loop shown in Figure 10.4.

The U.S.S.R. was already hypersensitive about its relationship with the Muslim world. This came about because (1) of the official atheism of the Communist Party of the Soviet Union, (2) of the Islamic faith of scores of millions of its own population inhabiting the Central Asian republics, (3) Iraq had been a profitable client-state of the Soviet military-industrial complex, and (4) the decade-long conflict with Muslim Afghanistan during the 1980s. The Soviets sought, from the outset of Iraq's *Anschluss* with Kuwait in August 1990, to adopt and maintain a special posture *vis-à-vis* both Iraq and the Arab-Islamic world in general.

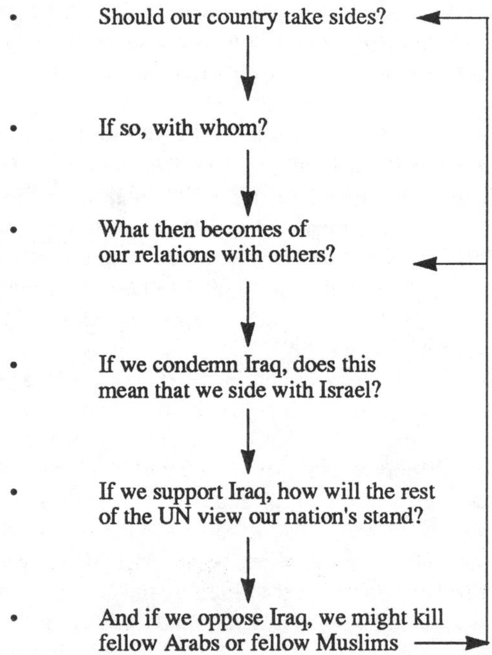

Figure 10.4 Feedback decision circuit of how a state envisages its possibly deteriorating relationships with others. In place of Iraq, on can read as easily "Peru", "United States", "Russia", etc... In the last step, the terms "Arabs" and "Muslims" are equally substitutable.

Within the general framework of the series of resolutions voted in 1990 by the Security Council of the UN (instruments numbered 660-662, 664-667, 669-670, 674, 677-678; see Figure 10.3), it was the Soviets especially who maneuvered bilaterally and multilaterally (and with fervor) during the autumn-winter of 1990-1991. Their aim: by lending their good offices in order to preclude and successfully reduce, if not eliminate, the intensity of the conflict (that finally broke out on 17 January 1991 between Iraq and the 28 opposing, coalesced countries of the UN system), they hoped to hold down Muslim/non-Muslim passions within the then existing Soviet borders.

The Environmental Dimension

We should briefly discuss the *problématique* or problem analysis of the environment, specifically the issues raised *à priori* (i.e. before the onset of the military campaign of January-February 1991). This was reflected through the repeated warnings made by the President of Iraq that hostilities would force him to incapacitate Kuwait's petroleum-producing economy.

The reader will recall that, in late February 1991, Iraqi forces destroyed refineries and set fire to some 600 wellheads during their withdrawal from occupied Kuwait. The resulting environmental cataclysm, according to a report published by the World Meteorological Organization (WMO) in May 1991, involved the combustion daily of 5 million barrels of oil and more than 70 million cubic meters of natural gas. This conflagration was glutting the atmosphere each day with more than 40,000 tonnes of sulfur dioxide, 3,000 tonnes of nitrogen oxides and 0.5 million tonnes of carbon dioxide, "the effects of which on human health in the region remain unknown" (WMO, 1991).

There was also generalized pollution by the production daily of 100,000 tonnes of sooty smoke, a minor calamity adding to the damage caused by the Allied forces to power and chemical plants, the desert floor, and national infrastructure in Iraq responsible for water supply and disease control.

The impact of war on the natural and artificial environments has become, indeed, an increasingly critical issue in policymaking as the world has belatedly recognized. War damage can be extensive (as in both Kuwait and Iraq) - and permanent. Policies are now being mulled at the international level to place intentional destruction in the category of punishable war crimes.

Policy and Strategic Limitations of the United States

The United States, already confronted by extensive economic and societal dysfunctions on its home soil, deemed itself - after intensive self-examination - obliged to take on a largely unwanted additional experiment. The American systemic instabilities evident at the opening of the century's final decade included:

- a large, and still growing, national debt
- a seriously negative international balance of payments
- a weakened dollar, internationally speaking
- poor performance of its educational and training systems
- loss of leadership in research performance
- unimproved societal status of the nation's African and Hispanic subcultures
- the "drug war."

Refer again to Figure 10.1. In our attempt to describe a scenario model, we must at this point adjust Figure 10.1: it requires a little amplification. We need to consider, for example, the motivations for Iraq's incursion upon its neighbor in August 1990.

The process involved becomes therefore one of maximization of system (revised or new legislation, changes in public perception of the effectiveness of various social systems - education, health, order and security), rather than a more passive optimization of mere process. The latter is what we discerned in the cases of the creation of the city of Washington and the revamping of Japanese police structure.

Iraqi Motives Are Varied, Complex

The motivations guiding Iraqi policy include:

• A search for an equitable sharing of the petroleum riches lying underground in both Iraq and Kuwait. The Middle Eastern nations sit atop about 70 per cent of the world's estimated petroleum reserves (Durix, 1991), and some prime consumers of petroleum (France, Germany, Italy and Japan) have virtually no such resources of their own.

• Non-acceptance of the shared boundaries between the two countries, and a sometimes deeply founded dubiety surrounding the legitimacy of the Kuwaiti state. This situation dates at least from the time of the fragmentation of the Ottoman empire at the end of the First World War. Since then, each decision-making stage of the question, "Who has the right to what, in which proportions, and for how long?" leaves more difficult questions in its aftermath.

• The contempt (or perhaps envy?) of Iraq's leaders, and others, for the ostentatiously opulent life-style of the Kuwaiti ruling clan. Newspaper accounts of the personal funds spent in Cairo, London and along the French Riviera by Kuwait's profligate elements abroad while Kuwaiti and other Coalition servicemen died on "the sands of Araby" left adverse images in the popular mind.

• A consuming drive for power by a national leader who wished to be seen as the savior of his people.

• Other, perhaps less compelling reasons (such as local boundary disputes and the details of possible future sharing of petroleum royalties).

Added to these pre-annexation conditions came:

1. - A post-annexation resolve on the part of Iraq's leader to circumvent or break the economic blockade imposed upon his country by UN sanctions.

2. - Systematic exploitation by each side of modern mass communications to present the position of Iraq and its adversaries, respectively, in the most favorable light possible.

A Standard View of Iraqi Motivations?

As with so many other facets of human relations, the morphology of a problem inevitably depends upon points of view and their interpretation. The war situation in the Gulf, and even the status of the so-called petroleum monarchies roundabout the Gulf itself, were not to be seen through compatible vision by the two sides concerned.

A letter submitted by the Iraqi government to the Arab League in July 1990 showed, according to Professor Mussama al-Mubarak (a political scientist at Kuwait University) and reported by Kuwaiti Television, revealed how determined Iraq was in its petroleum and frontier negotiations with Kuwait (Viorst, 1991, p. 65). "Many Kuwaitis think the government should have been more flexible in dealing with Iraq's demands during the negotiations" of Spring 1990, she commented, adding:

If the government had read Iraq's statements carefully - especially the letter to the Arab League.., it would have understood how serious Iraq was, and it probably would have settled."

There follow, to illustrate further points of view, typical passages published during the period by strongly pro-Iraq journalists from the Arab world.

Writing in the well established *Al Arab* (The Arabs) of London, Ahmed Al Hawny noted in this daily newspaper in mid-February 1991:

> The war is to continue, not to liberate Kuwait, but to destroy Iraq. Kuwait is only Osman's tunic [a pretext]. The Americans and their allies encouraged the [Kuwaiti] royal family to loot Iraq's petroleum, and they offered Iraq satellite imagery revealing this pillage. At the same time they urged the Kuwaiti to dump more oil on the markets so as to lower the revenue for Iraq and Iran (Al Hawny, 1991).

Or again, Diab Mokhadama made the following remarks in *Al Destour* (The Constitution), a daily published in Amman, Jordan:

> During his last visit in the region, the American Secretary of Defense [Richard Cheney] declared having underestimated the capacity of the Iraqi forces. What can this affirmation mean other than an avowal of powerlessness? As to the spokesman for the Israeli Ministry of Defense, he claims astonishment at the resistance shown by the Iraqi people to the thousands of bombings. What does this signify? (Mokhadama, 1991)

In parallel with these commentaries favorable to the Iraqi position, a Cairo centrist daily, *Al Ahram* (The Pyramids), maintained that one of the outcomes of the Gulf War must be an unequivocal redistribution within the region of the riches derived from petroleum. This exhortation was signed Ibrahim Nafeh (Nafeh, 1991).

Shortly after the invasion of Kuwait by Iraq, a highly placed official in the Department of State in Washington affirmed in a classified analysis that "no one was paying attention" to purchases by Iraq - all through the the 1980-1990 decade - "of western materiel that could be adapted to mass-destruction weapons." This statement by Assistant Secretary of State Richard A. Clarke was made available to the Congress of the United States in June 1992, leaked immediately to the press, and reproduced in daily media worldwide (Clarke, 1992).

Revelations of this kind were reinforced by the continuing efforts, throughout Spring 1992, of Congressman Henry Gonzalez of Texas to read into the *Congressional Record* (and thus make public) numerous official papers indicating the willingness of Vice-President and then President George Bush - throughout the 1980s - to support Saddam Hussein, mainly because of Saddam's opposition to Iranian leaders and policy.

It was unstated but official United States policy, in other words, to help Saddam arm and re-arm during the decade before August 1990.

Transcultural Political Error

In testimony presented in late March 1991 before the Senate Foreign Relations

Committee in Washington, former United States Ambassador April Glaspie stated that the President of Iraq and his cabinet, prior to the outbreak of fighting, had repeatedly assured the United States that Iraq would not attack Kuwait in the bilateral dispute. "The one thing that [the Iraqi] wanted more than anything else was medium-term credit... They were in debt to everybody... [Saddam Hussein] did not believe our clear and repeated warnings that we would support our vital interests... We didn't understand Saddam Hussein. That's absolutely true" (Glaspie, 1991).

In mentioning the name of the American ambassador serving in Iraq when the crisis began in mid-1990, the fact cannot be overlooked that the incumbent diplomat was a woman trying to operate in the man's world that is almost every Arab country, and virtually all the Islamic societies from Morocco to Indonesia. Despite her knowledge of Arabic, Ambassador Glaspie was left at an irreversible disadvantage when dealing with her opposite numbers in Baghdad, trying to communicate her country's position.

Diplomatic hierarchies are as formal - and, the nearer the top, perhaps as inflexible - as military bureacracies. One can only conclude, therefore, that Glaspie was faithfully reflecting policy instructions on Iraq that had been formulated in Washington during the closing years of the 1980s.

Congressional hearings have not succeeded, however, in establishing responsibility for the suspected lacuna in the political dialogue exchanged between American and Iraqi leaders in the months before the events of 2 August 1990. It would appear, then, that the Americans placed themselves in a no-win negotiating pattern with the Iraqi leadership during the pre-crisis months, partially by the appointment of a woman - albeit an informed and experienced diplomat - to their ambassadorship in all-male Baghdad. (Another no-win contribution was the U.S. role in helping Saddam Hussein arm throughout the 1980s, already mentioned).

It would not be surprising if historians determined that the Iraqi President received confusing, if not erroneous, signals from the American administration through its embassy in Baghdad. "Erroneous signals" in the sense that Saddam Hussein could have interpreted the American leadership's seeming insouciance, during the Iraq-Kuwait face-off in 1989-90 concerning petroleum supplies, in an otherwise serious tourney of high-pressure international politics. Indeed, writers Efraim Karsh and Inari Rautsi give credence to this view in their book published after the Gulf hostilities (Karsh and Rautsi, 1991).

The Israel Factor

A major concern shared by many nations, whether Occidental or Islamic, pivoted about the possible direct engagement of Israel in Mid-Eastern hostilities during the winter of 1990-1991. The 28-country group led by the American President sought to keep Israel out of combat so as to avoid breaking the quasi-solidarity of this bloc of nations and to contain the spread of possible battlegrounds.

The Iraqi President inflamed matters after 15 January 1991 by launching Scud surface-to-surface missiles against targets in Israel, and in Saudi Arabia too. The provocation of Israel was unsuccessful, despite the dead and wounded civil victims of the Scud campaign. In Saudi Arabia, the Iraqi missile aggression served only to envenom further the Saudi sentiments concerning Iraq's leader. See Ruina (1991), for a concise, balanced appraisal of the commitment to combat of the Patriot surface-to-air missiles to counteract the Scud threat.

During the year after the hostilities with Iraq, the commander of Israel's Air Force revealed that his country and the United States had shared a secret contingency plan - never used - to permit Israeli aircraft to retaliate on Iraq for the Scud raids. Major-General Avihu Bin-Nun, writing in a military journal, stated "at the last moment we were told to cancel the mission." He accused his government of wreaking chaos on Israeli armed forces by foot-dragging and wrong decisions (*Israel Air Force Magazine*, 1992). Concurrently, American military authorities evaluated severely, and publicly, the real performance of the U.S.-made Patriot countermissiles.

The Religious-Archeological Factor

In addition to the strongly religious appeals for unity and solidarity made persistently by Iraq's President to fellow Arabs and Muslims (which had their most telling reactions in Jordan and among Israel's Palestinian Arabs), there was also an element of profound concern for the preservation by the Coalition of Islamic holy sites and archeological monuments: mosques, sepulchers, pilgrimage destinations.

While there are few locales of this type of cultural heritage in Kuwait, they abound in Iraq. These include Babylon, Basra, Ctesiphon, Khorsabad, Najda, Nimrud, Nineveh, Samarra and Ur. It may be superfluous to add that locations such as these are equally cherished in the Occident as cradles of global civilization. To complicate things for military planners and commanders in the Coalition forces, Iraqi authorities had sited military airfields or nuclear installations close to some of these, or else valuable petroleum facilities lay nearby.

Decisions to attack various strategic and tactical targets, whether by aerial bombing or artillery at ground level, had to take into account all these concerns in order to (1) preserve a common cultural patrimony, and (2) conserve as many as possible productive oil wells (Dorfman, 1991). The historical sites were not damaged, as a matter of record, by Coalition actions.

How the Adversaries Manipulated Political Forces

First, there were the true motives of the United States to be considered, uppermost among which is the geo-economic factor of an uninterrupted and low-cost petroleum supply. In terms of *Realpolitik*, this overshadows in fact the proclaimed objective of restoring Kuwait to its full independence and legitimate governance, and its liberation from peremptory annexation through force of arms exercised by a totalitarian and despotic neighbor.

Second, between August and December 1990 there was the issue of hostages, a negotiating device used almost throughout the course of Arab history but one now scorned by countries of the European tradition. The retention of hostages in Iraq - which, in any event, was resolved when the Iraqi president ordered all of them freed in December 1990 - was thus perceived as conventional and normal in the cultural framework of one side, yet far from orthodox and humane in the moral view of the other.

Stated otherwise, Iraqi perception of the role of the temporary "state guests" contrasted sharply with the persecution of these same actors as it is traditionally perceived through the ethos of European norms (and now accepted, through political extension, by geographically removed actors such as China, Japan, India and Pakistan). How the assessment was made by the Iraqi leadership to release its foreign hostages remains enigmatic; it requires thorough analysis.

This factor is further perplexing, when trying to understand the dynamics of decision, especially since "Americans [play] into their enemies' hands by becoming neurotic when terrorists take hostages" (Fishlock, 1987), but historians know that this attitude dates from the days of the Barbary pirates in the early 19th century (Johnson, 1991).

Third, the use of the word "side" raises questions of meaning and interpretation of the fundamental *problématique*, or "what was really the issue?" The Gulf crisis has been characterized, it is true, as a bilateral stand-off between Iraq and "the rest of the world." But, in reality, this is so only as it was interpreted through the bi-level mechanism of the UN's Security Council and General Assembly.

This mechanism was skillfully manipulated by an element of one of the two sides in the conflict, the government of the United States (much as had been the case in 1950 on the occasion of the attack by the People's Democratic Republic of Korea against its southern neighbor), and almost as masterfully countermanipulated by the leadership in Iraq during the period August 1990-January 1991.

The Specter of Weapons of Mass Destruction

Coalition governments forecasted the likely use by Iraq, once open conflict would break out, of non-conventional arms: biological agents, toxic gases, possibly nuclear weapons.

Indeed, the President of Iraq maintained publicly that his side would use all arms available. The biological vectors which could be used by Iraq included bacterial, fungal, rickettsial and viral agents, and toxins. Suspected of being under development were also microorganisms capable of spreading anthrax, botulism, cholera and typhoid, and certain mycotoxins.

Iraq's poison-gas capability comprised, according to various Western sources, distilled "mustard" gas, a nasty blistering substance; hydrogen cyanide (very difficult to handle, however); and nerve gases such as tabun and VX. Any of these is capable of wreaking physiological and psychological havoc on an opponent.

During the pre-combat build-up of forces and tension on both sides, Iraq in fact showed no signs of preparing to deliver weapons of the kinds described. Once air and then ground hostilities had begun, Iraq still gave no evidence of exploiting stocks of these superweapons - whether deliverable by rocket transport, by airborne bombs and canisters or underwing tanks, or by artillery.

The record thus begs the question: How extensive, in fact, was Iraq's arsenal of such weapons? Data acquired during the summer of 1991 suggest that Iraq was, indeed, extensively equipped. We shall return to this subject, further ahead.

The Terrorism Variable

Terrorism as a form of political expression resurged on the world scene in the 1970s and 1980s, about a century after its last peak of activity (when its locales were mainly the Austro-Hungarian Empire, the United States of America, and Russia). International terrorism in our times began to draw the attention of the peace-research institutes in the 1980s, when specialists such as Daniel Hernant of the Institut de Polémologie in Paris began to study the problem. See, for example, Richard Allan's *Terrorism: Pragmatic International Deterrence and Cooperation* (Allan, 1990). It is also worth reading Jeff Hoover's brief report, "Terrorism in the 1990s: Prospects for A Safer 21st Century" (Hoover, 1991).

Despite warnings, given by some of the Coalition governments and in threats made by the Iraqi President (that terrorism would be accentuated during the crisis), few such events occurred during the five months beginning on 2 August 1990. Then in January 1991, four days after the air war against Iraq was launched, an operation designed to strike a Western embassy in Manila was mounted from Iraq's own embassy in Bangkok.

The attempt failed but led to the identification of Iraqi terrorists positioned in Africa, too. Security interrogations of these groups revealed that Iraq's President had readied a thousand such agents to strike throughout the world, using Iraq's diplomatic installations as bases. Unexplained incidents in Lima, Athens, several towns in Turkey, and in Beirut and London followed - all seemingly the work of local activists (Poutaut, 1991).

Whether or not Iraq's Ba'ath government was in fact behind these actions, the impact of repeated acts of Iraq-associated terrorism had little effect during hostilities and as late as Summer 1991.

Fidelity of Model

Now, back to our model. Votes cast at the UN's Security Council in New York do not reflect with exactitude all the political nuances shading the intended meaning of a "yes" or a "no" uttered in this international forum.

This is a typical shortcoming of any model; that is, the macro level does not reveal and may, indeed, mask greater variety at the micro level. A similar observation would apply to the present model's nontriviality, in the mathematical sense of not being related to immediate relevance. The intentions of Yemen, for instance, are not perceived within the Arab/Islamic world as identical with those of Egypt; nor those of the United Arab Emirates, again an example, with those of King Hassan II of Morocco; nor (yet another example) those of the Jordanese monarch with those of the Palestine Liberation Organization.

The nuances are made fuzzier still by the distinction that has to be drawn between votes cast by Arab nations (an ethnic bloc) and Muslim countries (a religious union), and within both these groupings by "integrationist" Muslims and those Muslims who do not share the fundamentalist view. The otherwise political polarization between perceived aggressor and the nation perceived to be aggressed then assumes an aspect not intended at the outset: a confrontation - in this case - of Arab nations/others, and, perhaps more grievously in terms of future international relations, an Islamic/non-Islamic face-off.

Both these dimensions were complicated further during the Gulf episode by the potential role of Israel - already listed as a factor in itself - as a disruption of Islamic religious solidarity. Would Israel, acting now politically and militarily, take sides? Would Israel become a target of Arab hostilities? Would Israel be constrained to enter active combat after weeks of bombing by Scud missiles? More questions of the same type were foremost in the minds of many observers.

Figure 10.5 shows that the optimization effort previously developed in Figure 1 becomes one of maximization of impact, tending to veer even further from game-theory optimization. Despite the high international stakes involved, the so-called Pareto-optimal win-win formula of game theory is here forsaken - at least, in part - in favor of a more discrete win/no win set of outcomes.

Tracing the Steps of the Model

In the case of Step 1, Figure 10.5, both Iraq and "others" endeavor primarily to appraise the situation in terms of national goals - given the emotional, practical and even intellectual limitations intrinsic to all evaluation. In paraphrase, this analytical step can be termed the statement of the problem.

Proceeding to Step 2, the same observation can be made, but one concretized with less of the inherent subjectivity that inevitably denatures (despite conscious effort to the contrary) the appreciation, and even the background, of the problem analysis undertaken in Step 1.

In elaborating Step 3, there need to be taken into account consequences of an economic nature or those involving the inviolability of sovereign national territory, domestic and foreign reactions of all sorts, and the effects on existing alliances or encumbering protocols.

Arriving at Step 4, we approach the critical alternatives (or final weighing of advantages and disadvantages) immanent in the case of the Iraq-Kuwait crisis, as viewed by either adversary:

- Deciding on suspending/breaking the UN blockade.

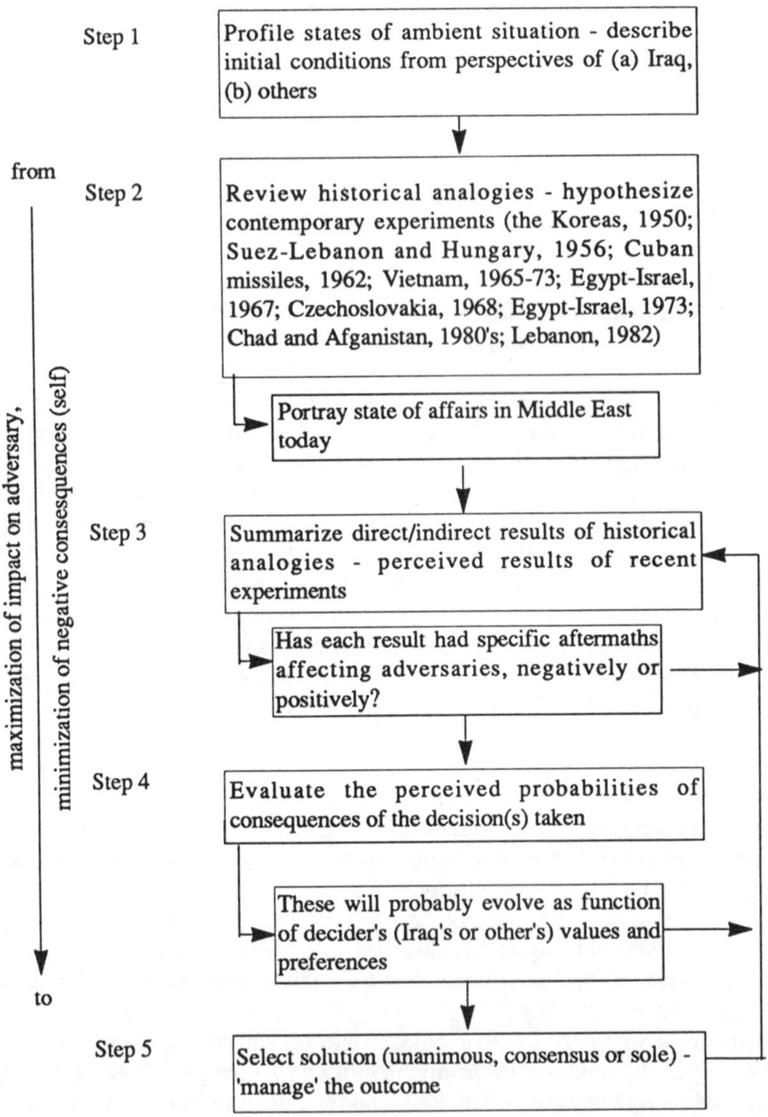

Figure 10.5. Decision-and-command management model applicable to the Gulf crisis of 1990-1.

- Deciding on a move to war, offensive/defensive, or not.
- Initiating novel political moves to resolve the conflict/settle the dispute.
- Temporizing further.
- Withdrawing from the contest - the Coalition by giving in to the status quo, or Iraq by offering its embroilment with Kuwait to perhaps the Arab League (Kodmani-Darwish, 1990) for fraternal resolution; the others by mandating the problem to the UN via a resolution for a mediation/peacekeeping force.

Step 4 implies a special iteration with regard to the first category listed immediately above, deciding on a move to war - meditation of a particularly humane kind, envisaging the "unique terror" associated with imminent battle 2,500 years ago by Greek hoplites (Hanson, 1989). Such contemplation seems to apply less to naval or aerial engagement than to war on the ground, although the history of war since 1914 reflects the broadened human risk involved - civil casualties included, in other than classical ground conflict. The increasingly high-technology nature of war, particularly since the Second World War, has led to a depersonalization and distancing from the effects of war.

A Sense of Reckoning

The Gulf war of January-February 1991, lasting 43 days, recalls certain peculiarities of the battle of Agincourt (northern France), fought on 25 October 1415 and about halfway through the Hundred Years' War. In the process of committing soldiers to the "supreme sacrifice" so familiar in epic letters, Shakespeare has King Henry V of England plead in the presence of an aide, Sir Thomas Erpingham,

> O God of battles! steel my soldiers' hearts;
> Possess them not with fear; take from them now
> The sense of reckoning, if the opposed numbers
> Pluck their hearts from them.
> Not to-day, O Lord,
> O! not to-day...
>
> *The Life of King Henry V*, Act IV, 1

A few hours later, Erpingham marshaled Henry V's archers for action at Agincourt (today called Azincourt), a devastating encounter that saw:

- an invading force of 5,000 English soldiers, taking on a French army of some 22,500 fighting on its own soil, lose only one soldier killed for every 85 French men of arms who perished in a battle lasting barely three hours. (This ratio of 1:85 may be as high as 1:5,000, or 1:10,000 if civilian casualties are counted, in the Gulf war. Figures from the Iraqi side are not available, however, and probably will never be calculated correctly.)
- the Armagnac cavalry and foot troops under France's King Charles VI, overwhelmed by a small infantry force using new technology, the massed use of longbows fired in rapid volleys.

(Hibbert, 1964)

The decision to commit Iraqi forces to combat was made by President Saddam Hussein, quite obviously, before the invasion of Kuwait on 2 August 1990. An analogous decision was mooted by the American and other Coalition leaders prior to (1) beginning the aviation and missile assaults against Iraq-occupied Kuwait before first light on 17 January 1991, and (2) the land attack which followed more than five weeks later on 24 February.

Next, Step 5 becomes the materialization ensuing from the weighing of the decisive factors representing the final phases of the procedure: the conversion of iteration to action - for the better, one trusts - and as specified in the previous paragraph. Either side will interpret, almost always, its own "action" as a positive measure, history reflecting extremely few choices that were made willfully on the grounds of a less desirable action.

Repeated defeat of the Gauls, for instance, in open battle with the Roman army continued during the second and first centuries B.C. (King, 1990); the Japanese historical landscape is dotted, too, with such reversals (Morris, 1975); the colleges of the University of Oxford have been called by some British conservatives "the home of lost causes"; and the white population of the Republic of South Africa may be voluntarily abandoning via the referendum process, during the 1990s, the policy of white supremacy.

Not shown in Figures 10.1 and 10.5, you may agree, is a Step 6: the management of the set of envisageable consequences to Step 5's decision(s). This would introduce, however, a new dimension of complexity, especially when plans fail to work as conceived and then managed. Nevertheless, this is an increasingly important area of policymaking in a world that is increasingly an unstable field of forces (see Chapter 4).

A former foreign minister of France and a specialist in developing-country problems, Claude Cheysson, stated in an interview with the press:

> ... If Shi'ite Muslim Fundamentalists gain control of southern Iraq, the millennial balance between Arabs and Persians will be upset, and a fundamentalist surge that starts with Iraq's holy cities will go on to threaten Mecca and Saudi Arabia. Before the war, the Arabs constantly told us that the destruction of Saddam would create this risk... [T]he Arab masses are suffering a frightening shock at Iraq's humiliating collapse. And Arabs feel it was imposed upon them... Commendably, the Americans have demonstrated their willingness to assume their responsibilities in the Gulf, as they did in seeking - rightly - to curtail Saddam's inordinate power. And if postwar problems are treated in the spirit suggested by President Bush, everything [should go well] (Cheysson, 1991).

During both the confrontation and combat periods of the Gulf episode, various elements within the Coalition favored or opposed not only the Shi'ite but also the Sunni and Kurdish factions in Iraq, or combinations of these. Conversely, the Iraqi leadership (of Sunni persuasion) was required to put in balance its amity or enmity for any or all of these as well as corresponding support or hostility outside Iraq.

Once the Decision Was Taken

The decision to transform confrontation into violent coercion was the consequence of the expiry of the deadline imposed on Iraq by Security Council Resolution 678. After 15 January 1991, the UN's members were authorized by this resolution "to use all necessary means" in order to compel Iraq to evacuate Kuwait. Cuba and Yemen, again, voted against such a move; China opted to abstain from this vote.

It is now a matter of history that the armed conflict beginning 17 January 1991 lasted about six weeks. The decision process involved, with last-minute discussions sponsored largely by the Soviet Union, also represents another analysis related to the survival of the military-industrial complex of the U.S.S.R., a subject beyond the scope of the present chapter. Although we know more and more that things were not all that they seemed to be during this extremely troubled period, disassembly in detail of the roles of all actors must await access to fuller information.

SIGNIFICANCE OF THE GULF EPISODE

In 1992, the Gulf Crisis/War and its origins are far from being a nearly closed chapter of history. There remain enormous gaps in our apperceptions of cause-and-effect, yet a number of useful observations can be made.

Systemic Implications

Not only is our third crisis model concerning the Gulf more immediately transparent and more structurally complex (in the vernacular sense) than the first two models offered in this chapter, it also has important human, institutional, environmental and symbolic implications far surpassing those involved in designing a new capital or choosing and a building a national police system. (See Figure 10.6.) The weighing of alternatives and the selection-rejection modes applied in the decision process, however, are similar enough to be nearly identical.

One would reasonably expect that simple models apply to simple times (if, in relative terms, "simple" can be used to describe any phase whatever of mankind's social evolution). Have I permitted myself to confound evolution, i.e. change over time, with complexity? Again, see Figure 10.6. The answer to the question must be: Possibly. This I indicated previously, at the outset of the chapter when we tried to sort "scenario" from "model" - a scenario being, after all, a kind of action or thought model.

The Gulf crisis of the 1990s involved, besides (1) the petroleum drama and (2) a bald political and military scenario of confrontation between Iraq and other nations:

- The ethnic rivalries already suggested, both real and presumed.
- Religious fervor and clash (occasionally of a conspicuously virulent order).
- Problems related to climate and supplies of food, water and energy.

Aim		Societal Effects		Implications
Design a capital	→	Benign; useful, and institution-building; relatively uncomplex	→	For national and local governance; symbolic both nationally and internationally
		↓ rising		↕ scalar
Create a viable, modern national police service	→	Neither benign nor malign; necessary in national polity; relatively complex	→	Law and order nationwide; symbolic enough to be emulated elsewhere
		↓ complexity		↕ relation
Resolve a conflict, (perhaps) settle a dispute at supranational level	→	May be anodyne; possibly benign; malign as 'worst scenario'; complexity may rise to utmost levels	→	Peace or war regionally or on broader scale; symbolic as evolving history; new model to be replicated or abjured

Figure 10.6. Acceleration of organizational complexity, immediate societal effects, and structural and symbolic implications of the three representative decision-and-command management models presented in this chapter.

- Perturbation of established alliances, treaties and protocols inherent to custom.
- Dysfunction in diplomatic communication.
- Different levels of scientific and technological capabilities (military engineering included).
- Other factors (e.g. the displacement *of* worker populations).

The Enigmatic Superweapons

In reviewing the pertinence of our model, we need to stress again one input, still a speculation as things have developed to date, but a basic constituent element of our analysis. This very element was used, too, as a main justification for taking the war to Saddam Hussein's front lines.

What is the importance to be given to Iraq's determined effort to acquire weapons of mass destruction (chemical, bacteriological and nuclear arms)? As of late February 1991, the mass media reported that thus far 24 nations had been involved in furnishing Iraq material and equipment associated with these forms of warfare. In one country alone (not an active belligerent in the Coalition), 134 different firms had been identified and another 15 were under investigation by that nation's authorities.

More pointedly, what might have been Iraq's intentions in making use of such resources, and when and against whom? All the world recalled vividly how Iraq's forces used chemical weapons in suppressing part of the country's Kurdish minority in 1988, leaving some 5,000 villagers dead.

But such resources were not used in 1990 and 1991, at least during the hostilities with Coalition forces. How is this explained? Some governments may already have partial replies to the question, but they are not publicly available. Be this as it may, the threat of using such weapons was fully exploited at the time by the Iraqi side, leaving the world terrified at the prospects of their use.

Percentage of GNP spent on defense and on health/education
(Figures are for 1988, the last year available)

Country	*Defense*	*Health/education*
Iraq	32.0	4.5
Saudi Arabia	16.5	14.6
Libya	12.9	13.1
Syria	10.9	6.5
India	3.5	4.3
Iran	20.0	5.3
Egypt	7.8	5.8
Israel	13.8	9.4
Afghanistan	(not available)	
Ethiopia	8.2	5.2
Algeria	4.2	5.6

N.B. Although Afghanistan imported about US$13 billion in arms during the period in question, it has not made available the data related to these two columns.

Table 10.2. Comparative expenditures by selected countries on defense and health and education (*source*: ACDA, Washington).

There was, to be sure, a near-consensus of non-governmental observers that Iraq's potential for nuclear warfare was (and remained) limited. K. R. Timmerman, in his *The Death Lobby: How the West Armed Iraq*, states that more than 450 occidental firms helped construct Saddam Hussein's nuclear capability (Timmerman, 1991). And the March 1991 issue of *The Bulletin of the Atomic Scientists* carried two articles germane to the subject. These concluded, in part:

> Making a bomb to be carried by an attack aircraft is relatively easy, but making a missile warhead may be impossible without full-scale nuclear tests or extensive outside help. Iraq could conceivably deliver a nuclear device by a ground vehicle or a ship (Albright and Hibbs, 1991a, 1991b).

By October 1991, inspectors of the IAEA had determined that upwards of 100,000 artillery shells of the Iraqi forces had been tipped with receptacles of sarin.

Complicating the picture further was the question of how much Iraq was spending on armaments overall. The World Bank and the U.S. Arms Control and Disarmament Agency (ACDA) both found that, among the world's nations spending one-fifth or more of their gross national product on military expenditures, the top three in 1988 (the latest year for which figures were available) were all in western Asia: Iraq, followed by Yemen and Jordan.

ACDA also determined comparative budgetary allotments (Table 10.2) among the nations nearest Iraq-Kuwait, based on their cumulative arms imports during the nine-year period, 1980-1988, comparing spending on defense with that on health and education.

An Enigma Becomes Less Perplexing

By early summer 1991, inspectors of the UN's International Atomic Energy Agency made public the fact that Iraqi scientists had resorted to a method almost 50 years old (and long since abandoned by the nuclear powers) to process sub-atomic fuels. Under the codename of Petrochemical Project 3, Iraqi authorities had been planning since before the advent of Saddam Hussein to produce their own supply of lithium-6, that is to say, an isotope of the light metal used to boost nuclear detonations to thermonuclear level.

IAEA enquiries within Iraq established furthermore that the principal manager of the Iraqi nuclear effort was British-educated Jaffar Dhia Jaffar, assisted by another western-trained physicist, Hussein Sharistani. As of October 1991, Jaffar was identified by the IAEA as combined deputy chairman of Iraq's commission for atomic energy, deputy minister (Industry and Minerals), and general manager in charge of rebuilding Iraq's electricity distribution network. (International Atomic Energy Agency, 1991).

More Puzzling Questions from All Sides

By the time the aerial phase of the Gulf war had lasted five weeks and been supplemented by heavy artillery barrages, analysts the world over envisaged a number of realistic outcomes. But they raised some difficult questions at the same time concerning the causes, conduct and consequences of a war that (almost) no one had really wanted. It is useful to recall these:

1. <u>Foreseeable developments</u>.

a. The Iraqi President loses the war but holds on to power. He retains the custody of at least his Republican Guard, as well as the personal image of a hero who resisted superior Western forces. In counterpoise, the United States would keep a presence (part of the Navy in all likelihood; part of the Air Force, too?) in the Middle East. This basic scenario seemed, even then, most probable.

b. The Iraqi President affirms that he has won, despite dreadful losses. The West, by then including Israel, is willing to refrain from demanding conquest, this being beyond the purview of the UN resolutions. (This model attaches to the previous one.)

c. Saddam loses his life, in one way or another. The Americans, Soviets and Iranians (perhaps with others) offer mediation among Iraqi survivors, reversing however the Iraqi *Anschluss* of its Kuwaiti neighbor. Although Saddam lives at the final review of our text - July 1992 - the balance of the forecast is what began to transpire during the Moscow discussions held in the third week of February 1991 and to which the public is not yet privy.

2. Significance of the survival of Iraq's President.

Pro-Saddam elements construe Iraq's defeat in Kuwait as a mortification, another in a long series of historical events in which the Arab or Islamic worlds (or both) have been harassed or overwhelmed by a modern, industrial and "infidel" West.

On 26 February 1991, the French mass media cited the chairman of the French Parliament's lower-house foreign relations committee, Michel Vauzelle. According to these reports, Vauzelle, while visiting Saddam Hussein a few days before the fateful deadline of 15 January, had been told by the Iraqi President that there would be war and that he would lose it militarily. The mass media of the world later quoted Iraq's Ambassador to the UN, Abdul Amir Al-Anbari, who claimed in Vienna in Spring 1992 that "the invasion of Kuwait was a mistake."

The attitude of Saddam Hussein has been confirmed by another source because two months later, the former Iraqi Foreign Minister (by then sidelined in the powerless post of Deputy Prime-Minister), Tariq Aziz, told an American interviewer in Iraq that the interventions of the American Ambassador (April Glaspie) before war broke out were "correct", that "we knew that she was acting on available instructions. She spoke in vague diplomatic language, and we knew the position she was in. Her behavior was a classic diplomatic response, and we were not influenced by it." (Viorst, 1991, p. 67)

The American columnist, William Safire, also quoted widely in the world press in April 1992, told of a conversation with Iraq's Tariq Aziz on 6 October 1989; the secret minutes revealed James Baker requesting that Iraqi authorities overlook Congressional and media suspicions of an arms buildup.

Cesare Marchetti, a physicist at the International Institute of Applied Systems Analysis in Austria, has done much work in analyzing human performance. This is what he has to say about the actions of the President of Iraq during the Gulf Crisis/War:

> Saddam Hussein functions, as do others, within the constraints of system mechanisms. At the same time that people pretend to be rational, they "run the show" by their individual and free decisions. In terms of the logic of holding

power, I would guess that Saddam's sanity is normal. Leaving his tail exposed to the Americans, as it were, he blocked them in a position permitting him to retain power. This is the tactic of the lizard - ever ready to lose its tail to an attacker. Consuming the lives of 100,000 soldiers (or 1 million) in the process is part of the game. (The tail of the lizard, by the way, contains more than a million cells.) Waves of "clever" or "stupid" actions on the part of an individual can thus be dealt with if we can measure them consistently (Marchetti, 1991).

The demonizing of an adversary, in the case of Saddam Hussein first by the Iranians during the 1980s and then by some of the Coalition during the period 1990-1991, is problematic in terms of mass social psychology. Stephen Graubard, historian and editor of *Daedalus*, has also stigmatized the American President in *Mr Bush's War: Adventures in the Politics of Illusion*. Graubard sees Bush's success in a "nothing war" mainly in terms of having manipulated television in order to assure mass support (Graubard, 1992).

It may be enlightening to the world when future personality analyses of Saddam Hussein, who has often been typified as a second Hitler, can be the object of study such as Wolfgang Hammer provided in his perceptive three-volume opus, *Adolf Hitler* (Hammer, 1970).

3. Significance of the shock effect of this crisis/war.

Layashi Yaker, a career diplomat from Algeria who has served as ambassador to the Soviet Union and to the United States, has strong views on the real cost of the Gulf crisis/war of 1990-1991. Yaker, who is a member of the Club of Rome and of the Brandt and South Commissions, estimates that the total cost of the war in the Gulf will approximate $400 billion for all the parties concerned. "This is eight times the cost annually of all foreign aid to countries of the South, and I should not be surprised to learn that all the civil and military casualties in Kuwait and Iraq total 200,000 people." Further: "As to whatever cultural heritage there was in Kuwait, this is gone forever," and, "all this frightful destruction of lives and resources that began in a South-South confrontation could have been avoided" (Yaker, 1991).

The casualty figures for the Iraqi dead, wounded and missing (civil as well as military) being unavailable, all estimates may indeed be horrendous. Moroccan university professor and systems specialist Mahdi Elmandjra, a member too of the Club of Rome and a frequent visitor to the Middle East, estimates that the grand total of Iraq's casualties will include 350,000 dead (Elmandjra, 1992).

4. Impacts on democracy and civil rights.

The coalition states made much, during the final months of 1990, of the need to bring democracy and civil rights to both Iraq and Kuwait. With the cessation of hostilities, Iraq seemed no less totalitarian than before despite new assurances from the Iraqi President that popular elections would eventually take place throughout Iraq. He also promised that the Kurds would ultimately have autonomy, although his actions after the Gulf War leave autonomy doubtful and a new status for Palestinian guest workers a subject to be renegotiated.

The Kuwaiti population seemed to fare little better: its leadership failed to evolve, most ministers returning to the portfolios they had held *ante bellum*; the franchise remained limited; and investigative excesses were reported against those who had collaborated with Iraq between August 1990 and February 1991.

5. Significance of hostilities on Kuwait's leadership.

After the storm in the desert had raged in January-February 1991, there were some changes made with the return of Kuwait's leadership (restored, to the dismay of many, virtually intact). Sabah al-Ahmad, Foreign Minister, and Ali al-Khalifa, Finance Minister, had both had diplomatic dealings with Iraq during the first seven months of 1990. They were removed from their posts. Nawaf al-Ahmad and Salim al Sabah, respectively Defense and Interior Ministers, had simply abandoned ship on 2 August 1990. They, too, were not given back their portfolios.

Were the Decisions Made Homogeneously?

At the risk of boring the reader with a little repetition, the term "others" that I have used with reservations (if not misgivings) - "others" than Iraq - may suggest that the world is blessed with a homogeneity of outlook and uniformity of response. In fact both outlook and response are no more than velleities, factors that are more wish or hope than determined volition.

This wishful thinking began to manifest its transparency with the appearance in Spring 1991 of the first "inside the crisis" books. The allegations proffered included major differences in point of view on combat in the Gulf between the American President and some of his seniormost military officers. The latter tended to favor prolonged economic sanctions against Iraq in order to force a withdrawal from Kuwait. The President, if one is to believe a book (*The Commanders*) prepared hurriedly by the journalist Bob Woodward of *The Washington Post*, rejected this strategy because of the time it might require (Woodward, 1991).

Another eventuality, revolt against Iraq's President, was thwarted - according to a staff paper compiled for the United States Senate's Foreign Relations Committee and released in May 1991. The revolt by disgruntled officers of Iraq's army had been scheduled for March 1991. These officers would have worked in collusion with the country's Kurdish minority. But, by their own estimate that the United States and other Coalition powers would not support their effort, the organizers of the incipient rebellion failed to follow through.

Thus, although the orchestration of riposte to Iraq's annexation of Kuwait in 1990 bore an apparent solidity of front among the Coalition nations, this firmness was fragile and must remain dubious, too, as to immediate and longer range effects. Among the latter, one thinks immediately of:

- The residual politico-military power remaining in Iraq (with or without the prominence of Saddam Hussein).
- The rivalry for supremacy among the Sunni, Shi'ite (or Shia) and Kurdish populations.

- The relationship between Iraq, its Arab neighbors and near-neighbors, and Iran.
- The real relationship between the Arab world and the Occident as a whole (Mahdi Elmandjra insists, for example, that the Occident rejects cultural diversity in the world).
- The effects on the relationship between the Islamic world and Israel.

These implicit fragilities may, in fact, exceed in terms of consequences the otherwise solidary appearance of the ranging of countries by sides. Policymaking and decisionmaking are today seldom straightforward.

A Repetition of the Past

Numerous observers have noted the parallel features of the Iraq-Kuwait episode of 1990-1991 with the Korean peninsular war of 40 years earlier. Figure 10.6 applies also to the Korean War, although four decades ago something went wrong with the totality of the response of the "UN" forces. Despite token military contingents provided by a surprising number of UN allies, the brunt of the campaign was borne by one country, the United States. (We add, however, that an allied nation experienced the wrecking in Korea of one of its most ancient fighting units, Britain's Gloucestershire Regiment.)

The offensive launched in June 1950 by North Korea led within 12 months to a truce, within 36 months to an armistice - the northern side succeeding, however, in keeping the two Koreas separated until the present. The decision processes at play in the Korean case have been scrutinized since, but we still have a poor understanding of how the North Koreans were provoked to attack their southern brothers and create, ultimately, what has become the Mother of Stalemates.

The Future Role of the UN

The UN, we have seen, played a maximal role in the period of the Gulf Crisis and War and their aftermath. The Secretary-General interceded as much as he was asked and permitted to do. "Permitted," because the UN system is the servant of its combined board of directors (who make policy decisions) and clientele (who develop requirements): the one-and-same member nations. The UN is not super-government.

The Americans, in the persons of their President and his Secretary of State (James Baker), resorted to the UN's good offices in tandem with Baker's direct involvement in negotiations with both the Iraqi side and third parties, notably the Soviet leadership. The Soviets, French and Iranians acted similarly.

All failed, however, to persuade the Iraqi leadership to withdraw its forces and other presence (an imposed government) from occupied Kuwait during the phase of the war ending with the launching of ground attacks in late February 1991. At the same time, the role of the UN was pushed somewhat to the wings in the military theater - although Security Council Resolutions 660 and 678 remained the supreme justification for the intercession of all these actors.

Reflections on the events leading to the January-February war in Kuwait and Iraq include the belief of some observers that Saddam Hussein was given insufficient time to iterate the various possibilities for responding to the 12 resolutions of the UN's Security Council and subsequent "calls to order" made by President George Bush. "Because of misinformation and misperceptions," reminds Robert McNamara, "there are misjudgments as to where a nation's interests lie and what can be accomplished" (McNamara, 1991). This perception is further complicated by accusations made in the United States in 1991 and 1992 (already mentioned) that the American President transformed his country's relationship with the Iraqi President into deformed foreign policy.

There remains some sentiment, especially among pacifist groups throughout the world, that President Bush manipulated the crisis (even if implicitly through the UN) when he uttered his ultimata to the adversary so as to leave the latter too little reasonable time for response to a threat of war and - especially - reaction to be prepared. In game theory, such tactics are an expression of the win-win formula; or, perhaps, more accurately if we consider the globally negative charisma of the Iraqi President, the expression of a desire for what game theory calls a "Pareto-optimal" solution to the problem of how best to deal with a refractory dictator.

A leading Jesuit theologian in Great Britain summed up the perplexity of anti-war believers, stating that the "disappointment" caused by the launching of ground operations in February 1991 "reflects the failure of society to resolve disputes by discussion and a readiness to resort" to the use of weapons (McDade, 1991). Both views from the pacifist side may be akin to what von Neumann and Morgenstern called n-person games in characteristic function form (von Neumann and Morgenstern, 1944). See also Chestnut's model, Chapter 3.

But the foreign minister of minuscule Luxembourg, Jacques Poos, speaking for the combined foreign ministers of the 12-nation European Community, invoked the authority of the UN system in such crises. He stated in an interview broadcast by Paris radio station France Inter:

> The UN must play a role, after the Gulf war as before, of direct involvement with whatever will be the consequences of the crisis (Poos, 1991).

The Coalition forces representing the UN were required to take into account a diversity of factors as they designed their war plans to implement the resolutions of the UN's Security Council. In Iraq, on the contrary, the planners may not have taken fully into account what the Coalition's comprehensive response might be.

Once the Coalition had begun to implement Resolution 678, one wonders (to select a dramatic, yet real, scenario) what the chief intelligence officer of the Iraqi President dared tell his superior concerning the gravity of Iraq's predicament day by day. And, if the truth was told, how the truth was accepted and with what consequences for the bearer of bad tidings?

A Shift in Political Paradigm?

Let me add a major impression that the Gulf Crisis/War leaves, the impression related to what Thomas Kuhn has called in reviewing the history of science a "paradigm shift" (Kuhn, 1970). By this, Kuhn means a suddenly and significantly altered manner of thinking or doing. The work of Descartes and Newton in the 17th century provided the evolution of physics one such paradigm shift. Much later, Albert Einstein, Niels Bohr and Werner Heisenberg developed another in our century (French and Kennedy, 1985).

We recalled earlier in this chapter a paradigm shift, in effect, in the development of weapons and their use: the calamitous conflict at Agincourt. The Gulf episode may prove to be a comparable paradigm shift in the political sphere, because it brought the sensitivities and rivalries inherent to the Middle East to the forefront of the global scene. The shift was a translation of the decisional processes applied by both politico-military actors. André Danzin, a member of the Club of Rome, alludes to such disruption in the North-South context as a "civilizational shift," a movement from "the logic of traditional obedience to the logic of initiatives" (Danzin, 1992).

Tariq Aziz, the Iraqi official cited earlier, attributes this historic displacement to a West unwilling under any circumstances to accept Iraq as an equal partner in a negotiation, or in any sort of relationship. More important, the West was not prepared to tolerate the rise of a strong, modern, assertive, developed country in the Arab world (Viorst, 1991, p. 67).

Tariq Aziz concludes his appraisal of such decisional motivation as follows. He had no choice, he reports, but to believe that what had happened to Iraq was what America wanted; in that sense, it was foreordained (Viorst, 1991, p. 67).

The Gulf events, in other words, propelled for the first time since the Arab conquests of Spain and in India one Arab nation into the role of major disrupter of regional (if not world) order. The military wherewithal that helped make this possible came, in large part, from allies-turned-antagonists. The arms build-up provided by one of these states, France, is expertly recounted by two French authors in their volume, *Notre allié, Saddam* (Angeli and Mesnier, 1992). France and much of the rest of the world, donning the umbrella of the UN, reacted sharply and en masse against the disrupter, with telling effects on this recalcitrant state.

A paradigm shift of the type postulated could take the form of a North-South polarization suggested by the director (until 1992) of the International Institute for Strategic Studies, François Heisbourg. He foresees this polarization involving western Europe and North America (possibly together with Japan) *vs* the Arab states. But Heisbourg believes that the determination of such a polarization by the Arab countries themselves would depend on the threat of hegemony they - alone - foresee on the part of Iran. Iran, in other words, could play a pivotal role in whether a polarization occurs, perhaps in the 1990s, in a classical North-South configuration or in an Iran *vs* the Gulf states mode (Heisbourg, 1992).

Such events may be part of a far more holistic process, one involving western

Asia, much of the entirety of Europe and North America as well as certain states in Africa and eastern Asia. What we need to determine next are the causes of a systemic breakdown of vast proportions - a sure case of systems theory applied to societal mechanics, a paradigm shift of global reach.

CONCLUSIONS

Complexity accumulates. To help bring peace to western Asia, former American Ambassador to Kuwait Nathaniel Howell believes that "there has been a remarkable stability of U.S. policy there. Although Iraq is wealthy because of its oil and agriculture, its lack of cash [in the late 1980s] came from overinvestment in the military. Future peaceful solutions to Iraq's problems with its neighbors may depend upon water-management arrangements, resolution of the problem of a Palestinian homeland, or even some sort of cultural-technological Marshall Plan for the entire region" (Howell, 1992). The diplomat could have added that future relations will depend on Iraq's relations with the United States. Political commentator Charles Krauthammer maintains that

> It was not the UN that reversed Saddam's conquest of Kuwait. It was the U.S. Army, based in Saudi Arabia, helped by Britain and France (Krauthammer, 1992).

The accrual of complexity marks the difficulty of much decisionmaking as time advances. Interrelationships between individuals and the social structures of which they are part, and between the social structures themselves and the interests and values they represent, become more intricate with the passage of time. Sensitivity to differences of opinion becomes more profound. The individual or group is thus often submerged inside large collective causes. Despite all this, awareness of new ways of resolving problems and disputes seems, in many cases, to improve.

The perfectibility of man and society may be an ideal, but history shows too that this ideal is founded on a reality that astonishes. Choosing a site for a national capital (those who did so came from countries with long established hub cities) was considered a complex but necessary process at the end of the 18th century. One can imagine how much more complicated such challenges would be in the 1990s.

Suppose that Kurdistan became a sovereign state. Which faction would dominate among Kurds who are pro-Iraqi, pro-Iranian, pro-Turkish, pro-Syrian, and where would they locate their capital in terms of cultural affinity, demographic concentrations, climate and natural resources, transport and communication?

Selecting a police structure at the national level in an Asian country emerging from feudalism faced innumerable obstacles in the late 19th century. Are we, here present a century later, better qualified to judge what makes the optimal police force?

Solving problems, avoiding clashes. As (or when) the next regional political confrontation threatens to be civil or international war (the former Yugoslavia is a case in point), with possible global repercussions, will the world have learned useful lessons from the Gulf crisis at the end of the 20th century? How to avoid conflict, render conflict superbly "efficient," or find non-belligerent substitutes?

There are, evidently, both yes and no responses to these questions - not to speak of "perhaps" or "we do not know." The point is that growing complexity makes the yes/no alternative to the answering of such questions all the more difficult as we tread the path of time. History has shown that changing national or ethnic substrates underlying political transformations are integral to the application of systemic analysis to these shifts.

Quick, decisive responses by others to such shifts may have lost their public acclaim as leaders and their advisors appreciate ever more the weight and ramifications of the response chosen. The termination of hostilities in Iraq in February 1991 may prove to have been, indeed, the beginning of a new test period for the acceptance of totalitarianism by democracies. (See, for example, Bulloch, 1991). This is perhaps the main lesson to be learned, therefore, from the analyses presented in this chapter.

Stability from perturbation. The geopolitical model used here can be replaced - with comparative ease - by decision simulations (whether in the decision-and-command mode or not) applicable to other spheres: economic policymaking; business strategy; planning natural-resource management, academic structures and their operation; and other society-of-scale challenges to management. The fashioning of models, this is to say, needs in future to encompass as many auxiliary factors as thought, energy and time permit. It needs to include as many variables in the "equation" as possible, extraneous as some of the variables might seem in their relation to the principal components of the model.

It may be fitting to end both chapter and book with reference again to the world-societal field discussed by De Greene in Chapters 1 and 4. In the case of the design of the city of Washington, the field was little perturbed at the time of initial layout but unforeseen perturbations later manifested themselves. Still, decision- and policymaking were relatively straightforward; there were few later regrets. In the case of the design of the Japanese national police force, careful comparison of alternatives led to system stabilization. But the case of the Gulf Crisis/War represents a mammoth perturbation of an unstable field of forces, and its aftermath is likely to include future surprises - not excluding further destabilization of the field.

Influence and decisions

The analysis of decisionmaking is one way of studying power relations. Decisions... do not reveal power directly. What they may show directly is influence, or the way in which power is translated into action. The relative power of contending forces is an inference... from a careful observation of the workings of influence. (Cox and Jacobson, 1977, p. 115)

REFERENCES

Albright, David, and Mark Hibbs, 1991a. "Iraq and the Bomb: Were They Even Close?". *The Bulletin of the Atomic Scientists*, 47 (3), pp. 12-15.

Albright, David, and Mark Hibbs, 1991. "Hyping the Iraqi Bomb", *The Bulletin of the Atomic Scientists*, 47 (3), pp. 26-29. See also this journal's report a year later by the same authors, "Iraq's Shop-Till-You Drop Nuclear Program", 48 (3), pp. 26-37.

Al Hawny, Ahmed, 1991. Cited in French translation, *Courrier International* (Paris), 21-27 February, p. 2.

Allan, Richard, 1990. *Terrorism: Pragmatic International Deterrence and Cooperation*, New York: Institute for East-West Security Studies, and Boulder: Westview Press.

Angeli, Claude, and Stéphane Mesnier, 1992. *Notre allié, Saddam*. Paris: Orban.

Badinter, Elisabeth, and Robert Badinter, 1988. *Condorcet (1734-1794), un intellectuel en politique* (2d ed. rev.). Paris: Fayard-Le Livre de Poche, p. 314 and note, same page; p. 335, n. 2.

Bulloch, Alan, 1991. *Hitler and Stalin, Parallel Lives*, London: Harper Collins.

Cheysson, 1991. Interviewed by Joseph Fitchett of the *International Herald Tribune*, 11 March.

Clarke, 1990. Reported by R. Jeffrey Smith of *The Washington Post*, as cited in the *International Herald Tribune* of 6 June.

Cox, Robert W., and Harold K. Jacobson, 1977. "Decision-making". *International Social Science Journal*, 19 (1), pp. 115-135.

Danzin, André, 1991. Speaking before the Association Futuribles Internationale, Paris, 26 March.

DeGreene, Kenyon, 1990 and 1991. Personal communications.

Department of the Army, 1984. *Staff Organization and Operations* (FM 101-5). Fort Monroe VA:Tradoc, May, pp. F4-F5.

Djalili, Mohammad-Reza, 1991. Analyse des conflits du Tiers Monde: éléments d'une typologie, *Revue internationale des sciences sociales*, 43 (1), pp. 163-171.

Dorfman, Andrea, 1991. Will Bombs Rock the Cradle? *Time* (International Ed.), 25 February, p. 62.

Durix, Philippe, 1991. Personal communication.

Elmandjra, Mahdi, 1992. Speaking before the Association Futuribles Internationale, Paris, 26 March.

Fisher, Roger, and William Ury, 1981. *Getting to Yes, Negotiating Agreement without Giving In*. Boston: Houghton, Mifflin.

French, A. P., and P. J. Kennedy, 1985. *Niels Bohr, A Centenary Volume*. Cambridge (Massachusetts): Harvard University Press.

Fukuda, Tokuzô, 1987. As cited in Westney, 1987, p. 35.
Glaspie, April, 1991. Quoted by *The New York Times*, as published in the *International Herald Tribune*, 21 March, p. 3.
Gould, Stephen Jay. 1983. *Hen's Teeth and Horse's Toes*. New York: Norton.
Graubard, Stephen, 1992. *Mr Bush's War: Its Political and Moral Consequences*. New York: Hill & Wang.
Hammer, Wolfgang, 1970, 1972, 1974. *Adolf Hitler* (3 vols.). Munich: Delp.
Hanson, Victor David, 1989. *The Western Way of War, Infantry Battle in Classical Greece*. New York: Knopf.
Hansson, P. A., 1991. "Chaos: Implications for Forecasting," *Futures*, (23), 1, pp. 51-2.
Heisbourg, François, 1992. Speaking before the Association de la Presse Etrangère, Paris, 8 April.
Hibbert, Christopher, 1964. *Agincourt*. London: Pan Books.
Hoover, Jeff., n.d. (probably 1991). "Terrorism in the 1990s: Prospects for A Safer 20th Century" (Meeting report, Security 1). New York and Prague: Institute for East-West Security Studies.
Howell, Nathaniel, 1992. Speaking before France-Amérique (a private association), Paris, 4 February.
International Atomic Energy Agency, 1991. The press releases of the IAEA from March through December record the findings and consequences of multiple missions by IAEA inspectors to various nuclear development sites in Iraq.
Israel Air Force Magazine, 1992. Portions of a full-length article appearing in the January issue were given wide dissemination throughout the world's mass media during this time.
Johnson, Paul, 1991. *The Birth of the Modern, World Society 1815-1830*. London: Oxford University Press.
Karsh, Efraim, and Inari Rautsi, 1991. *Saddam Hussein: A Political Biography*. Oxford, Pergamon-Brassey.
Kennedy, Roger G., 1989. *Orders from France: The Americans and the French in A Revolutionary World*. New York: Knopf.
King, Anthony, 1990. *Roman Gaul and Germany*. Berkeley: University of Southern California Press.
Kodmani-Darwish, Bassma, 1990. Que faire de l'Irak? (What to Do with Iraq?) *Politique étrangère*, 55 (4), p. 775-792.
Kremenyuk, Victor A. (Ed.), 1991. *International Negotiation: Analysis, Approaches, Issues*. San Francisco: Jossey-Bass.
Kuhn, Thomas, 1970. *The Structure of Scientific Revolutions* (2d ed.). Chicago: University of Chicago Press.
Kuwaiti Television, 1991. Originally broadcast at any unspecified time and hour, Professor al-Mubarak's letter was reproduced in the world's mass media in October and November.

Makower, M. S., and E. Williamson, 1967. *Operational Research - Problems, Techniques, and Exercises*. Sevenoaks (Great Britain): Hodder& Stoughton.

Marchetti, Cesare, 1991. Personal communication.

McDade, John, 1991. Interview by BBC Radio 4, 7:00 A.M. news of 24 February.

McNamara, Robert, 1991. Interviewed by Carl Bernstein, *Time* (International edition), 11 February.

McNeill, William H., 1982. *The Pursuit of Power, Technology, Armed Force and Society Since A.D. 1000*. Chicago: University of Chicago Press.

Mokhadama, Diab, 1991. Cited in French translation, *Courrier International* (Paris), 21-27 February.

Morison, Samuel Eliot, 1965. *The Oxford History of the American People*. New York: Oxford University Press.

Morris, Ivan, 1975. *The Nobility of Failure: Tragic Heroes in the History of Japan*. New York: Knopf.

Mumford, Lewis, 1961. *The City in History, Its Origins, Its Transformations, and Its Prospects*. New York: Harcourt, Brace & World.

Nafeh, Ibrahim, 1991. Cited in French translation, *Courrier International* (Paris), 21-27 February.

Poos, Jacques, 1991. Interviewed after the 8:00 A.M. news, France Inter, 20 February.

Poutaut, Jean-Marie, 1991. *Le Point*, 11 February.

Raiffa, Howard, 1982. *The Art and Science of Negotiation*. Boston: Harvard University Press-Belknap.

Richardson, Jacques G. (Ed.), 1988. *Windows on Creativity and Invention*. Mt Airy (Maryland): Lomond Publications.

Riley-Smith, Jonathan, 1991.*The Atlas of the Crusades*. London: Times Books.

Ruina, Jack, 1991. "Comments on the SDI, Patriot and ABM Treaty." *Pugwash Newsletter*, 29 (1), pp. 21-22.

Schellenberg, James A., 1982. *The Science of Conflict*. New York: Oxford University Press.

The Economist, 1991. "World Body Seeks Head, A More Effective United Nations Needs A New Sort of Leader to Run It". 27 April.

Timmerman, Kenneth R., 1991. *The Death Lobby: How the West Armed Iraq*. Boston: Houghton- Mifflin.

Unesco, 1990. *Guidelines for the Planning and Implementation of Impact Evaluations* (CEU/Work.Series/2.Rev.3). Paris: Unesco,Central Programme Evaluation Unit, November.

Viorst, Milton, 1991. "Report from Baghdad". *The New Yorker*, 24 June, pp. 55-73.

Von Neumann, John, and Oskar Morgenstern, 1944. *Theory of Games and Economic Behavior*. New York: Wiley.

Weber, Max, 1947.*The Theory of Social and Economic Organization*. Oxford: Oxford University Press.

Westney, D. Eleanor, 1987. *Imitation and Innovation, The Transfer of Western Organizational Patterns to Japan*. Cambridge (Massachusetts): Harvard University Press.

Woodward, Bob, 1991. *The Commanders*. New York: Simon and Schuster.

World Meteorological Organization (WMO), 1991. Press release no. 465, 1 May.

Yaker, Layashi, 1991. Speaking before the Association Futuribles Internationale, Paris, 6 May.

INDEX